T0204304

Longman Mathematical Texts

Electricity

Longman Mathematical Texts

Edited by Alan Jeffrey

Elementary mathematical analysis I. T. Adamson
Set theory and abstract algebra T. S. Blyth
The theory of ordinary differential equations J. C. Burkill
Random variables L. E. Clarke
Electricity C. A. Coulson and T. J. M. Boyd
Waves C. A. Coulson and A. Jeffrey
A short course in general relativity J. Foster and J. D. Nightingale
Integral equations B. L. Moiseiwitsch
Functions of a complex variable E. G. Phillips

Longman Mathematical Texts

Electricity

Second edition

The late **C. A. Coulson**
and **T. J. M. Boyd**

Professor of Applied Mathematics in the University of Wales at University College of North Wales, Bangor

Longman
London and New York

Longman Group Limited London

Associated companies, branches and representatives throughout the world

Published in the United States of America by Longman Inc., New York

First published by Oliver and Boyd Ltd. 1948
Published as second edition by Longman Group Ltd. 1979

British Library Cataloguing in Publication Data

Coulson, Charles Alfred
Electricity. – 2nd ed. – (Longman mathematical texts).
1. Electricity
I. Title II. Boyd, Thomas James Morrow
537 QC522 78-40510
ISBN 0-582-44281-8

Printed in Great Britain by Richard Clay (The Chaucer Press), Ltd, Bungay, Suffolk

Contents

Preface

In preparing a new edition of **Electricity** my aims have been twofold. In the first place it seemed to me vital to preserve the essence of the original book, with its clarity and economy of style, which has served successfully to introduce a generation of students to the subject. Thus the first eight chapters are based firmly on Professor Coulson's book—though with some re-ordering and some change on emphasis—and in these the subject is developed along conventional lines. Alternative treatments, such as those in which Maxwell's equations are introduced at an early stage, are not without appeal but I have eschewed these since, on balance, I believe that the traditional development is still the best. Perhaps its most serious shortcoming lies in the fact that many students find a prolonged discussion of the physics leading up to Maxwell's equations rather lacklustre. In lecturing on the subject I try to counter this by appealing to contemporary applications of the basic theory and I have adopted the same approach in this book. Thus in Chapter 5, and again in Chapter 8, some simple illustrations from plasma physics are introduced to highlight ideas in magnetostatics and in electromagnetic induction.

In the second place in the interests of providing a balanced account of classical electrodynamics I felt that it was necessary to extend the coverage of the subject beyond that of the first edition to include a full discussion of electromagnetic waves and radiation. These, together with a short introduction to relativistic electrodynamics, make up the remainder of the book.

In preparing this edition I have adopted another feature of courses I have given on classical electrodynamics, namely the use of computer graphics as an aid to understanding. I am certain that many lecturers would agree that one of the difficulties in presenting a theoretical account of the subject to the undergraduates of today compared, say, with those for whom Professor Coulson wrote, stems from their lack of direct experience in carrying out many of the simple basic experiments in electromagnetism which were once an essential part of any school physics curriculum. It is now commonplace to find undergraduates with little feel for the magnitudes typical of electric and magnetic quantities and with only the haziest ideas of the electric or magnetic fields from simple configurations of charges or currents. I have felt for a long time that one of the best antidotes to some of these problems is to make as full use as possible of computer graphics. For the past few years I have taught courses on the subject with the aid of a graphics system and some specimen programs have been included in the book by way of illustration of what can be done. Others have been used to plot charged particle trajectories (Chapters 5 and 12), field lines of TE and TM modes in a rectangular waveguide (Chapter 10) and radiation patterns (Chapter 11). Since sophisticated graphics facilities are now widely available in most computing laboratories—and indeed in many university physics and engineering departments—there are few, if any, technical obstacles to adopting this teaching aid. Used imaginatively, I have found that the interest of students has been stimulated in innumerable ways.

The last four chapters of the book extend the subject beyond the coverage in the first edition. Chapter 9 includes a full discussion of electromagnetic waves and examples from plasma physics are used to illustrate the basic concepts of wave motion. This is followed in Chapter 10 by a treatment of wave propagation in bounded media which includes an illustration on dielectric waveguides from modern optical communications theory. Chapter 11 presents a discussion of electromagnetic radiation including both radiation from antennas and from accelerated point charges. The book ends with a chapter on relativistic electrodynamics without which no introductory account of the subject from a theoretical standpoint can be considered complete. There are of course very many topics such as the relation of the equations of macroscopic electrodynamics to the microscopic description, together with much of what is embraced by classical electron theory that have been omitted. These require more sophisticated discussion and are best left in my view to an advanced course in the subject.

Professor Coulson, in his original preface, wrote that no one really understands the subject until he or she has worked a good many examples in it. That is no less true today, and most of the examples in the original edition have been preserved and others added. Moreover I have added a number of worked examples in several chapters.

In giving thought to a revised edition of his book, Professor Coulson had decided to change from the Gaussian-cgs system of units of the earlier edition to SI units. I have followed his wishes in this matter though I believe it essential for students of the subject to be familiar with both systems and to this end a short appendix on units is included.

I am conscious of my debt to a number of people who have helped me very considerably by commenting on the various drafts of the book and by working examples. These include a number of research students at Bangor notably Gareth Humphreys-Jones, Ivan Moshkun and David Cooke. I have had invaluable help on computer graphics from Dr. G. A. Gardner and Philip Range and particularly from Terry Hewitt who has tested all the graphics programs used in the book, whether listed or not. I am also indebted to Mrs. M. Walker for her expert preparation of the typescript and to Professor Alan Jeffrey and Longman for their patience over the many delays which the work has suffered due to other commitments.

T. J. M. BOYD
Bangor, 1979.

1

Preliminary survey

§1.1 Electrostatics

The fact that a piece of amber, when rubbed, will attract small particles of matter was known 2 500 years ago by Thales of Miletus. From this simple experimental fact has developed the whole science of electrostatics, which deals with the properties of electricity at rest. Indeed the very word electricity is derived from the Greek word for amber, η'λεκτρον. Since the beginnings of physics with the Milesian school of philosophers in the sixth century B.C., a great deal of experimental knowledge of electricity has accumulated, especially in the last 200 years. This knowledge has seldom been obtained in the most systematic order, so our approach in this book will not be to report the various experimental findings as they were first made, but rather to start with a general survey of the whole subject. Readers interested in a full history of the classical theory of electricity and magnetism are referred to [1].† The subject itself is properly described as **classical electrodynamics** and concerns all classical – as opposed to quantal – aspects of the subject, from electrostatics to magnetic effects and relativity.

Electricity – in this general sense – is the study of positive and negative charges and their interactions. Like charges repel each other, unlike charges attract. They do so with a force that is enormous compared with gravitation and is responsible for holding our material universe together. The smallest negative charge is that of the **electron** – the first of the elementary particles – whose existence had been anticipated by Larmor in 1894 and which was discovered and measured for the first time by J. J. Thomson in 1897. The name itself had been introduced by Stoney in 1881 in the sense of an elementary unit of electricity, either positive or negative, which he used to interpret Faraday's classical experiments on electrolysis carried out in 1833.

† See the Bibliography on p. 383

The smallest positive charge has the same magnitude as that of the electron and is found on two additional elementary particles, the **proton** and the **positron**. The proton (Greek for "first") was discovered and named by Rutherford in 1919 and is of course the nucleus of the hydrogen atom. The existence of atomic nuclei had been previously discovered by Rutherford in 1911. The positron, or positive electron, had its origins in Dirac's theory of the electron which showed that an electron can exist in a negative energy state. Dirac interpreted these negative energy states as positive electrons. Their existence was established by experiment in 1933 by Anderson and they were named positrons. Classical electrodynamics is not concerned with positrons since they are produced in the free state only under special conditions. On colliding with electrons in the presence of a nuclear field both particles suffer annihilation with the production of very short wavelength electromagnetic radiation.

All charges are integral multiples of these fundamental units, the magnitude of which is 1.6×10^{-19} coulomb. The smallness of the charge on the electron in relation to ordinary measurements is reflected in the fact that in a 60-watt lamp at 200 volts approximately 2×10^{18} electronic units of charge flow along the filament per second. In addition to the charge on the electron and proton the other property of these elementary particles which concerns us in classical electrodynamics is their mass. The mass of the electron is 9.109×10^{-31} kg; the proton is approximately 1 836 times heavier, at 1.673×10^{-27} kg. Of course both particles possess other properties such as spin and statistics. The spin of the proton for example is revealed by a study of the spectrum of the hydrogen atom. Associated with spin is a magnetic moment which will concern us, at least superficially, in Chapter 6, when we refer to the interpretation of ferromagnetism in terms of electron spin.

In classical – and in quantum – electrodynamics the electron is a point charge, i.e. it has no physical extension. Neither the electron nor the proton can be a point in the strict mathematical sense of the word. Indeed we shall meet a quantity in Chapter 11 which is called the classical radius of the electron and has dimensions of the order of 10^{-15} metre. In classical electrodynamics we are not concerned with phenomena on this scale length and so will take the elementary charges to be point particles. We must recognize however that this does involve a basic inconsistency in the theory (divergence of the self-energy of the electron, cf. §3.4) which lies outside the scope of this book but which is discussed in theories of classical charged particles [2].

Matter is composed of atoms built from protons and neutrons (yet another elementary particle of mass comparable to the proton but devoid of electric charge) which form the nuclei, and of electrons which form their outer structure. Atoms are typically of dimension 10^{-10} metre and their physical behaviour is described by quantum mechanics. In a cubic metre of an ordinary solid there are about 10^{29} atoms; in diamond, for example, there are 4×10^{29}. We know from elementary physics that as we heat matter to higher temperatures it usually passes first to a liquid phase and from this to a gaseous state. What happens when we continue to heat the gas is that the bonds between the nucleus and some of the outer electrons are broken and we obtain matter in a fourth state known as a **plasma**. In the most general sense a plasma is that state of matter in which enough free charged particles (i.e. those not bound to an atom) are present for its dynamical behaviour to be dominated by electromagnetic forces. As we continue to heat the matter, now in plasma form, more and more of the remaining electrons are stripped from the nucleus until eventually the atoms are fully ionized. Our sun and stars in general are hot enough for the matter to be almost completely ionized; interstellar gas is also ionized due to the action of stellar radiation. Indeed virtually all matter in the universe is in the plasma state. Moreover since its behaviour is in very many cases classical we shall make reference to it throughout this book to illustrate points in the theory but without discussing details of plasma physics [3].

A study of the forces which positive and negative charges exert on one another forms the content of Chapters 2 and 3. This may be said to represent the science of electrostatics proper.

§1.2 Electric currents

The study of electrostatics, or charges at rest, leads naturally to a study of electric currents, or charges in motion. The current may be caused by movement either of the positive or negative charges, or of both. Thus in a plasma positive ions carrying a positive charge move in one direction, and electrons carrying negative charge move in the opposite direction, so that the total current has two components. On the other hand in metals such as copper, the charge is carried entirely by electrons. It makes no difference to our formulation of the laws of current flow, developed in Chapter 4, which type of carrier bears the charge, for in all cases the current is measured by the rate at which the charge flows, i.e. the net amount crossing unit area in unit time. The direction of the current is determined by the sign of the product of the charge carrying the current and the mean drift velocity.

The definition of current in terms of a mean or average value of particle drift is an instance of a vital distinction between **microscopic** and **macroscopic** quantities. On a microscopic scale the charges are moving with a distribution of velocities. On a macroscopic scale however we consider only the average motion of charges within a small volume to obtain a mean drift velocity, which, when multiplied by the average number of charges and their respective charge, measures the electric current. For purposes of discussing the flow of current, materials may be placed in one of two categories – **insulators** or dielectric materials and **conductors**. An insulator (e.g. glass) is a substance in which it is extraordinarily difficult to cause electric current to flow. The explanation is simple, since in these materials all the electrons are firmly bound to the positive charges. As we cannot easily separate them it follows that no net flow of charge can take place. A conductor on the other hand is a substance in which a certain number of electrons (or negative ions in some instances) are easily separated from their associated positive charges and one or both can move under the influence of a force of the right kind. Thus in metals such as copper there are electrons (those in the outermost structure of the copper atoms) known as **free** or **conduction** electrons, which are able to flow freely through the material, giving rise to a current while the positive charges remain fixed. In a plasma the positive charges – positive ions – also move and contribute to the current, though their contribution is usually negligible in comparison to that of the electrons. This is due to the inertia of the ions relative to that of electrons so that they acquire a much smaller drift velocity.

In **electrolytes**, on the other hand, each molecule of electrolyte separates spontaneously into positive and negative ions and the total current is the sum of these two currents.

There is a further very important class of conductors – on which a whole technology is based – known as **semiconductors**.

They are so called because their ability to conduct electricity is much less than metals since they possess many fewer conduction electrons. For example, whereas copper has on average one or two conduction electrons per atom a semiconductor such as germanium might have about one per 10^8 atoms. The current in a semiconductor is carried not only by the electrons but by positive carriers, apparent positive charges, known as **holes**.

§1.3 Magnetostatics

Associated with the flow of electric current are magnetic fields. In particular for the steady currents considered in Chapter 4 the magnetic fields are constant in time and these effects are described in Chapter 5 on magnetostatics. Of course magnetic effects had been known long before electric currents were understood. For example, Lucretius mentions that certain mineral ores such as lodestone have the power of attracting small pieces of iron placed near them, and one of the earliest attempts at a perpetual motion machine makes use of this attraction. These forces were called magnetic forces, and the attracting materials were called permanent magnets. From this beginning there developed, quite independently of electrostatics, a branch of physics, magnetism, which gained great practical importance when it was realized that the earth itself behaved like a large permanent magnet. It was the discovery of Oersted (1820) and Ampère (1825) that the same magnetic effects are produced by electric currents as by permanent magnets, which related the two hitherto distinct sciences. Indeed the name "electrodynamics" is due to Ampère.

§1.4 Electric and magnetic materials

In the theory of electricity, in contrast to atomic physics, we are not primarily concerned with the forces exerted by one atom, or one electron, on another atom, but rather with the average effect of large numbers of particles. The distinction is that between **microscopic** and **macroscopic** points of view. From the microscopic point of view we deal with individual atoms and electrons and their dynamics, which is usually quantum mechanical. In the macroscopic description we average those forces over a large number of atoms in a small volume; for that purpose it is often quite fair to neglect the peculiar individual atomic and interatomic effects of a microscopic description. The relation between the two levels of description, microscopic and macroscopic, is rather subtle. There is no adequate elementary treatment; at a more advanced level the relationship is discussed in [4].

If we are not to take into account the detailed structure of matter, but are to use averages in which we have effectively smoothed it out to become homogeneous and continuous, we might wonder what advantage there is in introducing the microscopic point of view in the first place. Essentially the microscopic viewpoint throws light on the fundamental physical processes. This enables us to view our subject as one whole and means that we shall not have to introduce from time to time apparently unrelated physical assumptions, for we shall see how the macroscopic equations arise quite naturally from simple microscopic properties of the atom and electron. This is particularly the case in discussing the relation between electric currents and magnetism.

It was Ampère who proposed the hypothesis that each atom was really a minute electric current. We know of course – though Ampère did not – that, in an atom, electrons move in orbits about the nucleus. They also possess spin. We know too that charges in motion mean that an electric current is flowing. Combining these two facts it follows that in each atom there are indeed tiny electric currents, both orbital and spin. It may happen – and in fact very often does – that in each atom there are pairs of electrons moving so that the net current vanishes and the atom possesses no net magnetic moment. It may equally well happen that although each atom is itself magnetic, these microscopic magnets are arranged in a given block of material in random directions; then again the substance is not a permanent magnet. But it may happen that a majority of the microscopic currents are oriented in the same direction and then we do have a permanent magnet. In the case of the substances which are not permanent magnets, it is possible, by using the right kind of force, to alter the orbits of the electrons and in this way we may induce a temporary magnetism which vanishes when the perturbing force is removed. This is the phenomenon of induced magnetism. Thus the difference between substances which are, or are not, permanent magnets is not that they are made of essentially different material, but rather that with permanent magnets we have no simple way of destroying the cooperative effect of the separate atoms. With non-permanent magnets, on the other hand, this cooperation is solely the result of forces exerted from outside, and automatically disappears when the force is removed.

Chapter 6 examines electrostatics and magnetostatics when electrical and magnetic materials are present. There is a close parallelism between electrostatics and magnetostatics so that the same mathematical analysis can be employed to solve problems in either field. We therefore break off, in Chapter 7, to discuss in detail a number of such problems, and to illustrate techniques – analytic and numerical – required in their solution.

§1.5 Electrodynamics

We have so far been mainly concerned with steady current and static fields. When currents do not remain constant we have to distinguish two cases, depending on whether the change is slow (when the currents are said to be quasi-steady) or rapid. In the first of these, associated particularly with Faraday, Henry and Lenz, we find that if by any means we try to change the current flowing in a circuit then the system behaves as if it were trying to prevent such change. That is to say, new systems of currents are set up, the effect of which is to counteract the original change. This is the phenomenon of **electromagnetic induction**, discussed by Faraday, and independently by Henry, in 1831. Chapter 8 is devoted to a study of quasi-steady currents in general terms and to just two of the many important aspects of induction, namely an introduction to the theory of alternating current circuits and to the hydromagnetics of electrically conducting fluids.

At this point we have brought most of the common electrical effects within the power of our equations, but not all. It took the genius of Maxwell to recognize an omission. We have seen that under certain circumstances an atom may be deformed so that the positive and negative charges are slightly separated. While this separation is taking place there is a very small flow of charge, i.e. a microscopic current is created. If the change is extremely rapid and the negative charge fluctuates about the position of the positive charge then an oscillating current may be said to flow. This is a part of Maxwell's **displacement current**, and we shall see in

Chapter 9 that although for quasi-steady systems this new term is not effective, for sufficiently rapidly varying currents it often becomes dominant. When it is included and combined with another part, the vacuum displacement current, we are led to the general equations of classical electrodynamics, known as **Maxwell's equations**. Upon these equations all subsequent development rests. These equations predicted the possibility of **electromagnetic waves**, which were detected by Hertz in 1887. Electromagnetic waves are discussed in Chapter 9 together with some examples of their propagation characteristics in plasmas. The study of electromagnetic waves is continued in Chapter 10 with an examination of effects at material boundaries. The laws of reflection and refraction in classical optics are derived and applied to treat the propagation of laser light in optical wave guides, which is a topic of great potential importance for communications. In Chapter 11 we turn from an examination of the propagation characteristics of electromagnetic waves to the generation of electromagnetic radiation by studying the inhomogeneous wave equations. Both radiation from antennas and from individual accelerated point charges are treated.

No account of classical electrodynamics can be considered complete without a discussion of the subject within the framework of the **special theory of relativity**. Chapter 9 reveals that within electrodynamics there is an intrinsic velocity, that of light, which is a maximum velocity for the propagation of electromagnetic disturbances. From a Galilean standpoint the velocity of light must appear different to observers in motion relative to one another. If an observer could move with the speed of light in a vacuum, c, then a light wave would appear as a spatially oscillating electromagnetic field at rest. This Einstein could not accept. It appeared to him "intuitively clear that, judged from the standpoint of such an observer, everything would have to happen according to the same laws as for an observer who, relative to the earth, was at rest". Einstein postulated that the velocity of light is the same when viewed by any observer. The main ideas of relativistic electrodynamics are introduced in Chapter 12, leading to the final synthesis in which electric and magnetic fields, scalar and vector potentials, charge and current densities are all seen as the time-like and space-like components of various entities.

2

Electrostatics

§2.1 Law of force

The study of electrostatics is based primarily upon the experiments of Cavendish using a charged sphere, and of Coulomb using a torsion balance. These show that if two charges e_1 and e_2 are a distance r apart the force between them is proportional to $e_1 e_2 / r^2$. If the charges are of like sign the force is repulsive; if unlike, it is attractive. This is known as the **inverse square law** of electrostatics.

It is impossible to prove that the law of force is exactly that of the inverse square. The original experiments of Cavendish showed that if the force was proportional to $1/r^n$ then n lay between 2 ± 0.02. Plimpton and Lawton (1936) were able to show that the exponent in Coulomb's law differs from 2 by not more than one part in 10^9. We may therefore use the inverse square law with complete confidence, provided that we do not apply it in conditions where the dimensions differ essentially from those under which it has been determined. The experiments mentioned were concerned with the dependence of the electrostatic force on distance for distances of the order of centimetres. There have however been more recent experiments assuring the validity of the law over much smaller distances. In particular measurements made by Lamb and Retherford (1947) on the relative positions of the energy levels of the hydrogen atom confirmed that the exponent is again correct to one part in 10^9 over atomic scale lengths, i.e. at distances of the order of an angstrom unit (10^{-10} metres). Moreover if we appeal for evidence from nuclear, rather than atomic, physics then we find that at nuclear scale lengths (about 10^{-15} metres) there are electrostatic forces still varying approximately according to the inverse square law. Since our interest lies in macroscopic rather than microscopic electrical properties we may accept the validity of the law with complete confidence throughout this book.

Coulomb's law may be expressed as

$$\mathbf{F}=C\frac{e_1e_2}{r^3}\mathbf{r} \tag{2.1}$$

in which $\mathbf{F}$ is the force exerted by e_1 on e_2 and $\mathbf{r}$ the vector displacement from e_1 to e_2. The constant of proportionality C depends on the system of units adopted. In this book we shall use SI units in which the unit of force is the newton, the unit of charge the coulomb and the unit of length the metre. The **coulomb** is defined in terms of the ampere, the unit of current which in turn derives from magnetic experiments. Consequently since these units are defined independently of (2.1), this law is used to evaluate the constant of proportionality C. Clearly the dimensions of C are newton-metres²/coulomb².

In the interests of simplifying matters which come later in the development of electrodynamics we replace C by

$$C=\frac{1}{4\pi\varepsilon_0}\text{ newton-m}^2/\text{coulomb}^2$$

where the new constant ε_0 is called the **permittivity of free space** and has a value 8.854×10^{-12} coulomb²/newton-m². Thus the interaction between two point charges at rest in a vacuum takes place according to **Coulomb's law**

$$\mathbf{F}=\frac{1}{4\pi\varepsilon_0}\frac{e_1e_2}{r^3}\mathbf{r}$$

If we place the charge e_1 at some point $\mathbf{r}_1$ in space with e_2 at $\mathbf{r}$ then the force exerted by e_1 on e_2 is now

$$\mathbf{F}=\frac{1}{4\pi\varepsilon_0}\frac{e_1e_2(\mathbf{r}-\mathbf{r}_1)}{|\mathbf{r}-\mathbf{r}_1|^3}.$$

The question of the law of force in other media is left till Chapter 6.

§2.2 Electric field

Implicit in the formulation of the law of force is the fact that the magnitude of the force is proportional to the product of the charges e_1e_2. Thus if we double e_1, say, we also double the force. Coulomb's torsion balance experiments show one further very important fact. Suppose we have an additional charge present. Then the charges e_1 and e_2 exert on this third charge a force equal to the sum of the forces which each would separately exert on it. These additional facts of nature constitute the **principle of superposition** which states that for any given system of charges the total force acting on a **test charge** e is simply that obtained from a superposition of the separate forces acting on e. Thus if we have a set of N point charges $\{e_j\}$, the force acting on e due to these is

$$\mathbf{F}=\frac{e}{4\pi\varepsilon_0}\sum_{j=1}^{N}\frac{e_j(\mathbf{r}-\mathbf{r}_j)}{|\mathbf{r}-\mathbf{r}_j|^3} \tag{2.2}$$

This leads us to the concept of the **electric field**. The test charge e experiencing the force $\mathbf{F}$ is said to be in a field of force and we can define the strength or **intensity** of this electric field, denoted by the vector $\mathbf{E}$, by the ratio

$$\mathbf{E}=\frac{\mathbf{F}}{e}=\frac{1}{4\pi\varepsilon_0}\sum_{j=1}^{N}\frac{e_j(\mathbf{r}-\mathbf{r}_j)}{|\mathbf{r}-\mathbf{r}_j|^3} \tag{2.3}$$

There is a difficulty, however, raised by a definition of $\mathbf{E}$ as the ratio of force to a finite test charge since this test charge will itself cause the set of charges to redistribute themselves. The definition using (2.3) is sometimes replaced by one which adopts a limiting process to ensure that the test charge e does not materially affect the charge distribution which is the source of $\mathbf{E}$, i.e. (2.3) is replaced by

$$\mathbf{E}=\lim_{e\to 0}\frac{\mathbf{F}}{e}$$

This would be an acceptable procedure were the limit accessible! This is not the case since the minimum charge occurring in nature is that of the electron. Alternatively a consistent definition of the field could be had without resort to a limiting process by using one of the set $\{e_j\}$ as our test charge.

It will often be convenient to think of electric charges not as separate and distinct points but rather as smeared in a continuous distribution throughout a given volume or over prescribed surfaces. We may then introduce a volume charge density $\rho(\mathbf{r})$ such that the charge in an element $d\mathbf{r}$ of a volume V is $\rho(\mathbf{r})\,d\mathbf{r}$ and likewise a surface charge density $\sigma(\mathbf{r})$ such that on an element dS of a surface S the charge is $\sigma(\mathbf{r})\,dS$. Proceeding as in the case of the set of discrete charges we now have an electric field defined by

$$\mathbf{E}(\mathbf{r})=\frac{1}{4\pi\varepsilon_0}\int_V \frac{\rho(\mathbf{r}')(\mathbf{r}-\mathbf{r}')}{|\mathbf{r}-\mathbf{r}'|^3}\,d\mathbf{r}'+\frac{1}{4\pi\varepsilon_0}\int_S \frac{\sigma(\mathbf{r}')(\mathbf{r}-\mathbf{r}')}{|\mathbf{r}-\mathbf{r}'|^3}\,dS' \quad (2.4)$$

Knowing the distributions ρ and σ allows us – at least in principle – to determine the electric field. The problem is that in practice these distributions are rarely known!

Note that the expression for the field in (2.4) is quite general and includes (2.3). By using the **Dirac delta function** which has the properties

$$\begin{aligned} \delta(\mathbf{r}-\mathbf{r}')&=0 \quad \text{for } \mathbf{r}'\neq\mathbf{r} \\ \int \delta(\mathbf{r}-\mathbf{r}')\,d\mathbf{r}'&=1 \quad \text{if } \mathbf{r}'=\mathbf{r} \text{ is included in} \\ & \qquad \text{the region of integration} \\ &=0 \quad \text{otherwise} \end{aligned}$$

and

$$\int f(\mathbf{r}')\delta(\mathbf{r}-\mathbf{r}')\,d\mathbf{r}'=f(\mathbf{r})$$

we can represent a distribution of N charges e_j situated at points $\mathbf{r}=\mathbf{r}_j$ by

$$\rho(\mathbf{r})=\sum_{j=1}^{N} e_j\delta(\mathbf{r}-\mathbf{r}_j) \quad (2.5)$$

so that from (2.4), with $\sigma=0$,

$$\begin{aligned} \mathbf{E}(\mathbf{r})&=\frac{1}{4\pi\varepsilon_0}\int_V \sum_{j=1}^{N} e_j\delta(\mathbf{r}'-\mathbf{r}_j)\cdot\frac{(\mathbf{r}-\mathbf{r}')}{|\mathbf{r}-\mathbf{r}'|^3}\,d\mathbf{r}' \\ &=\frac{1}{4\pi\varepsilon_0}\sum_{j=1}^{N} e_j\cdot\frac{(\mathbf{r}-\mathbf{r}_j)}{|\mathbf{r}-\mathbf{r}_j|^3} \quad (2.3) \end{aligned}$$

Since the electric field was introduced as force per unit charge it has units of newton coulomb^{-1} or, more commonly, volt m^{-1} (cf. §2.3).

§2.3 Electrostatic potential

Consider again our expression for the electrostatic field

$$\mathbf{E}(\mathbf{r})=\frac{1}{4\pi\varepsilon_0}\int\rho(\mathbf{r}')\frac{(\mathbf{r}-\mathbf{r}')}{|\mathbf{r}-\mathbf{r}'|^3}\,d\mathbf{r}'$$

and note that we can rewrite the integrand as a function of $\mathbf{r}$ by observing that

$$\frac{(\mathbf{r}-\mathbf{r}')}{|\mathbf{r}-\mathbf{r}'|^3}=-\boldsymbol{\nabla}\left(\frac{1}{|\mathbf{r}-\mathbf{r}'|}\right)$$

Since the operator $\boldsymbol{\nabla}$ involves $\mathbf{r}$ but not $\mathbf{r}'$ we may write

$$\mathbf{E}(\mathbf{r})=-\frac{1}{4\pi\varepsilon_0}\boldsymbol{\nabla}\int\frac{\rho(\mathbf{r}')}{|\mathbf{r}-\mathbf{r}'|}\,d\mathbf{r}' \tag{2.6}$$

from which it follows that

$$\boldsymbol{\nabla}\times\mathbf{E}=0 \tag{2.7}$$

since the effect of operating with the curl operator on the gradient of any scalar function is identically zero. We define the **electrostatic potential** $V(\mathbf{r})$ by

$$V(\mathbf{r})=\frac{1}{4\pi\varepsilon_0}\int\frac{\rho(\mathbf{r}')}{|\mathbf{r}-\mathbf{r}'|}\,d\mathbf{r}' \tag{2.8}$$

so that

$$\mathbf{E}=-\boldsymbol{\nabla}V \tag{2.9}$$

Note that if we add a constant to (2.8) the electric field is unchanged so that to this extent the potential is arbitrary. To make clear the physical significance of the electrostatic potential let us suppose that a unit test charge is moved from A to B under the influence of an electric field $\mathbf{E}(\mathbf{r})$. Now the work we do against the field in moving the charge (which is just (-1) times the work done by the field) is

$$\begin{aligned}W&=-\int_{\mathrm{A}}^{\mathrm{B}}\mathbf{F}\cdot d\mathbf{s}=-e\int_{\mathrm{A}}^{\mathrm{B}}\mathbf{E}\cdot d\mathbf{s}\\&=e\int_{\mathrm{A}}^{\mathrm{B}}\boldsymbol{\nabla}V\cdot d\mathbf{s}\\&=e(V_{\mathrm{B}}-V_{\mathrm{A}})\end{aligned} \tag{2.10}$$

In other words eV may be interpreted as the **potential energy** of the test charge in the electric field. Note that in (2.10) the work done depends only on the positions of A and B and not on the path connecting them. Whichever path we follow from A to B involves us in the same expenditure of work. If this were not so and a path 1 existed which involved us in doing less work than some other path 2, then by going out along 1 and back via 2 we should recover more work than we expended. By continuing this process we could obtain an indefinite amount of work or energy. This is impossible in all natural phenomena, in which the fields of force are said to be **conservative**.

From (2.10)

$$\int_A^B \mathbf{E} \cdot d\mathbf{s} = -(V_B - V_A)$$

and so for a closed path the line integral vanishes, a result which follows directly from (2.7) and Stokes' theorem, i.e.

$$\int_S \nabla \times \mathbf{E} \cdot d\mathbf{S} = \oint \mathbf{E} \cdot d\mathbf{s} = 0 \tag{2.11}$$

One final point worth noting about the potential is that in practice it is often preferable to work in terms of V rather than $\mathbf{E}$. It is clear that determining $\mathbf{E}$ by first calculating V using (2.8) and then taking the gradient will in general be more straightforward than trying to calculate $\mathbf{E}$ directly from (2.4).

The SI unit of potential is the **volt**. A potential difference of 1 volt between two points exists if 1 joule of work per coulomb is done in moving charge from one point to the other. We remarked in the previous section that the electric field is measured in units of volt m^{-1}.

§2.4 Equipotentials and field lines

Consider all points for which the potential has the same value V. As a rule these will be on a surface which we call an **equipotential surface**. By choosing different values of V we generate a family of equipotentials. Since the potential at each point is unique, there is one and only one equipotential through any point P. If the potential at any point $\mathbf{r}$ is given by $V(\mathbf{r})$ then the family of equipotentials is described by the equation

$$V(\mathbf{r}) = \text{constant}$$

By choosing different numerical values for the constant we obtain the various members of the family of surfaces.

For example, consider the special case of a solitary charge e at the origin. Then from (2.8)

$$V(\mathbf{r}) = \frac{e}{4\pi\varepsilon_0} \int \frac{\delta(\mathbf{r}')}{|\mathbf{r}-\mathbf{r}'|} d\mathbf{r}' = \frac{e}{4\pi\varepsilon_0 r} \tag{2.12}$$

so that the equipotentials described by (2.12) are the surfaces $r = \text{constant}$, i.e. concentric spheres with centre at the origin.

By definition V is constant along any equipotential surface. It follows from (2.10) that if A and B are distinct points on an equipotential

$$\int_{\mathrm{A}}^{\mathrm{B}} \mathbf{E} \cdot d\mathbf{s} = 0$$

i.e. either $\mathbf{E} \equiv \mathbf{0}$ along the path or $\mathbf{E}$ is perpendicular to $d\mathbf{s}$. We conclude therefore that the direction of $\mathbf{E}$ at any point is normal to the equipotential surface through that point.

This result suggests a way of mapping the electric field. Suppose we generate a family of lines such that at any point on one of these lines the direction of $\mathbf{E}$ coincides with the tangent to the line at this point. These lines, known as **field lines** or **lines of force**, were introduced by Faraday as a means of picturing the electric field. Through any given point there is usually just one field line and the family of lines so obtained are the orthogonal trajectories of the equipotential surfaces; that is, every field line cuts every equipotential surface at right angles.

It is straightforward to write down the differential equations of the field lines. Since the direction of the line is identical with that of $\mathbf{E}$, the lines must be defined by $dx : dy : dz = E_x : E_y : E_z$, i.e. by

$$\frac{dx}{E_x} = \frac{dy}{E_y} = \frac{dz}{E_z} \tag{2.13}$$

In the particular case of a solitary charge at O the field lines are evidently radii drawn from O, represented in Fig. 2.1 by a

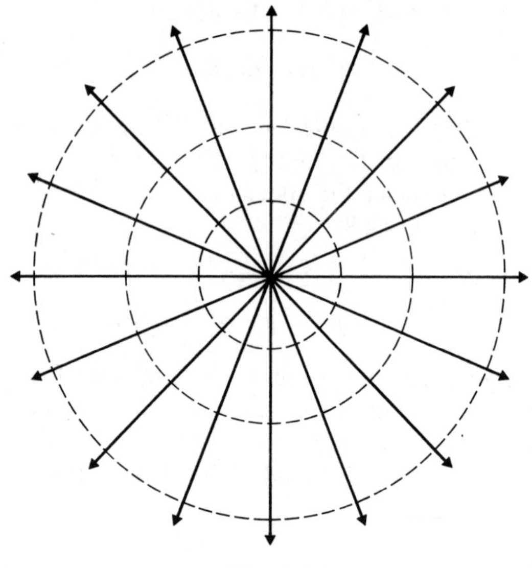

Fig. 2.1

section of the three-dimensional bundle of lines. Arrows indicate the direction of $\mathbf{E}$ which is outwards in this case as we have chosen a positive point charge. The radii all intersect the equipotentials (concentric spheres with O as centre) at right angles. In Fig. 2.1 the equipotentials are represented by the dotted lines. We see from the figure that the position of the charge is a singular point at which the direction of $\mathbf{E}$ is indeterminate.

Another exception to the statement that at any given point there is usually just one field line, crops up whenever $\mathbf{E} = \mathbf{0}$. From (2.13) it is clear that at such points $dx : dy : dz$ is indeterminate so that the field lines themselves are indeterminate. If $\mathbf{E} = \mathbf{0}$ at a point P it follows that a charge placed at P experiences no resultant force and is in a state of equilibrium. Such points are referred to as **neutral points** or **points of equilibrium** of the field.

A simple example of an electric field with a neutral point is that provided by two equal and like point charges. Figure 2.2 represents the field lines for two equal positive charges, symmetrically placed about the origin. From considerations of symmetry we expect to find a neutral point at the origin.

Although this is a simple system physically it illustrates that solving (2.13) has already become tedious. If we write down $\mathbf{E}(\mathbf{r})$

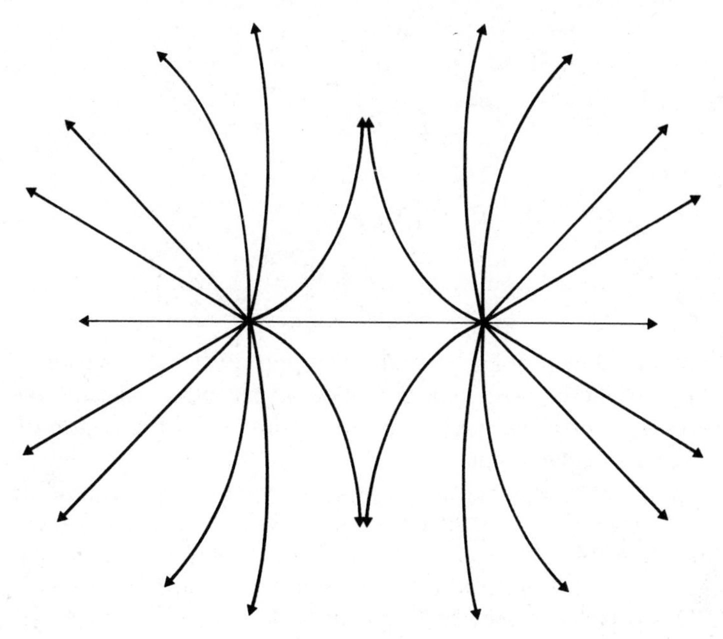

Fig. 2.2

from (2.3) with $\mathbf{r} = x\hat{\mathbf{i}} + y\hat{\mathbf{j}}$, $\mathbf{r}_1 = -a\hat{\mathbf{i}}$, $\mathbf{r}_2 = a\hat{\mathbf{i}}$ then

$$\mathbf{E} = \frac{e}{4\pi\varepsilon_0}\left[\frac{(x+a)\hat{\mathbf{i}} + y\hat{\mathbf{j}}}{((x+a)^2 + y^2)^{3/2}} + \frac{(x-a)\hat{\mathbf{i}} + y\hat{\mathbf{j}}}{((x-a)^2 + y^2)^{3/2}}\right] \tag{2.14}$$

giving expressions for E_x and E_y to use in the differential equation (2.13). It is possible to solve this differential equation (problem 7) and use the solution to plot the field lines. An alternative approach would be to try to infer the essential properties of the field lines **directly** from the differential equation. Such a qualitative approach can give us worthwhile information about the topology of the field which is often all we want to know, leaving the detailed mapping of **E** to be done by computer, as described in §2.9.

In this case it is easy to see for example from (2.13) and (2.14) that for $y = 0$, $\dfrac{dy}{dx} = 0$ and for $x = 0$, $\dfrac{dy}{dx} = \infty$ giving the x and y axes as field lines. At the origin therefore, $\dfrac{dy}{dx} = \dfrac{E_y}{E_x}$ is indeterminate and it follows that we must have $E_x = 0 = E_y$, i.e. that the

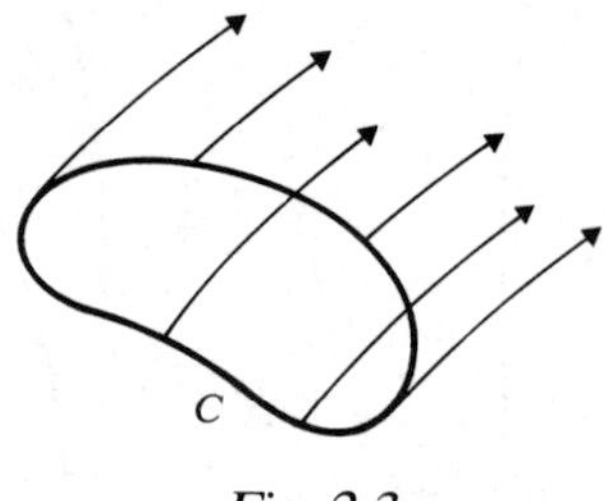

Fig. 2.3

origin is a neutral point. It is also straightforward to determine the asymptotic behaviour of the field, i.e. far away from the two sources at $(-a, 0)$ and $(a, 0)$, and to examine it in the region of these two singular points.

Next let us take any small closed contour C (Fig. 2.3) and draw the field lines that pass through every point of C. These lines form a thin tube, the cross-section of which may vary along its length but whose generators at any point are all effectively parallel. This is called a **tube of force** and is a concept due to Faraday who used it with great skill to picture the behaviour of the field. We define the **strength** of a tube of force as the product of the magnitude of $\mathbf{E}$ and the normal sectional area. We shall see in §2.7 that the strength of a tube remains constant along its length. A unit tube is one whose strength is unity.

Tubes of force, like field lines, are everywhere parallel to the direction of $\mathbf{E}$, and the potential falls continuously as we move along them in the positive direction. A normal cross-section of a tube of force forms part of an equipotential surface, since it cuts all field lines through its boundary at right angles.

§2.5 Flux

It follows from our definition of tube strength that the number of unit tubes crossing any small area $d\mathbf{S}$ is $\mathbf{E} \cdot d\mathbf{S}$. This quantity is called the **flux** of $\mathbf{E}$ across $d\mathbf{S}$. If we have a surface S bounded by a closed curve C we may write

$$\text{flux of } \mathbf{E} \text{ across } S = \int \mathbf{E} \cdot d\mathbf{S} \tag{2.15}$$

If S is a closed surface we speak of the flux of $\mathbf{E}$ out of, or into,

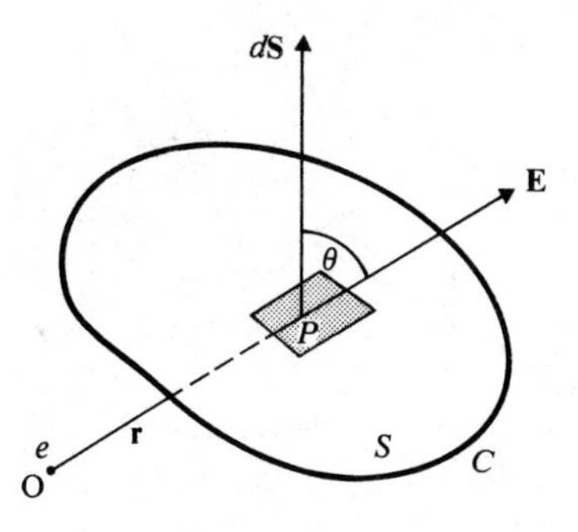

Fig. 2.4

S. If the direction of dS is that of the outward normal, then

$$\text{flux of } \mathbf{E} \text{ out of } S = \int \mathbf{E} \cdot d\mathbf{S} \tag{2.16}$$

In (2.15) the integration is over an open surface, in (2.16) it is over a closed surface. These formulae represent the total number of unit tubes of **E** crossing the open surface bounded by *C*, and emerging from the closed surface *S*, respectively.

Let us next consider the flux of **E** across a given surface *S* (Fig 2.4) when the field is due to a single charge *e* at O. Then if θ is the angle between an element $d\mathbf{S}$ and **E**

$$\begin{aligned}\mathbf{E} \cdot d\mathbf{S} &= E\, dS \cos\theta = \frac{e}{4\pi\varepsilon_0 r^2}\, dS \cos\theta \\ &= \frac{e\, d\omega}{4\pi\varepsilon_0}\end{aligned} \tag{2.17}$$

where $d\omega$ is the element of solid angle subtended at O by the surface element dS. It follows from (2.15) that the flux of **E**

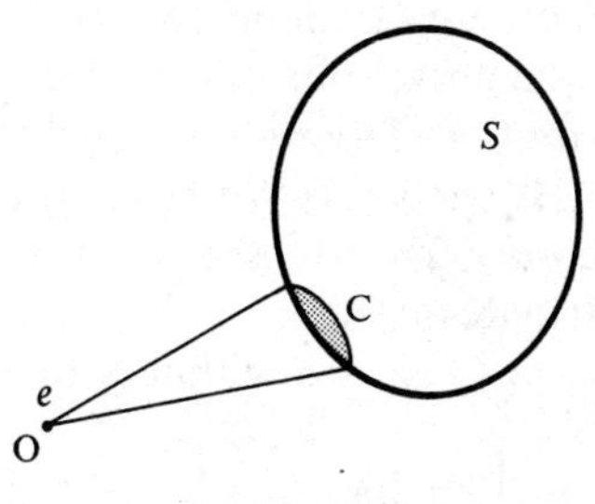

Fig. 2.5

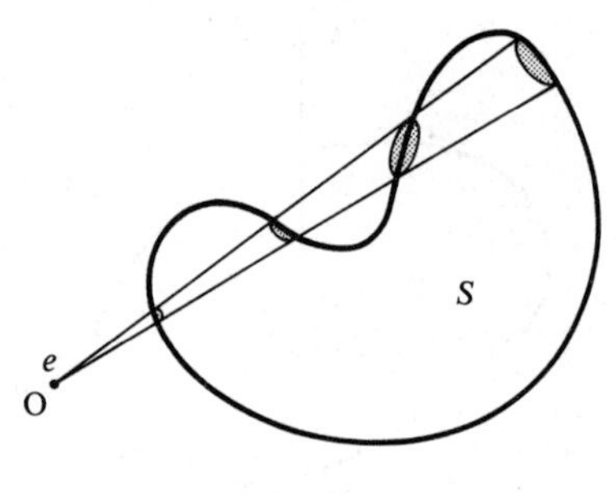

Fig. 2.6

across the whole surface S is simply $(e\omega/4\pi\varepsilon_0)$ where ω is the solid angle subtended at O by the boundary curve C.

If the surface S is a closed surface surrounding O, then $\omega = 4\pi$ so that the flux of $\mathbf{E}$ out of any closed surface containing a charge e is e/ε_0. This can be expressed by saying that e/ε_0 unit tubes of force leave any charge e; alternatively, $1/\varepsilon_0$ unit tubes of force leave unit charge.

What happens in the case of a closed surface not surrounding O? A convenient way of picturing this, illustrated in Fig. 2.5, is to view it as the limit of an open surface on a boundary curve C which shrinks to zero. This shows that when the surface is closed, $\omega = 0$, and hence the flux out of any closed surface not containing the charge e is zero.

These conclusions are not affected if the surface S is re-entrant as in Fig. 2.6.

§2.6 Gauss' Law

This brings us to an extremely important result known as Gauss' law. Suppose we have a set of N charges $\{e_j\}$ and we draw a closed surface S which may surround some or all of them. Then, since the principle of superposition shows that the total field $\mathbf{E}$ is the sum of the contributions from each separate charge, the same is true for the flux of $\mathbf{E}$ out of S. We have just seen that charges outside S make no contribution to the flux out of S, whereas each charge e_j inside contributes e_j/ε_0. Thus if charges $e_1, e_2 \ldots e_i$ lie outside S and $e_{i+1}, e_{i+2} \cdots e_N$ are within S then

$$\int \mathbf{E} \cdot d\mathbf{S} = \frac{1}{\varepsilon_0} \sum_{j=i+1}^{N} e_j \tag{2.18}$$

This is **Gauss' law**:

The total flux out of any closed surface S is equal to $\frac{1}{\varepsilon_0}$ times the total charge enclosed by S.

We have proved Gauss' law on the hypothesis that the charges are discrete, but the result may be easily extended to a continuous charge distribution characterized by a charge density ρ. Regarding the volume distribution as a set of discrete charges $\rho\,d\mathbf{r}$ the result is equally applicable in this case. The enclosed charge is now simply $\int_V \rho\,d\mathbf{r}$ where V is the volume of the space enclosed by S, so that $\int E \cdot dS = \frac{1}{\varepsilon_0}\int_V \rho\,d\mathbf{r}$. If we now use Gauss' theorem

$$\int \mathbf{E}\cdot d\mathbf{S} = \int_V \nabla\cdot\mathbf{E}\,d\mathbf{r}$$

then

$$\int_V \left(\nabla\cdot\mathbf{E} - \frac{\rho}{\varepsilon_0}\right) d\mathbf{r} = 0$$

Since this result holds for arbitrary V it follows that

$$\nabla\cdot\mathbf{E} = \frac{\rho}{\varepsilon_0} \tag{2.19}$$

which is the differential form of Gauss' law.

We have now fully specified the electrostatic field since earlier we showed that $\nabla\times\mathbf{E}=0$. The fact that $\nabla\times\mathbf{E}=0$ in no way depends on the inverse square nature of electrostatic interaction, but requires only that the force be central. Gauss' law, on the other hand, demands an inverse square law as well as invoking the principle of superposition and so is simply an alternative expression of Coulomb's law. We may combine the two differential equations for the field since we know that the vanishing curl of the electrostatic field means that $\mathbf{E}=-\nabla V$ which, together with (2.19), gives

$$\nabla^2 V = -\rho/\varepsilon_0 \tag{2.20}$$

This is known as **Poisson's equation** and in the particular case of $\rho=0$ it reduces to **Laplace's equation**

$$\nabla^2 V = 0 \tag{2.21}$$

Equations (2.20) and (2.21) are alternative expressions of Gauss' law; our method of deriving this result showed that it depended only on the validity of the inverse square law and the principle of superposition. It follows therefore that (2.20) and (2.21) are equivalent to a statement that the law of force between charges is the inverse square, and that electric fields and potentials are additive.

§2.7 Deductions from Gauss' law

There are a number of important consequences of Gauss' law. An immediate deduction is that the potential cannot have a maximum or minimum value except at points where there is respectively a positive or negative charge. Suppose the potential **has** a maximum at some point P so that V decreases in all directions away from P. If we picture a small sphere with centre at P, the lines of $\mathbf{E}$ cross this surface everywhere from inside to outside. Thus the flux of $\mathbf{E}$ out of the surface is necessarily positive. It follows from Gauss' law that if the flux $\int_S \mathbf{E} \cdot d\mathbf{S} > 0$, then $\int \rho \, d\mathbf{r} > 0$, i.e. there must be a positive charge enclosed within the surface, however small the sphere. This charge must therefore be at P which establishes the result. An analogous argument applies to minimum values of V.

It is important to realise that the above argument does not necessarily imply a **point** charge at P. The conditions are equally well satisfied by a charge cloud in which the density at P is not zero. For at a point where V is a maximum (but finite) $\partial^2 V/\partial x^2$, $\partial^2 V/\partial y^2$, $\partial^2 V/\partial z^2$ must all be negative. So $\nabla^2 V < 0$ and hence, by (2.20), $\rho > 0$.

A second important deduction from Gauss' law concerns equilibrium in an electrostatic field. Suppose we place four identical negative charges at the corners of a square and add a positive charge at the centre. The force on the positive charge is clearly zero but is this equilibrium position stable? The answer to this question may be found from Gauss' law and the result is known as **Earnshaw's theorem**:

A freely movable charge cannot be in stable equilibrium in an electrostatic field except at a point already occupied by charge.

For with a given system of fixed charges, the potential energy of a movable charge e is eV. Stable equilibrium of e at some point P requires that eV be a minimum at P. But we have already seen that this is impossible **unless** there is charge at P. The importance of Earnshaw's theorem lies in the fact that it shows that no purely stationary system of charges can be in stable equilibrium under their own influence. The theorem was invoked in early models of the atom to show that electrons in an atom or molecule can never be at rest so that these configurations must be in a state of dynamic, rather than static, equilibrium. A cold plasma (i.e. one in which the thermal or random energies are negligible) is a further example of a system of charges which cannot exist in stable equilibrium under the influence of purely electrostatic forces, no matter how complex the structure of the field.

Another deduction from Gauss' law is that **tubes of force can only begin or end on charges**. For suppose a tube of force begins at a point P. If we surround P by a small closed surface there must then be more tubes leaving this surface outwards than there are entering it. This implies there is a net flux of $\mathbf{E}$ out of the surface and hence, from Gauss' law, that there is a positive charge at P. Similarly tubes of force can only end on negative charge.

It is now possible to prove the statement in §2.4, that the strength of a tube of force remains constant along its length. To see this, consider a tube of force as in Fig. 2.7 with fields $\mathbf{E}_1$ and $\mathbf{E}_2$ across the ends of the section. From the result just proved, since the tube clearly does not either begin or end between dS_1 and dS_2 there can be no charge contained within the section of the tube. Hence there is no net flux of $\mathbf{E}$ out of the surface of the tube. Obviously no other tubes may cross the walls of the one we are considering and so the flux across one end must equal that across the other, i.e. $\mathbf{E}_1 \cdot d\mathbf{S}_1 = \mathbf{E}_2 \cdot d\mathbf{S}_2$. Now we defined the strength of a tube to be $\mathbf{E} \cdot d\mathbf{S}$ and hence the strength does remain invariant along the length of a tube.

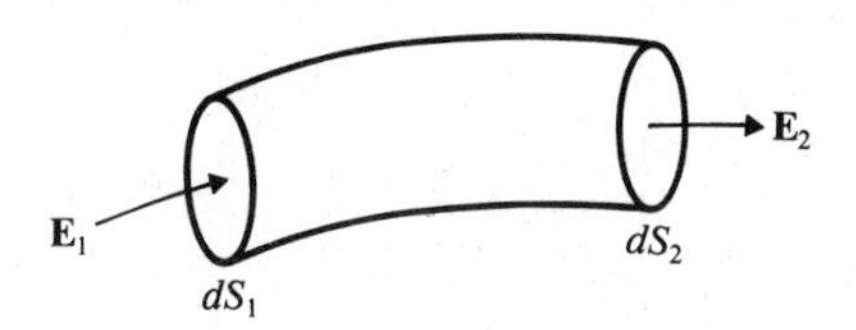

Fig. 2.7

As another example of Gauss' law, consider the field due to an infinite layer of positive charge with constant surface density σ, lying on the plane ABCD in Fig. 2.8. By symmetry we know that the field lines must be directed normally away from the layer of charge. Consider therefore the flux of $\mathbf{E}$ out of the right cylinder PQR, of cross-section dS, whose generators are all normal to ABCD. If we now apply Gauss' law to this cylinder we find, since there is no flux across the curved surface,

$$(E_1 + E_2)\, dS = \frac{\sigma}{\varepsilon_0}\, dS$$

i.e.

$$E_1 + E_2 = \frac{\sigma}{\varepsilon_0}$$

Thus the field is everywhere constant in magnitude and directed away from the layer of charge. If there are no other charges present then by symmetry

$$E_1 = E_2 = \frac{\sigma}{2\varepsilon_0} \tag{2.22}$$

Surfaces such as the cylinder PQR which we use in applying Gauss' law are often called **Gaussian surfaces**. It is important to recognize that (2.22) applies **only** to the field due to the charge layer in the absence of **all** other charges.

As a final example consider the equations of the field lines when we have a set of N collinear charges $\{e_j\}$. Let PQ (Fig. 2.9)

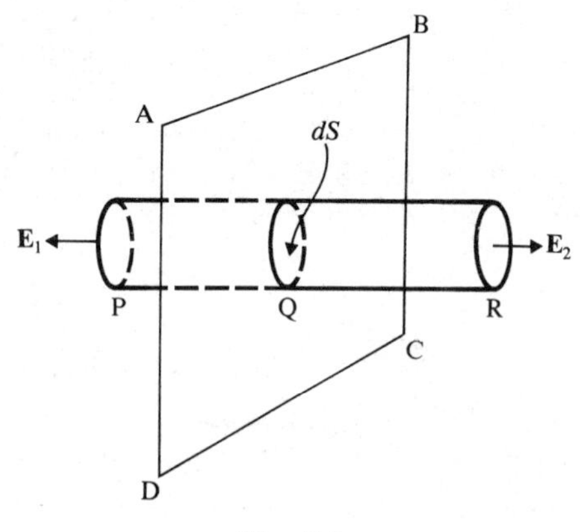

Fig. 2.8

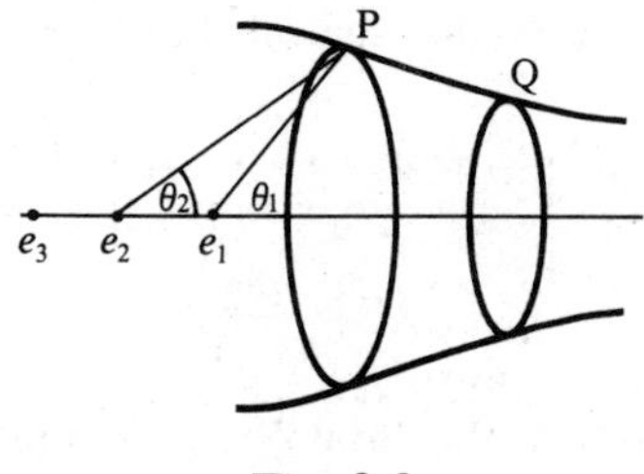

Fig. 2.9

be an element of a field line. Since the field must have cylindrical symmetry about the line of charges we may rotate PQ about this line and generate a closed surface. Clearly no field lines can cross its curved part and so the flux of **E** across the plane circle at P is the same as that across the plane circle at Q. From §2.5 we know that the total flux from left to right across P is $\sum_{j=1}^{N} e_j\omega_j/4\pi\varepsilon_0$ where ω_j are the solid angles subtended at e_j by the circle. Now the solid angle ω_j is given in terms of the angles θ_j by $\omega_j = 2\pi(1-\cos\theta_j)$ and so $\dfrac{1}{2\varepsilon_0}\sum_{j=1}^{N} e_j(1-\cos\theta_j)$ has the same value at P and at Q. Therefore the field lines are defined by the equation

$$\sum_{j=1}^{N} e_j \cos\theta_j = \text{constant} \tag{2.23}$$

In the above argument we assumed that none of the charges lie between P and Q. We leave it as an exercise to the reader to show that our result is still valid if this restriction is removed, so that (2.23) defines the field lines in their entirety.

§2.8 Examples

(A). Calculate the electric field due to a sphere of charge of uniform density and radius a.

Since the distribution of charge is symmetric, **E** is radial. We choose for our Gaussian surface a sphere of radius r concentric with the sphere of charge and consider two cases: $r>a$ and $r<a$.

Over this surface S, by Gauss' law in the case $r>a$

$$\int_S \mathbf{E}\cdot d\mathbf{S}=4\pi r^2 E=\frac{1}{\varepsilon_0}\int_V \rho\, d\mathbf{r}=\frac{4\pi a^3\rho}{3\varepsilon_0}$$

$$\therefore \quad E=\frac{\rho a^3}{3\varepsilon_0 r^2} \qquad r>a$$

Note that if we introduce the **total** charge contained within our sphere of charge $Q=\frac{4}{3}\pi a^3\rho$ we may write

$$E=\frac{1}{4\pi\varepsilon_0}\frac{Q}{r^2} \qquad r>a$$

i.e. the field is the same as if we had a **point charge** Q at the origin.

For the case $r<a$ we again use an appropriate Gaussian surface to which we apply Gauss' law

$$\int_S \mathbf{E}\cdot d\mathbf{S}=4\pi r^2 E=\frac{1}{\varepsilon_0}\int_V \rho\, d\mathbf{r}=\frac{4\pi r^3\rho}{3\varepsilon_0}$$

$$\text{i.e. } E=\frac{\rho r}{3\varepsilon_0} \qquad r<a$$

The two forms for E match, as they should, at $r=a$. The form of the electric field is shown in Fig. 2.10.

(B). Sketch the field lines for a system of two charges $4e$ and $-e$.

We saw in §2.4 that a direct approach to problems of this type (i.e. solving the differential equation for the field lines) is best

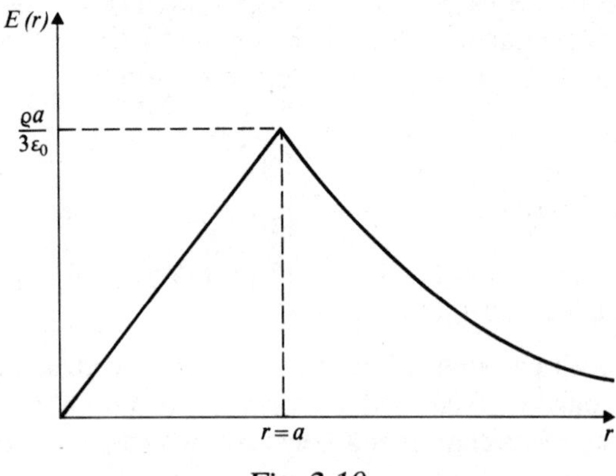

Fig. 2.10

avoided if possible. However, we can get this information directly by applying (2.23). Suppose we put the charge $4e$ at the origin, $-e$ at $(a, 0)$ and let the coordinates of the neutral point P be $(x, 0)$. Then x is determined from

$$\frac{4e}{x^2} - \frac{e}{(x-a)^2} = 0$$

i.e.

$$x = 2a$$

The field line from the origin to P will be a separatrix; in other words, lines lying inside this will end on the charge $-e$ while those outside will go to infinity.

To sketch the separatrix it would be helpful to know something of its behaviour close to the origin. We are now able to find this very simply by applying (2.23), which is the equation of the family of field lines. We are interested in one particular line, namely that through the neutral point, for which the value of the constant in (2.23) is, say, C

$$\text{i.e. } 4e \cos \theta_1 - e \cos \theta_2 = C$$

Now since P lies on this line we know that $\theta_1 = 0 = \theta_2$ so that $C = 3e$. We know too that the origin lies on this field line so that from (2.23)

$$4e \cos \theta_1 - e \cos \pi = 3e$$

$\therefore$

$$\theta_1 = \cos^{-1} \left(\tfrac{1}{2}\right)$$

allowing us to sketch the separatrix and field lines enclosed by it and without.

§2.9 Computer mapping of field lines and equipotentials

The example just considered shows how we may determine the general features of the electric field from two charges. To map the field in detail or to plot field lines and equipotentials for more complicated charge configurations is a task for a computer. A program which does this is listed below and Figs. 2.11 and 2.12 show plots of field lines (dashed lines) and equipotentials for the example considered in §2.8 and for three equal charges set at the corners of an equilateral triangle.

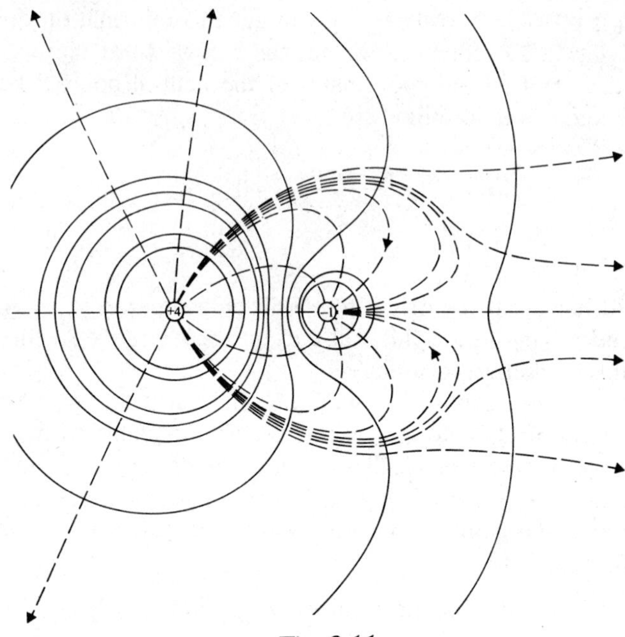

Fig. 2.11

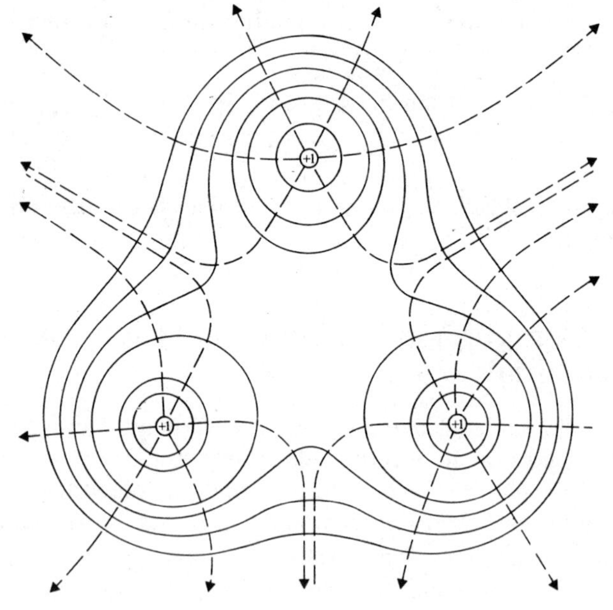

Fig. 2.12

```
      PROGRAM FIELDS
C     FIELD LINES AND EQUIPOTENTIALS
C
      DIMENSION X(100),Y(100),Q(100)
      DATA PI,KYES,KNO/3.141592654,1HY,1HN/
      DATA KEX,KEQ,KFC,KFP/1HX,1HE,1HC,1HP/
      DATA XMIN,XMAX,YMIN,YMAX/-2.0,2.0,-2.0,2.0/
      DATA E1,E2,D1,D2/.05,5.0E-3,.05,.05/
      DATA NTTY/5/
C
C     CLEAR SCREEN AND INITIALIZE DISPLAY
      CALL INITAL
C
C     SCALE SCREEN FROM XMIN TO XMAX,YMIN TO YMAX
      CALL SCAL(XMIN,XMAX,YMIN,YMAX)
C
C     READ IN DATA ABOUT CHARGES
      WRITE(NTTY,9000)
      READ(NTTY,9010)NCHARG
C
      DO 10 ICHARG=1,NCHARG
      WRITE(NTTY,9020)ICHARG
      READ(NTTY,9030)Q0,X0,Y0
C
C     PUT POINT ON SCREEN AND STORE DATA
      CALL POINT(X0,Y0)
      CALL NUM(Q0)
      Q(ICHARG)=Q0
      X(ICHARG)=X0
      Y(ICHARG)=Y0
10    CONTINUE
C
C
C     GET A COMMAND FROM THE TERMINAL
      WRITE(NTTY,9040)
20    WRITE(NTTY,9050)
      READ(NTTY,9060)ICOM
      IF (ICOM .EQ. KEX) GOTO 900
      IF (ICOM .EQ. KEQ) GOTO 100
      IF (ICOM .EQ. KFC) GOTO 200
      IF (ICOM .EQ. KFP) GOTO 300
      GOTO 10
C-----------------------------------------------------------------
100   CONTINUE
C     SET TO DRAW EQUIPOTENTIALS
      KFLD=KNO
      D1=5.0E-3
      GOTO 350
C-----------------------------------------------------------------
200   CONTINUE
C     DRAW FIELD LINE FROM A CHARGE
      KFLD=KYES
      D1=.05
      WRITE(NTTY,9070)
      READ(NTTY,9080)ICHARG,THETA
      THETA=PI*THETA/180.
      Q0=Q(ICHARG)
      X0=E1*COS(THETA)+X(ICHARG)
      Y0=E1*SIN(THETA)+Y(ICHARG)
      CALL POINT(X(ICHARG),Y(ICHARG))
      CALL VECT(X0-X(ICHARG),Y0-Y(ICHARG))
      GOTO 400
```

```
C-----------------------------------------------------------------
300   CONTINUE
C     DRAW FIELD LINE FROM ANY POINT
      KFLD=KYES
      D1=.05
350   WRITE(NTTY,9090)
      READ(NTTY,9100)X0,Y0,Q0
      CALL POINT(X0,Y0)
      ICOUNT=-1
      GOTO 400
C-----------------------------------------------------------------
400   CONTINUE
C     GENERAL PURPOSE SECTION
C     STORE STARTING POINTS
C     A 'GOTO 20' MEANS AN END TO THE LINE BEING DRAWN
      XSTART=X0
      YSTART=Y0
C     ASSUME FIELD IS ZERO
410   EX=0.
      EY=0.
C
C     WORK OUT FIELD COMPONENTS EX,EY AT X0,Y0
      DO 420 I=1,NCHARG
      X1=X0-X(I)
      Y1=Y0-Y(I)
      RCUBED=(X1*X1+Y1*Y1)**1.5
      EX=EX+Q(I)*X1/RCUBED
420   EY=EY+Q(I)*Y1/RCUBED
C
      EX=EX*SIGN(1.,Q0)
      EY=EY*SIGN(1.,Q0)
      IF (KFLD.EQ.KYES)GOTO 430
C     INTERCHANGE EX,EY FOR EQUIPOTENTIAL
C     (I.E. PERPENDICULAR TO FIELD LINE)
      S3=EX
      EX=EY
      EY=-S3
C     NEXT NORMALIZE AND CALCULATE DX,DY(D1=STEP)
430   S1=AMAX1(ABS(EX),ABS(EY))
      IF (S1.LT.E2)GOTO 20
      DX=EX*D1/S1
      DY=EY*D1/S1
C     IF IT IS A FIELD LINE PLOT A POINT
      ICOUNT =ICOUNT+1
      IF (KFLD .EQ.KYES)GOTO 440
      IF ((ICOUNT/20)*20.NE.ICOUNT)GOTO 450
440   CALL POINT(X0,Y0)
450   IF (KFLD.EQ.KYES)CALL VECT(DX,DY)
C     UPDATE CURRENT POINT
      X0=X0+DX
      Y0=Y0+DY
C
C     TEST IF OUT OF SCREEN
      IF (X0.GT.XMAX .OR. X0.LT. XMIN)GOTO 20
      IF (Y0.GT.YMAX .OR. Y0.LT.YMIN)GOTO 20
      IF (KFLD.EQ.KYES)GOTO 480
C
C     TEST IF CLOSING A CONTOUR
C     THIS PIECE IS ONLY EXECUTED FOR EQUIPOTENTIAL (KFLD='NO')
      IF (ICOUNT .LT. 10)GOTO 410
      R=SQRT((X0-XSTART)**2+(Y0-YSTART)**2)
      IF (R.LE.D2)GOTO 20
```

```
      GOTO 410
C
C     NOW TEST IF WE ARE WITHIN E1 OF ANY CHARGE
480   DO 490 I=1,NCHARG
C     NO USE TESTING FOR CHARGE ICHARG , IT WAS THE START POINT
      IF (ICHARG.EQ.I) GOTO 490
      IF (ABS(X0-X(I)).LT.E1 .AND. ABS(Y0-Y(I)).LT.E1)GOTO 500
490   CONTINUE
C     NOT NEAR A CHARGE SO TRY AGAIN WITH NEW X0 AND Y0
      GOTO 410
C     DRAW A VECTOR TO FINISH.
500   CALL VECT(X(I)-X0,Y(I)-Y0)
      GOTO 20
C-----------------------------------------------------------------
900   CALL FRAME
        CALL GREND
        STOP
C-----------------------------------------------------------------
9000  FORMAT(1H1,30HHOW MANY POINT CHARGES(<100): ,$)
9010  FORMAT(I2)
9020  FORMAT(1H ,40HGIVE THE CHARGE,XCOORD,YCOORD OF CHARGE ,
     1I2,1H:,$)
9030  FORMAT(3F8.2)
9040  FORMAT(1H ,51HTYPE ONE OF THE FOLLOWING X(EXIT),
     1,E(EQUIPOTENTIAL),/,1H ,54HC(FIELD LINE FROM A CHARGE),
     2P(FIELD LINE FROM A POINT))
9050  FORMAT(1H ,8HCOMMAND:,$)
9060  FORMAT(A1)
9070  FORMAT(1H ,45HTYPE CHARGE REF. NUMBER AND ANGLE(IN DEGREES),$)
9080  FORMAT(I2,F7.2)
9090  FORMAT(1H ,45HTYPE X,Y OF START POINT & DIRN OF LINE(+1,-1),$)
9100  FORMAT(3F7.2)
C-----------------------------------------------------------------
      END
```

§2.10 Problems

1. Estimate the relative strength of the electrostatic and gravitational interactions between a proton and an electron. [The relevant masses and charge are given in Appendix 1 and the gravitational constant $G=6.67\times10^{-12}$ newton$-$m^2/kg^2.]

2. A point charge $e_1=+10^{-7}$ coulomb is placed at the origin and a charge $e_2=-75.10^{-9}$ coulomb at the point (0.4 m, 0). Calculate the electric field at a field point A, (0.2 m, 0) and at a second field point, B, in the xy plane a distance 0.32 m in from e_1 and 0.24 m from e_2. Calculate the work done in moving a charge $e_3=+10^{-10}$ coulomb from B to A.

3. An electron of charge e(<0) and mass m is free to move in the plane of the paper. There is also a fixed charge $Z|e|$ at the origin. Show that if the only force acting on the moving charge is

the electrostatic attraction of the fixed charge then it may describe a circle of any chosen radius r about the origin provided its velocity v satisfies $mv^2 = Ze^2/4\pi\varepsilon_0 r$.

4. If the electric field is zero everywhere within a uniformly charged spherical shell, show that the electrostatic law of force must vary as the inverse square of the distance. Can you establish this without using (2.19)?

5. Two charges of opposite sign are placed in given positions. Show that the equipotential surface for which $V=0$ is spherical, whatever numerical values the charges may have. Discuss the apparently exceptional case when the two charges are equal in magnitude.

6. Electric charge is distributed on an infinite periodic lattice with density given by $\rho(x, y, z) = \rho_0 \cos lx \cos my \cos nz$. Determine the potential due to this charge distribution.

7. Find the equations of the electric field lines corresponding to charges e at $x = \pm a$. Sketch the field lines.

[*Note:* A transformation is needed to separate the variables in the differential equation for the field lines $dx/E_x = dy/E_y$. Substitute

$$u = \frac{x+a}{y}, \qquad v = \frac{x-a}{y}\Bigg]$$

8. A certain distribution of electric charge is spherically symmetric about the origin, and the total charge inside a sphere of radius r is $Q(r)$. Show that the potential $V(r)$ is given by

$$V(r) = \frac{1}{4\pi\varepsilon_0}\int_r^\infty \frac{Q(r)}{r^2}\,dr$$

Show that this may be written in the equivalent form

$$V(r) = \frac{Q(r)}{4\pi\varepsilon_0 r} + \frac{1}{\varepsilon_0}\int_r^\infty r\rho(r)\,dr$$

where $\rho(r)$ is the charge density at distance r from the origin.

9. In the ground state of a hydrogen atom the electronic charge $e(<0)$ has a density distribution given by $\rho(r) = \dfrac{e}{\pi a_0^3}\exp(-2r/a_0)$

where $a_0 = 0.529 \times 10^{-10}$ m is the Bohr radius. Show that the potential due to the electronic charge in the atom, $V_e(r)$, is

$$V_e(r) = \frac{e}{4\pi\varepsilon_0 r}(1 - e^{-2r/a_0}) - \frac{e}{4\pi\varepsilon_0 a} e^{-2r/a_0}$$

Calculate the corresponding electric field strength and also the **total** potential and electric field in the atom. Sketch $V(r)$ and $\mathbf{E}$ as functions of r.

10. A fixed circle of radius a has a charge e placed at a distance $3a/4$ from its centre, on a line through the centre perpendicular to the plane of the circle. Show that the flux of $\mathbf{E}$ through the circle is $e/5\varepsilon_0$. If a second charge e', similarly placed at a distance $5a/12$ on the opposite side of the circle, results in no net flux through the circle show that $e' = 13e/20$.

11. An infinite plane slab of thickness d is filled with charge of uniform density ρ_0 with no charge elsewhere. Find the potential V and the electric field strength $\mathbf{E}$ everywhere.

12. Show that at all finite distances from an infinite plane charged on each side with uniform surface density σ the electric field is of constant magnitude σ/ε_0.

13. Use Gauss' flux theorem to show that there is a change σ/ε_0 in the normal component of $\mathbf{E}$ on crossing a layer of charge of density σ. Deduce that if an electric field line crosses a positive layer of charge it is refracted towards the normal.

14. Obtain (2.22) for the field due to an infinite plane layer of charge of density σ, by direct summation of the contribution from each element.

15. Electric charge is distributed with constant density σ on the surface of a disc of radius a. Show that the potential at distance x away from the disc along the axis of symmetry is $\frac{\sigma}{2\varepsilon_0}\{(a^2 + x^2)^{1/2} - x\}$. Deduce the value of the electric field and, by letting a tend to infinity, reproduce (2.22) for the field due to an infinite layer of charge.

16. Electric charge is distributed on an infinite plane surface with surface density σ. P is a point distant a from the plane, and dS is a surface element whose distance from P is r. Prove that the

electric field at P has a component away from the plane equal to $\frac{a}{4\pi\varepsilon_0}\int\frac{\sigma\,dS}{r^3}$. For constant σ show that this gives a value $\sigma/2\varepsilon_0$ and deduce that in such a case one half of the field arises from those points of the plane that are less than $2a$ from P.

17. Show that the potential at the centre of a square sheet uniformly charged with charge density σ is given by

$V_0=\frac{\sigma d}{\pi\varepsilon_0}\ln(1+\sqrt{2})$ where d is length of a side of the square.

18. Electric charge is distributed at constant density e per unit length of an infinite straight line, i.e. a **line charge** of strength e. Apply Gauss' law to a cylinder of unit length and radius r coaxial with the line charge and deduce that the intensity of the electric field at distance r from the line is $(e/2\pi\varepsilon_0 r)$. Show that the potential is $-\frac{e}{2\pi\varepsilon_0}\ln r+\text{constant}$. Why is it not possible to determine the constant without further information?

19. Obtain an alternative proof of (2.23) which gives the electric field lines for a system of collinear charges, as follows. Let P and Q be two adjacent points along a certain field line so that the direction of the field at P is PQ. Write down the condition that there is no resultant field perpendicular to PQ and show that this is equivalent to the relation $\delta\left[\sum_j e_j\cos\theta_j\right]=0$.

20. For systems consisting of a charge $+ne$ at the origin and either $\pm e$ at some other point, show that in either case there is just one neutral point. Show also that with $(+ne, -e)$ the field lines which separate those lines that go from $+ne$ to infinity from those that go to $-e$ leave the positive charge at an angle

$\cos^{-1}\left(\frac{n-2}{n}\right)$ to the line joining the charges.

For $n=2$ sketch the electric field lines and a section of the equipotential surfaces.

21. Use (2.23) to show that the equation for the electric field lines due to the system of charges $-e$, $+2e$, $-e$ arranged collinearly is $r=C(\sin^2\theta|\cos\theta|)^{1/2}$ provided r is large compared with

the charge separation. Sketch the electric field lines for this configuration.

22. Use the program provided in §2.9 (or a variation of it) to plot the electric field lines for the arrangement of charges in problem 21 (known as a linear quadrupole).

23. Equal charges $+e$ are fixed at the four corners of a square of side $\sqrt{2}a$. A fifth charge $+e$, whose mass is m, is placed at the centre of the square and is free to move. Show that it is in equilibrium at this point and that the equilibrium is stable for all small displacements in the plane of the charges, the period of small oscillations in any direction in this plane being $2\pi\sqrt{2\pi\varepsilon_0 ma^3/e^2}$, but that it is unstable with respect to motion perpendicular to the plane.

24. Using the charge density for the electron in a hydrogen atom given in problem 9 verify that the electron–proton interaction energy U is given by $U = -e^2/4\pi\varepsilon_0 a_0$ where a_0 is the Bohr radius.

25. In example A in §2.8 we showed that the electric field $E(r)$ due to charge distributed symmetrically throughout a sphere of radius a is the same for $r > a$ as if all the charge were concentrated at the origin. Is this a consequence of the spherically symmetric distribution or the inverse square nature of Coulomb's law?

26. Use the program provided in §2.9 (or a variation of it) to plot electric field lines and equipotentials for charge distributions of your own choosing.

3

Electrostatics of conductors and dipoles

§3.1 Conductors and dielectrics

We know from the elementary physics of electricity that materials, as far as their electrical properties are concerned, form two main categories – **conductors** and **insulators** or **dielectrics**. A conductor is a substance in which the carriers of charge are essentially free; in the context of this book they will almost invariably be electrons. An insulator or dielectric material, on the other hand, is one in which the charges are bound, more or less strongly, to the molecules composing the dielectric. If we apply an electric field to a conductor we produce a steady drift of charge through the material; this constitutes an electric current and will be discussed in Chapter 4. Repeating this with a dielectric produces a different picture in which the charges, while suffering some displacement on account of the electric field, remain attached to their parent molecules; dielectric effects will be discussed in Chapter 6. One may think of matter, in terms of its electrical properties, distributed over a spectrum. At one end there are perfect conductors which offer no resistance to the movement of charge. The phenomenon of **superconductivity** has been recognized since 1911 when experiments by Onnes showed that the resistance of mercury vanished at 4.2 K. Since then many substances, both elements and compounds, have been found to behave as superconductors, with a range of critical temperatures. Recently, for example, superconducting behaviour at temperatures of around 60 K has been claimed for an organic conductor TTF–TCNQ. At the other end of the spectrum we find very strong dielectrics or perfect insulators. Between these two extremes lies a great variety of electrical material including **semiconductors** on which a whole technology is based. A discussion of superconductors and semiconductors demands more than classical electrodynamics and so falls outside the scope of this book. We begin by discussing electrostatic fields near charged conductors.

§3.2 Electrostatics of conductors

In the last chapter we learned how to examine the electrostatics of **prescribed** distributions of charge. We now go on to see how Gauss' law enables us to obtain certain properties of the charge distribution in conductors. Since a perfect conductor is one in which there is no resistance to the movement of charge it follows that if there is any charge at a point inside such a conductor there cannot at the same time be any field at that point. Otherwise there would be an electrostatic force on the charge, causing it to move in the direction of the field. We can soon see from Gauss' law that there cannot be any charge at rest inside the conductor. For if $\rho \neq 0$, since $\nabla \cdot \mathbf{E} = \rho/\varepsilon_0$, $\nabla \cdot \mathbf{E} \neq 0$. In consequence $\mathbf{E} \neq 0$ and the distribution of charges would move under the influence of this field. We conclude that if a conductor is charged in electrostatics (i.e. no current is flowing) then the charge must reside on the surface where strong forces act to prevent it from leaving.

Furthermore, the surface of the conductor must be at the same potential everywhere otherwise charges would flow from regions of high potential to those of low potential. This would continue until the entire surface was an equipotential.

In this condition there can be no tubes of force inside the conductor; if there were, since there is no charge inside, the tubes would have to start and end on the surface. This is impossible because the potential drops continuously along a tube of force and we have just shown that the potential is the same at all points of the surface. Since there are no tubes of force inside, there is no field inside. So $\mathbf{E} = \mathbf{0}$ inside a conductor and the potential maintains the same constant value at all points of the conductor.

Since the surface of the conductor is an equipotential, it follows that the direction of the field just outside the surface is perpendicular to the surface, i.e. the electric field lines leave the surface along the normal $\hat{\mathbf{n}}$. Thus if E_s is the component of $\mathbf{E}$ in any direction tangential to the surface, $E_s = 0$. The normal component $E_n \equiv \mathbf{E} \cdot \hat{\mathbf{n}}$ may be calculated in terms of the surface charge density σ by applying Gauss' law to the Gaussian surface in Fig. 3.1. The total charge contained within this surface is $\sigma\, dS$. There is no flux of $\mathbf{E}$ across dS_1 or out of the sides of the cylinder so that the total flux of $\mathbf{E}$ out of the volume is $E_n\, dS_2$.

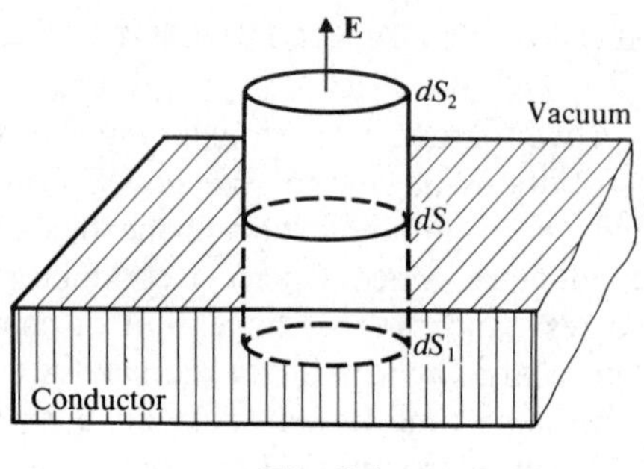

Fig. 3.1

Hence by Gauss' law, $E_n\, dS_2 = E_n dS = \dfrac{\sigma}{\varepsilon_0}\, dS$ so that the electric field just outside the surface of a conductor is

$$E_n = \sigma/\varepsilon_0$$

Denoting the field just outside the conductor by $E_{n2}\hat{\mathbf{n}}$ and that just inside by $\mathbf{E}_1(\equiv 0)$ then

$$E_{n2} - E_{n1} \equiv [\hat{\mathbf{n}} \cdot \mathbf{E}] = \frac{\sigma}{\varepsilon_0} \tag{3.1}$$

where [ ] denotes the step in the quantity within. We know that $[\hat{\mathbf{n}} \times \mathbf{E}] = 0$ (since $E_{s2} = 0$) and so it follows that

$$[\mathbf{E}] = \frac{\sigma}{\varepsilon_0}\hat{\mathbf{n}} \tag{3.2}$$

We see from this that a sheet of charge **on a conductor** gives rise to an electric field twice as big as that due to a sheet of charge on its own (which we considered in §2.7) and it is instructive to see why this is so. The charges in the neighbourhood of a point P on the surface of a conductor do give rise to a field $\sigma/2\varepsilon_0$ both inside and outside the surface. But since $\mathbf{E} = \mathbf{0}$ inside a conductor we conclude that the residual charges (i.e. those other than charges in the neighbourhood of P) are the source of an additional field at P, equal in magnitude to $\sigma/2\varepsilon_0$. The resultant field inside the conductor vanishes while that just outside becomes σ/ε_0. When we considered a sheet of charge on its own in §2.7 we specified that no other charges be present in deriving (2.22).

We have seen that $\mathbf{E} = \mathbf{0}$ inside a conductor. Now suppose that we have a cavity within the conductor as in Fig. 3.2 and that a

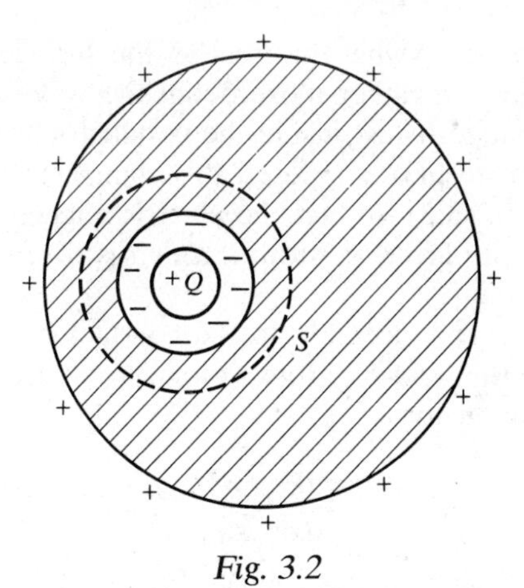

Fig. 3.2

positive charge Q is placed within the cavity. Choose a Gaussian surface S within the conductor and enclosing the cavity. Since $\mathbf{E}=\mathbf{0}$ over S it follows from

$$\int_S \mathbf{E}\cdot d\mathbf{S}=0=\frac{1}{\varepsilon_0}\times\text{enclosed charge}$$

that a charge $-Q$ has been induced on the surface of the cavity. If the conductor was initially uncharged then the net charge on it must still be zero and so a positive charge Q must exist on the outer surface of the conductor.

If the cavity does not contain any charge then $\mathbf{E}=\mathbf{0}$ within it and a uniform potential exists throughout both conductor and cavity. This property, that electric fields are excluded from regions surrounded by conducting material, is important in practice. No **static** charge distribution outside a conductor can produce electric fields within and so electrical equipment in a laboratory may, on occasions, be enclosed by a copper screen to shield it from the influence of external fields. The fact that the field vanishes is a direct consequence of Coulomb's law; if the law of force was not inverse square there would be an electric field within the cavity.

§3.3 Force on a charged conducting surface

Since a field exists at the surface of a charged conductor the charge on this surface will experience a force; this force, like the field itself, will be everywhere normal to the surface. To calculate it we must consider in rather more detail the way in which the field changes from zero just inside the conductor to σ/ε_0 just outside. Hitherto we have supposed this change to take place discontinuously, as in (3.2); in fact the electric charge on the surface occupies a very thin layer of width t which is of the order of a few atomic diameters.

Suppose that at distance x from the conductor the charge density is ρ and the field $\mathbf{E}$. From Poisson's equation, since $\mathbf{E}$ is normal to the conductor

$$\frac{dE}{dx}=\frac{\rho}{\varepsilon_0} \tag{3.3}$$

Now consider the charge at different distances inside a small cylinder of unit cross-section shown dotted in Fig. 3.3. The amount of charge between the planes $x, x+dx$ is $\rho\,dx$ so that the force on this charge is, using (3.3),

$$E\rho\,dx=\varepsilon_0 E\frac{dE}{dx}\,dx$$

The total force on the charge in this cylinder is then

$$\int_0^t \varepsilon_0 E\frac{dE}{dx}\,dx=\varepsilon_0\int_0^{\sigma/\varepsilon_0} E\,dE=\frac{\sigma^2}{2\varepsilon_0} \tag{3.4}$$

Hence the required force is $\sigma^2/2\varepsilon_0$ per unit area of surface. But since the charges cannot normally escape from the conductor, it is communicated by them to the surface itself.

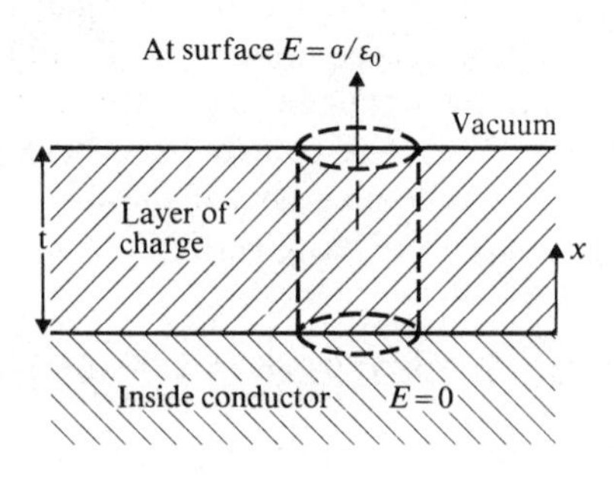

Fig. 3.3

§3.4 Energy of the electrostatic field

Consideration of the force on a charged conducting surface leads us to the concept of energy to be associated with an electric field. For electrostatic fields this energy must be solely the potential energy arising from the interaction between the charges. Our next problem is to calculate the mutual potential energy between a given set of charges $\{e_j\}$, $j=1\ldots N$. Denoting $|\mathbf{r}_i-\mathbf{r}_j|$ by r_{ij}, the mutual potential energy is simply

$$\begin{aligned}W&=\frac{1}{2}\sum_{i\neq j}\sum\frac{e_ie_j}{4\pi\varepsilon_0 r_{ij}}\\&=\frac{e_1}{8\pi\varepsilon_0}\left\{\frac{e_2}{r_{12}}+\frac{e_3}{r_{13}}+\ldots+\frac{e_N}{r_{1N}}\right\}+\frac{e_2}{8\pi\varepsilon_0}\left\{\frac{e_1}{r_{21}}+\frac{e_3}{r_{23}}+\ldots+\frac{e_N}{r_{2n}}\right\}+\ldots\\&=\tfrac{1}{2}e_1V_1+\tfrac{1}{2}e_2V_2+\ldots+\tfrac{1}{2}e_NV_N\end{aligned}\tag{3.5}$$

where V_1, $V_2\ldots V_N$ are simply the potentials at the positions of the charges $e_1, e_2\ldots e_N$ due to all the other charges.

If the charges are not localized as point charges but are distributed with volume density ρ and surface density σ the equivalent statement to (3.5) is

$$W=\frac{1}{2}\int\rho V\,d\mathbf{r}+\frac{1}{2}\int\sigma V\,dS\tag{3.6}$$

Replacing ρ using $\boldsymbol{\nabla}\cdot\mathbf{E}=\rho/\varepsilon_0$ the first term may be rewritten

$$\begin{aligned}\frac{1}{2}\int\rho V\,d\mathbf{r}&=\frac{\varepsilon_0}{2}\int V(\boldsymbol{\nabla}\cdot\mathbf{E})\,d\mathbf{r}\\&=\frac{\varepsilon_0}{2}\int[\boldsymbol{\nabla}\cdot(V\mathbf{E})-\mathbf{E}\cdot\boldsymbol{\nabla}V]\,d\mathbf{r}\\&=\frac{\varepsilon_0}{2}\int_S V\mathbf{E}\cdot d\mathbf{S}+\frac{\varepsilon_0}{2}\int\mathbf{E}\cdot\mathbf{E}\,d\mathbf{r}\end{aligned}$$

The surface integral is taken over the sphere at infinity (where a consideration of orders of magnitude shows that it vanishes for any finite set of charges) and over the surface of each conductor where

$$\frac{\varepsilon_0}{2}\int V\mathbf{E}\cdot dS=\frac{\varepsilon_0}{2}\int VE_n\,dS$$

E_n being the normal component of **E out** of V **into** each conductor. Applying (3.1) and allowing for the change in the direction of E_n (i.e. $E_n = -\sigma/\varepsilon_0$) we find

$$\frac{1}{2}\int \rho V\,d\mathbf{r} + \frac{1}{2}\int \sigma V\,dS = \frac{\varepsilon_0}{2}\int \mathbf{E}\cdot\mathbf{E}\,d\mathbf{r}$$

Hence the total electrostatic energy is

$$W = \int \frac{\varepsilon_0 E^2}{2}\,d\mathbf{r} \qquad (3.7)$$

This integration is to be performed over all space except that occupied by conductors. But since $\mathbf{E}=\mathbf{0}$ inside any conductor it may equally well be extended over all space.

Equation (3.7) is interesting because it shows that we may regard the energy of the charges as being distributed throughout all space, the **energy density** being $\frac{1}{2}\varepsilon_0 E^2$ per unit volume. This quantity is known as the energy density of the field. In the earlier work of Faraday and Maxwell, who could not understand how charges could exert forces on each other across a vacuum and who, in consequence, sought to dispense with the idea of action-at-a-distance, it was supposed that there was a medium, the aether, pervading all space. This transmitted the inverse square law from one charge to another by becoming stressed and it was argued that such a stress would require energy to create it; this was the energy $\frac{1}{2}\varepsilon_0 E^2$. Alternatively we may think of the tubes of force as being in a state of tension. Note the similarity between the energy density just obtained and the mechanical force per unit area on the surface of a conductor, $\sigma^2/2\varepsilon_0 = \frac{1}{2}\varepsilon_0\left(\frac{\sigma}{\varepsilon_0}\right)^2 = \frac{1}{2}\varepsilon_0 E^2$ from (3.2).

Although the concept of the aether is no longer valid, it is still convenient to regard the energy of this kind of electrostatic system as in some way residing in the medium with density $\frac{1}{2}\varepsilon_0 E^2$.

Suppose we use (3.7) to compute the electrostatic energy associated with a single point charge e. Substituting $\mathbf{E} = \dfrac{e\mathbf{r}}{4\pi\varepsilon_0 r^3}$ and taking $d\mathbf{r}$ to be a spherical shell, thickness $d\mathbf{r}$, gives

$$W = \frac{\varepsilon_0}{2}\int E^2\,d\mathbf{r} = \frac{\varepsilon_0}{2}\int_0^\infty \left(\frac{e}{4\pi\varepsilon_0}\right)^2 \frac{1}{r^4}\cdot 4\pi r^2\,d\mathbf{r}$$

$$= -\frac{e^2}{8\pi\varepsilon_0}\left[\frac{1}{r}\right]_{r=0}^{r=\infty}$$

We see that in this case (3.7) predicts an infinite **self-energy** for a point charge. This is a physically absurd result arising from the point-like nature of the charge. So we are obliged to conlude that the concept of an energy density of the electrostatic field is incompatible with that of point charges. An obvious way out of the difficulty over self-energy might appear to lie in the direction of a finite radius for the electron (problem 7). However, this escape route involves problems in electrodynamics when moving charges are considered.

§3.5 Electric dipoles

It frequently happens that we have a pair of equal and opposite charges $\pm e$ at a constant small separation l. Such a combination is called an **electric dipole** and $\mathbf{p}=e\mathbf{l}$ the **electric dipole moment** where $\mathbf{l}$ is drawn as in Fig. 3.4 from $+e$ to $-e$. Most molecules, e.g. H_2O and NH_3 are electric dipoles and the theory of such systems is of great importance in physical chemistry. In characterizing the separation of the charges as "small" what we mean is that we will be interested in the electric fields from dipoles only over distances r such that $l \ll r$. To calculate the field from the dipole in Fig. 3.4 we start from the potential $V(\mathbf{r})$,

$$V(\mathbf{r})=\frac{e}{4\pi\varepsilon_0}\left[\frac{1}{\left|\mathbf{r}-\frac{\mathbf{l}}{2}\right|}-\frac{1}{\left|\mathbf{r}+\frac{\mathbf{l}}{2}\right|}\right]$$

and expand the denominators, retaining only linear terms since $l/r \ll 1$. To this approximation

$$V=\frac{e}{4\pi\varepsilon_0}\frac{\mathbf{l}\cdot\mathbf{r}}{r^3}=\frac{1}{4\pi\varepsilon_0}\frac{\mathbf{p}\cdot\mathbf{r}}{r^3} \tag{3.8}$$

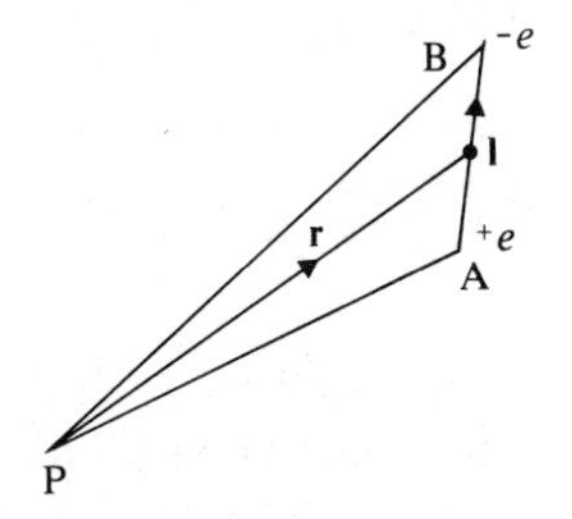

Fig. 3.4

i.e. the potential due to a dipole decreases with distance as the inverse square whereas that due to a single charge (monopole) decreases as $1/r$. It is physically reasonable that the dipole potential should fall more rapidly since as r increases the dipole appears increasingly as a null charge. The expression for **E** follows directly from

$$\mathbf{E} = -\frac{1}{4\pi\varepsilon_0}\nabla\left(\frac{\mathbf{p}\cdot\mathbf{r}}{r^3}\right)$$

$$= -\frac{1}{4\pi\varepsilon_0}\left[\frac{1}{r^3}\nabla(\mathbf{p}\cdot\mathbf{r}) + (\mathbf{p}\cdot\mathbf{r})\nabla\left(\frac{1}{r^3}\right)\right]$$

Since $\nabla(\mathbf{p}\cdot\mathbf{r}) = \mathbf{p}$, $\nabla\left(\frac{1}{r^3}\right) = -\frac{3\mathbf{r}}{r^5}$, this gives

$$\mathbf{E} = \frac{1}{4\pi\varepsilon_0 r^3}\left[\frac{3(\mathbf{p}\cdot\mathbf{r})}{r^2}\mathbf{r} - \mathbf{p}\right] \tag{3.9}$$

i.e. the electric dipole field varies as the inverse cube of the distance from the dipole. The concept of the **point dipole** is often useful. This is a fiction in which we picture $l \rightarrow 0$ in such a way that **p** remains finite.

It is often convenient to express the electric field at P in terms of its spherical polar components. In terms of the polar angle $\theta = \cos^{-1}(\hat{\mathbf{p}}\cdot\hat{\mathbf{r}})$, $V = \frac{1}{4\pi\varepsilon_0}\frac{p\cos\theta}{r^2}$ and hence

$$E_r = -\frac{\partial V}{\partial r} = \frac{p\cos\theta}{2\pi\varepsilon_0 r^3}$$

$$E_\theta = -\frac{1}{r}\frac{\partial V}{\partial\theta} = \frac{p\sin\theta}{4\pi\varepsilon_0 r^3} \tag{3.10}$$

$$E_\phi = -\frac{1}{r\sin\theta}\frac{\partial V}{\partial\phi} = 0$$

It is an easy matter in this case to write down and solve the differential equation for the electric field lines (problem 20). The field lines and equipotentials are sketched in Fig. 3.5.

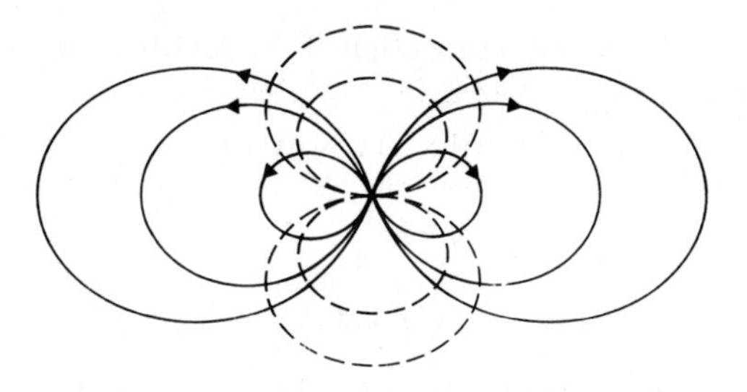

Fig. 3.5

§3.6 Forces on dipoles

Suppose that a dipole $\mathbf{p}$ is free to rotate about its centre in the presence of a constant field $\mathbf{E}$. Figure 3.6 shows a section in the plane through the dipole. The field acts from right to left and therefore exerts a force $+e\mathbf{E}$ on the charge $+e$, and $-e\mathbf{E}$ on $-e$. These are equivalent to a couple of magnitude $eEl \sin \theta$, i.e. $pE \sin \theta$. Since this is about an axis perpendicular to $\mathbf{p}$ and $\mathbf{E}$ it may be written as $\mathbf{p} \times \mathbf{E}$.

If W is the potential energy of the dipole in the field, then

$$-\frac{dW}{d\theta} = \text{couple tending to increase } \theta = -pE \sin \theta$$

So if we choose the zero of potential energy at $\theta = \pi/2$,

$$W = -pE \cos \theta = -\mathbf{p} \cdot \mathbf{E} \tag{3.11}$$

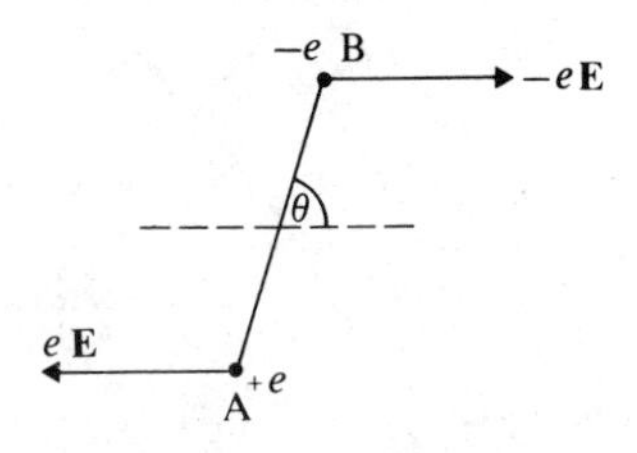

Fig. 3.6

Alternatively we may argue that if V denotes the potential due to the field $\mathbf{E}$, then

$$\begin{aligned} W &= eV_B - eV_A \\ &= el\frac{\partial V}{\partial s} \\ &= \mathbf{p}\cdot\nabla V = -\mathbf{p}\cdot\mathbf{E} \end{aligned}$$

Our next problem is to calculate the mutual potential energy of two given dipoles, $\mathbf{p}_1$ and $\mathbf{p}_2$. We may do so directly following the same method that gave us the potential due to a single dipole or, preferably, since we have done the work already, use (3.9) in (3.11). Then $W = -(\mathbf{p}_1 \cdot \mathbf{E})$, where $\mathbf{E}$ is the field due to $\mathbf{p}_2$, so that

$$W = \frac{1}{4\pi\varepsilon_0 r^3}\left[\mathbf{p}_1\cdot\mathbf{p}_2 - \frac{3(\mathbf{p}_1\cdot\mathbf{r})(\mathbf{p}_2\cdot\mathbf{r})}{r^2}\right] \qquad (3.12)$$

§3.7 Electrical double layers

In many biological and colloid problems it appears that on a given surface there are two layers of charge, of opposite sign, one just outside the other. These may cover part or the whole of the surface and together form what is known as an **electric double layer**. The strength p of the layer is defined as the product of the charge per unit area σ and the separation between the layers, t, i.e. $p = \sigma t$. The two charges $\pm\sigma\, dS$ shown in Fig. 3.7 are equivalent to a dipole of moment $\sigma t\, dS$, i.e. $p\, dS$. Thus instead of thinking in terms of layers of positive and negative charge we may

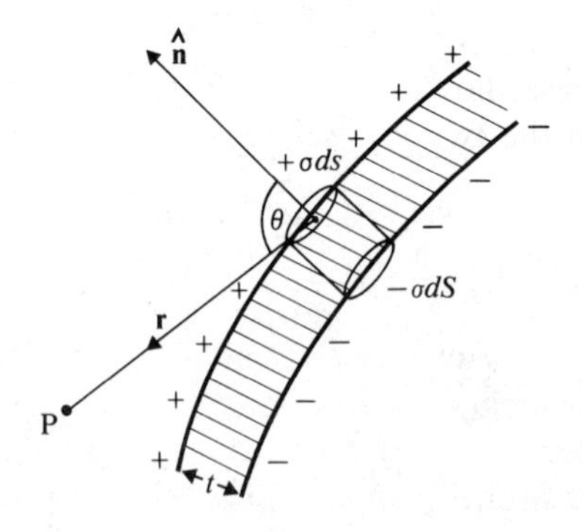

Fig. 3.7

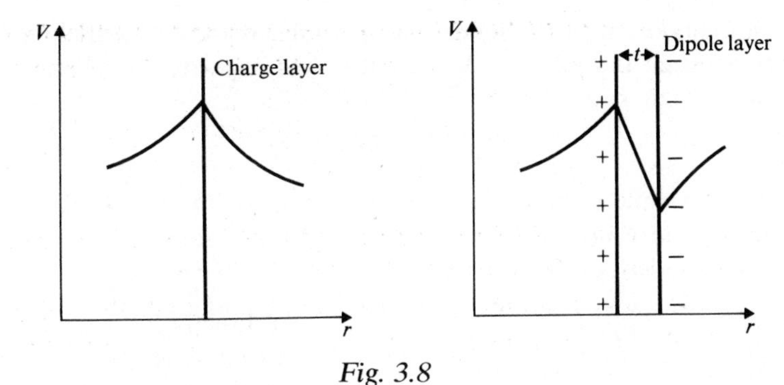

Fig. 3.8

equally well picture a layer of dipoles, all pointing normally to the surface, rather like the bristles of a hair-brush, such that the dipole moment per unit area is p.

It is straightforward to write down the potential at a point P due to an electric double layer of strength p, namely

$$V = \frac{p}{4\pi\varepsilon_0} \int \frac{\mathbf{r} \cdot d\mathbf{S}}{r^3} \tag{3.13}$$

i.e.

$$V = \frac{p\omega}{4\pi\varepsilon_0} \tag{3.14}$$

where ω is the solid angle subtended at P by the boundary curve of the double layer. The solid angle subtended by an open surface suffers a discontinuity of 4π as P crosses the surface. Consequently there is a potential discontinuity of magnitude p/ε_0 in the case of an infinitely thin double layer. We recall that in §2.7 when we considered a charge layer of density σ in the absence of all other charges, there was a discontinuity in the field amounting to σ/ε_0. In this case, however, there is no potential discontinuity. The behaviour in the two cases is illustrated in Fig. 3.8.

§3.8 Condensers (capacitors)

We conclude this chapter with a brief discussion of condensers. A condenser – also widely known as a capacitor – is a device for storing charge and consists, in its simplest form, of two plates which are conductors of arbitrary shape. The positive plate carries a charge $+Q$ and the negative plate a charge $-Q$ so that there is no net charge on the condenser as a whole. If the potential difference between the plates is V we define the **capacity** (also known as the capacitance) C by the ratio

$$C = Q/V \tag{3.15}$$

By the principle of superposition if we increase Q in a certain ratio, V is also increased in the same ratio, so that C is a constant independent of the charge on the plates. To calculate C for a given condenser we must first solve the potential equation, as in the following simple examples.

Parallel plate condenser. Consider two equal plates of area A a distance t apart as in Fig. 3.9. If we now place charges $+Q$ on one plate, $-Q$ on the other. These will be attracted by one another and will spread uniformly over the inner surfaces of the plates, establishing surface charge densities $+\sigma$, and $-\sigma$. Apart from a small correction which is needed to account for fringe effects (Fig. 3.10), the tubes of force will go straight from M to N. The electric field follows immediately from

$$\nabla \cdot \mathbf{E} = \frac{\rho}{\varepsilon_0} = \frac{Q-Q}{\varepsilon_0 A t} = 0,$$

i.e. $\mathbf{E} = \text{constant}$. The constant is given by (3.2) so that $\mathbf{E} = \dfrac{\sigma}{\varepsilon_0}\hat{\mathbf{x}}$.

To calculate the drop in potential between the plates we use this result with $\mathbf{E} = -\nabla V$, i.e.

$$\frac{dV}{dx} = -\frac{\sigma}{\varepsilon_0}$$

$$V = V_{x=0} - \frac{\sigma x}{\varepsilon_0}$$

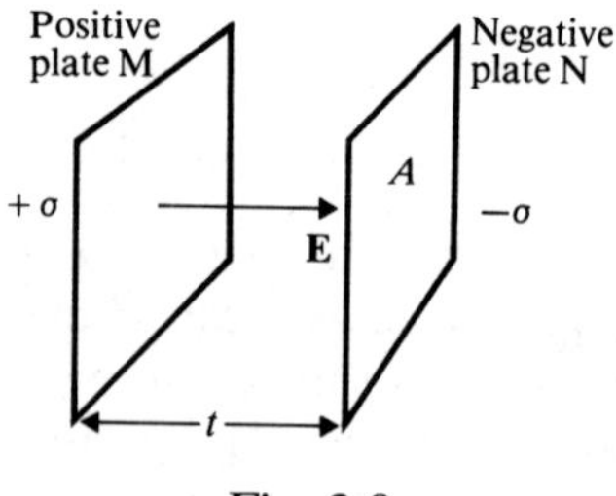

Fig. 3.9

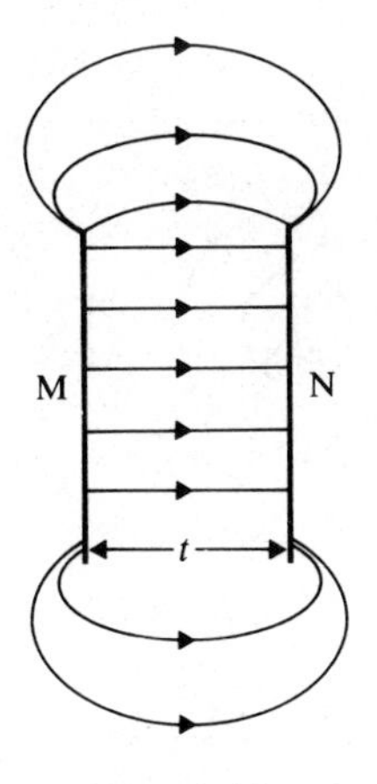

Fig. 3.10

giving the potential difference between the plates as $\frac{\sigma t}{\varepsilon_0}$. The total charge Q is simply σA and hence the capacity is $C = \frac{\varepsilon_0 A}{t}$. Evidently the smaller we make t for a given A, the larger C becomes. However in practice there is a limit on t imposed by the need to avoid electrical breakdown. High-voltage breakdown occurs when the electric field strength is such that a stray electron in the gap MN picks up enough energy to enable it to ionize a gas molecule, thus creating another electron. One gets very quickly a cascade of electrons giving rise to a spark across the gap.

Two concentric spheres. By symmetry the field is everywhere radially outwards (Fig. 3.11) and so by applying Gauss' law to the surface of radius r (shown dotted) we find

$$\mathbf{E} = \frac{Q}{4\pi\varepsilon_0 r^3}\mathbf{r} = -\nabla V$$

$$\therefore \quad V = \frac{Q}{4\pi\varepsilon_0 r} + \text{const.}$$

The difference in potential between the plates

$$V_a - V_b = \frac{Q}{4\pi\varepsilon_0}\left(\frac{1}{a} - \frac{1}{b}\right)$$

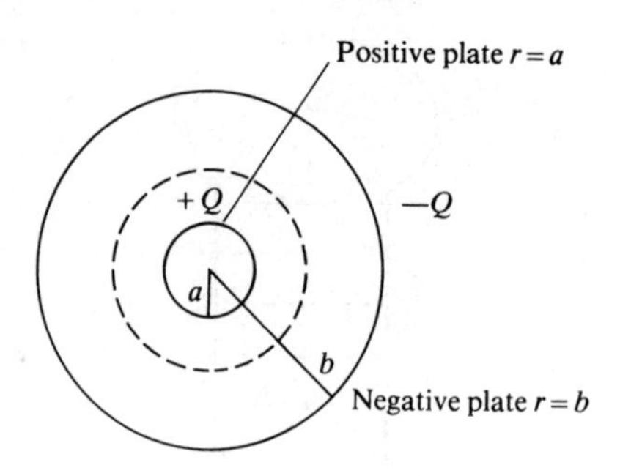

Fig. 3.11

and therefore

$$C=\frac{Q}{V_a-V_b}=4\pi\varepsilon_0\left(\frac{ab}{b-a}\right) \tag{3.16}$$

Note that if we make b very large we have a condenser, one of whose plates is a sphere of radius a and the other is effectively removed to infinity. In this limit

$$C=\lim_{b\to\infty}4\pi\varepsilon_0\left(\frac{ab}{b-a}\right)=4\pi\varepsilon_0 a$$

Condensers in series and in parallel. Arrays of condensers with their plates connected as in Fig. 3.12 are said to be **in series**. The capacity of their array is easily seen to be given by

$$\frac{1}{C}=\frac{1}{C_1}+\frac{1}{C_2}+\ldots+\frac{1}{C_n} \tag{3.17}$$

The arrangement in Fig. 3.13 on the other hand is of condensers

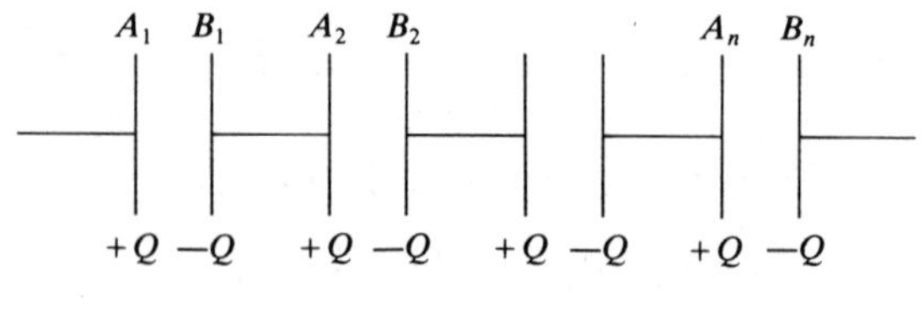

Fig. 3.12

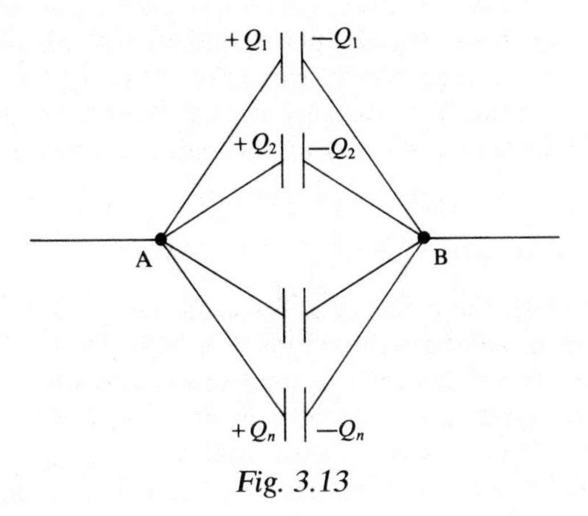

Fig. 3.13

in parallel with C now given by

$$C = C_1 + C_2 + \ldots + C_n \tag{3.18}$$

A condenser whose plates carry charges $\pm Q$ at potentials V_1 and V_2 will be a reservoir of electrical energy. The amount of energy stored is, from (3.5),

$$\sum_j \tfrac{1}{2} e_j V_j = \tfrac{1}{2}(QV_1 + (-Q)V_2) = \tfrac{1}{2}CV$$

where $V = V_1 - V_2$. This may be put in any of the equivalent forms

$$\text{Energy} = \tfrac{1}{2}QV = \tfrac{1}{2}CV^2 = \frac{Q^2}{2C} \tag{3.19}$$

One may also obtain (3.19) for simple geometries starting from (3.7) (problem 14). If the two plates are connected by a thick wire this energy will be dissipated as heat in the form of a spark at the moment of connection.

From the definition (3.15) we see that the unit of capacity is 1 coulomb volt^{-1} which is known as a **farad**. We saw that capacity is given in terms of $\varepsilon_0 \times$ a distance, so that the unit most often used for ε_0 is farad metre^{-1}. The farad is an inconveniently large unit; in practice condensers with capacities ranging from picofarads (10^{-12} farads) up to about 0.1 farad are used.

Consider again the parallel plate condenser of Fig. 3.9. The energy W may be regarded as a function either of the charge Q and plate separation t, or of the potential V and t. Calling these functions $W(Q, t)$ and $W(V, t)$ respectively, we have, from (3.19):

$$W(Q, t) = Q^2/2C = Q^2 t/2\varepsilon_0 A$$
$$W(V, t) = CV^2/2 = \varepsilon_0 AV^2/2t$$

Now, from (3.4), the mechanical force on the negative plate is $\sigma^2 A/2\varepsilon_0$ which may be written either as $+\partial W(Q, t)/\partial t$ or as $-\partial W(V, t)/\partial t$. Since this force is tending to decrease t, we may say that the force tending to increase t is either $-\partial W(Q, t)/\partial t$ or $+\partial W(V, t)/\partial t$. It is possible to show that if the electrostatic energy W can be calculated as a function of the charges or the potentials, together with a coordinate t which defines the position of one of the conductors, the mechanical force tending to increase t is always given by the two expressions we have quoted.

§3.9 Examples

(A). Find the force between the plates of a parallel plate condenser by applying the principle of virtual work.

Let F be the force between the plates whose separation is z. The electrostatic energy $U = Q^2/2C$ so that, for constant Q,

$$\delta U = \frac{Q^2}{2}\,\delta\left(\frac{1}{C}\right)$$

i.e.

$$F\delta z = \frac{Q^2}{2}\,\delta\left(\frac{1}{C}\right)$$

Since

$$C = \varepsilon_0 A/z, \qquad \delta\left(\frac{1}{C}\right) = \delta z/\varepsilon_0 A$$

$$\therefore \quad F = Q^2/2\varepsilon_0 A$$

as in §3.8.

(B). Two condensers of capacities C_1 and C_2 carry charges Q_1 and Q_2. Their negative plates are now connected to earth and their positive plates are joined. Show that the final potential of the positive plates is $(Q_1+Q_2)/(C_1+C_2)$ and that there is a loss of energy.

After connecting the plates we now have two condensers in parallel so that the new configuration has capacity $C=C_1+C_2$; also $Q=Q_1+Q_2$. The electrostatic energy of the system before connecting the plates was

$$U_i=\frac{Q_1^2}{2C_1}+\frac{Q_2^2}{2C_2}$$

and subsequently,

$$U_f=\frac{(Q_1+Q_2)^2}{2(C_1+C_2)}$$

i.e.

$$\Delta U=U_i-U_f=(Q_1C_2-Q_2C_1)^2/2C_1C_2(C_1+C_2)$$

What happens to the energy?

§3.10 Problems

1. Show that with a system of charged conductors in which the total net charge is zero, at least one conductor must be everywhere charged positively, and one everywhere negatively.

2. Point charges e_1, $e_2 \ldots$ at $A, B \ldots$ are such that the potential at A due to all the charges except e_1 is V_1, and at B due to all except e_2 is V_2, and so on. A second set of charges e_1', $e_2' \ldots$ at $A, B, \ldots$ gives potentials V_1', $V_2' \ldots$ Prove Green's Reciprocal Theorem which states that

$$e_1V_1'+e_2V_2'+\ldots=e_1'V_1+e_2'V_2+\ldots$$

Show that the same result is true if A, $B, \ldots$ are a series of conductors on which the total charges are e_1, $e_2, \ldots,$ and whose potentials are V_1, $V_2, \ldots$

Now make all the charges zero except e_1 and e_2', and reduce the second conductor to a point P. Deduce the following theorem: the potential of an uncharged conductor under the influence of a unit charge at P is the same as the potential at P due to a unit charge placed on the conductor. Hence show that a unit charge placed a distance f away from the centre of an uncharged conducting sphere raises the latter to potential $1/f$.

3. Consider a point charge e placed between two concentric conducting spheres at a distance r from the centre such that $a<r<b$ where a, b are the radii of the spheres. If the outer sphere is earthed show that the charge e' induced on the inner sphere is

$$e' = -ea(b-r)/r(b-a).$$

4. A spherical soap bubble is blown on the end of a thin tube, and is then given an electric charge Q. The tube is now made open to the air, so that the air pressure is the same inside and outside. It is known that if the radius is r, the effect of surface tension is to give an inward pressure $2T/r$ per unit area. If this is balanced by the mechanical force due to the charge, so that the bubble is in equilibrium, show that $Q^2 = 64\pi^2\varepsilon_0 Tr^3$.

5. S is a closed surface drawn in such a way that it encloses no electric charges. Show that according to Maxwell's theory the energy stored in that part of the field lying inside S may be written

$$\frac{\varepsilon_0}{2}\int V\frac{\partial V}{\partial n}\,dS$$

the integration being over the surface of S.

6. Verify (3.7) for the case of a conducting sphere of radius a which receives a charge Q.

7. A conducting sphere of radius a receives a charge Q. Verify that the potential is $Q/4\pi\varepsilon_0 r$, $r>a$ and $Q/4\pi\varepsilon_0 a$, $r\leq a$. Determine the electric field $\mathbf{E}$ at any point and by evaluating $\frac{1}{2}\varepsilon_0\int E^2\,d\mathbf{r}$ verify that the energy of the condenser formed by the given sphere at infinity is $Q^2/8\pi\varepsilon_0 a$. Notice that this implies that a genuine point charge ($a\to 0$) would have infinite self-energy.

8. Three concentric hollow conducting spheres have radii a, b, c. The inner and outer spheres are connected together by a fine wire, and form one plate of a condenser; the middle sphere is the other plate. By regarding this system as two separate condensers in parallel, or otherwise, show that the capacity is $b^2(c-a)/(b-a)(c-b)$.

9. Calculate the capacity of an orbiting satellite of diameter 3 m.

10. A conducting sheet of thickness d is placed between (and parallel to) the two plates of a parallel plate condenser, which are

a distance t apart. Show that this has the effect of increasing the capacity of the condenser by a factor $t/(t-d)$ per unit area. Consider the case in which $d \simeq t$.

11. A spherical condenser is made of two conductors radii a and b. The inner sphere carries a constant charge Q and the potential of the outer is zero. Under internal forces the outer conductor contracts from a radius b to a new value c. Show that the work done by the electric field is $Q^2(b-c)/8\pi\varepsilon_0 bc$.

12. A condenser consists of two coaxial cylinders of radii a and b each of length l. Show that its capacity is $2\pi\varepsilon_0 l/\ln(b/a)$.

13. S_1 is a fixed hollow conducting cylinder of radius b, and S_2 is a smaller solid conducting cylinder of radius a coaxial with S_1 S_2 is free to slide along its axis and a constant potential difference V is maintained between the two cylinders. Show that when a length l of S_2 is inside S_1, the electrostatic energy is $\pi\varepsilon_0 lV^2/\ln(b/a)$, and hence deduce that the movable cylinder experiences a force $\pi\varepsilon_0 V^2/\ln(b/a)$ drawing it inside the fixed cylinder.

14. Confirm (3.19) by a direct calculation using (3.7) for parallel plate and spherical condenser geometries.

15. Obtain equations (3.10) for the field due to a dipole by superposing the fields due to the two component charges, without calculating the potential as an intermediary. Notice how much easier it is to work in terms of the potential rather than the field.

16. An electric dipole of moment $\mathbf{p}$ is placed in an inhomogeneous electric field $\mathbf{E}$. Show that in addition to the couple of moment $\mathbf{p}\times\mathbf{E}$, it experiences a resultant force $(\mathbf{p}\cdot\nabla)\mathbf{E}$.

17. An electric dipole of moment $\mathbf{p}$ is free to rotate about its centre which is fixed, in the presence of a constant electric field $\mathbf{E}$. If its moment of inertia about the centre is I, and if we may assume that the only force on the doublet is the electrostatic couple due to the field, prove that small oscillations about its equilibrium position are simple harmonic, with period $2\pi\sqrt{(I/pE)}$.

18. Two equal charges e are at opposite corners of a square of side a, and an electric dipole of moment $\mathbf{p}$ is at a third corner

pointing towards one of the charges. If $p = 2\sqrt{2}ea$, show that the field strength at the fourth corner of the square is $\sqrt{\frac{17}{2}}\frac{e}{4\pi\varepsilon_0 a^2}$.

19. A and B are two electric dipoles such that the direction of A passes through B, and the direction of B is perpendicular to that of A. Show that the actual force exerted by A on B is not in the same direction as the actual force exerted by B on A. Explain why this does not contradict Newton's third law of motion, that action and reaction are equal and opposite.

20. An electric dipole is at the origin and its direction is that of the polar axis. Show that the electric field lines are given by $\sin^2\theta/r = \text{constant}$. The electric field lines and the equipotentials are sketched in Fig. 3.5.

21. A system of charges consists of $+2e$ at the origin and $-e$ at the two points $(\pm a, 0, 0)$. Show that at distances from the origin much greater than a, the potential may be written in the approximate form

$$V = -\frac{ea^2}{4\pi\varepsilon_0 r^3}(3\cos^2\theta - 1)$$

This is a quadrupole configuration, and is important in studying the forces between atoms and molecules. Refer back to problems 21 and 22 in §2.10.

22. An electric double layer of uniform strength p is distributed over the surface of a sphere. Show that the potential is zero at all external points, but has the value p/ε_0 at all internal points.

23. An electric double layer of uniform strength p is spread over one side of a circular disc of radius a. Show that at points distant x from the layer along the line of symmetry perpendicular to the plane of the disc, the electric field is $pa^2/2\varepsilon_0(a^2+x^2)^{3/2}$.

24. If the law of force between charges e_1 and e_2 was $e_1e_2/4\pi\varepsilon_0 r^n$ show that the potential due to a charge e would be $e/4\pi\varepsilon_0(n-1)r^{n-1}$ and the potential due to an electric dipole of moment $\mathbf{p}$ would be

$$(\mathbf{p}\cdot\mathbf{r})/4\pi\varepsilon_0 r^{n+1}$$

25. If a prescribed total charge is distributed over a set of fixed conducting surfaces show that the electrostatic energy in the reg-

ion bounded by the surfaces is minimum when the charges are distributed in such a way that each surface is an equipotential. [This result is known as **Thomson's theorem.**]

26. If a prescribed total charge is distributed over a set of fixed conducting surfaces show that the introduction of an uncharged conductor into the region bounded by the surfaces lowers the electrostatic energy.

27. One type of transmission line consists of two parallel conducting cylinders of circular cross-section, radius b, centres a distance $2h$ apart. Show that the capacity per unit length of this two-wire transmission line is given by

$$C = \pi\varepsilon_0/\cosh^{-1}(h/b)$$

28. In a parallel-plate diode a constant potential difference is maintained between the cathode (which emits electrons) and the anode which collects them. The cathode-anode separation is d. If the electrostatic potential between the plates is given by $V = Kx^{4/3}$ (K constant and x the distance from the cathode) calculate (a) the charge density $\rho(x)$, $0<x<d$ and (b) the surface charge density on the cathode and (c) the surface charge density on the anode.

4

Steady currents

§4.1 The current vector

Up to this point we have considered only static conditions, in which electric charges were at rest. If charges are in motion which results in a net drift of charge, we say that there is a conduction current or simply an **electric current**. In metals the conduction current is carried entirely by the conduction electrons. In semiconductors it is carried by electrons and holes. In plasmas and electrolytes there are both positive and negative charges (positive ions and electrons in plasmas, positive and negative ions in electrolytes) so that the total conduction current is the algebraic sum of the individual currents. However in a plasma, since the ions are massive compared with the electrons and therefore correspondingly sluggish, virtually all the current is carried by the electrons. In this chapter we shall only be concerned with **steady currents**, that is, situations in which the current flowing is independent of time, though it may vary spatially.

Current, which we will denote by I, is defined as the rate at which charge crosses an area A. Thus

$$I = \frac{dQ}{dt} \qquad (4.1)$$

The unit in the SI system is the **ampere** (named after the French physicist Ampère whose great contributions to electrodynamics will be described in Chapter 5) and, from (4.1), 1 ampere = 1 coulomb second^{-1}.

Suppose that at a point P the charges (one species only) are moving with mean velocity $\mathbf{v}$; then if there are N per unit volume, each carrying charge e, we define the **current density j** at P by

$$\mathbf{j} = Ne\mathbf{v} \qquad (4.2)$$

The direction of $\mathbf{j}$ is that in which the current flows; its magnitude is a proper measure of current density, for if $d\mathbf{S}$ is an element of area, all the charge within the cylindrical element with $d\mathbf{S}$ as base and generators of length v will cross $d\mathbf{S}$ in unit time. The volume of the cylinder is $\mathbf{v} \cdot d\mathbf{S}$, so that the total charge contained in it is $Ne\mathbf{v} \cdot d\mathbf{S} = \mathbf{j} \cdot d\mathbf{S}$. As in electrostatics (see §2.5) we call this the flux of $\mathbf{j}$ across $d\mathbf{S}$. Across a surface S, the flux of $\mathbf{j}$ will be given by $\int_S \mathbf{j} \cdot d\mathbf{S}$. It is important always to distinguish clearly between the current density $\mathbf{j}$ and the current I. From (4.1)

$$I = \frac{dQ}{dt} = \frac{d}{dt}\int_S Ne\mathbf{v}t \cdot d\mathbf{S} = \int_S \mathbf{j} \cdot d\mathbf{S} \tag{4.3}$$

since $\mathbf{v}$ is steady. The unit of $\mathbf{j}$ is amperes metre^{-2}.

Since $\mathbf{j}$ is a vector quantity we shall have lines and tubes of $\mathbf{j}$ just as we have discussed lines and tubes of $\mathbf{E}$ in §2.4. We describe these as **lines and tubes of current flow**. The differential equation of these lines is, by analogy with (2.13),

$$\frac{dx}{j_x} = \frac{dy}{j_y} = \frac{dz}{j_z} \tag{4.4}$$

Next let us draw any closed surface S and consider the balance-sheet of charge inside S. If ρ is the volume density of charge the total included charge is $\int_V \rho d\mathbf{r}$ and the rate at which this is decreasing is $-\frac{d}{dt}\int_V \rho \, d\mathbf{r}$. Now we suppose the volume V to be constant and so $\frac{d}{dt}$ operates solely on ρ which is a function of position and time, i.e.

$$I = -\int_V \frac{\partial \rho}{\partial t} d\mathbf{r}$$

This decrease is due to the outward flow of charge from S, so that

$$-\int_V \frac{\partial \rho}{\partial t} d\mathbf{r} = \int \mathbf{j} \cdot d\mathbf{S}$$

Using Gauss' theorem we get the flux equation

$$\int_V \left(\nabla \cdot \mathbf{j} + \frac{\partial \rho}{\partial t}\right) d\mathbf{r} = 0$$

and since this is for arbitrary V, the integrand must vanish identically. This result

$$\frac{\partial \rho}{\partial t}+\nabla \cdot \mathbf{j}=0 \tag{4.5}$$

is the **equation of continuity**. It is based on the law of conservation of charge which states that no net charge can be created or destroyed within the surface S. There is an analogous equation in fluid dynamics which expresses the physical fact that mass is conserved.

Since we are only concerned with steady conditions in this chapter, $\frac{\partial \rho}{\partial t}=0$ and the continuity equation becomes

$$\nabla \cdot \mathbf{j}=0 \tag{4.6}$$

Clearly the flux equation needs modification if our surface S encloses any source of current. Places at which current enters or leaves a conductor are known as positive and negative **electrodes**. If we draw S to surround an electrode, the total flux out of S must equal the current I flowing from the electrode, sometimes called the **strength** of the electrode, i.e. $\int \mathbf{j} \cdot d\mathbf{S}=I$.

§4.2 Electrical conductivity

We must now consider in more detail the means whereby a current flows. There are various ways of causing charges to move. We might arrange, for example, different concentrations of charge in different regions. Alternatively if we apply an electric field to a conductor the charge carriers will respond to the field and an electric current will flow.

The physics of current flow in conductors under certain conditions may be summarized in an experimental result, **Ohm's law:** $I=V/R$, where I is the current, V the potential difference which causes the current to flow and R the resistance, characteristic of the conductor and constant provided the temperature is held constant. We saw in §4.1 that **j**, rather than I, is to be used to develop the theory of current flow. If therefore we consider a conductor of cross-section S and length l we may recast Ohm's law in terms of j and E (since $V=El$), i.e. $j=\frac{l}{RS}E$, or introducing a new constant of proportionality σ in place of l/RS, $j=\sigma E$.

Moreover if the mean velocity $\mathbf{v}$ of the charges lies in the same direction as the field $\mathbf{E}$ causing the motion, $\mathbf{j}$ and $\mathbf{E}$ are parallel vectors and we may express Ohm's law in the form

$$\mathbf{j} = \sigma \mathbf{E} \tag{4.7}$$

where σ is the **electrical conductivity**. An alternative form for Ohm's law is

$$\mathbf{E} = \eta \mathbf{j} \tag{4.8}$$

η being the **resistivity** of the conductor. For a homogeneous isotropic conductor σ and η are constants independent of position though they are of course temperature dependent. In anisotropic conductors such as a plasma in a magnetic field or certain crystals, the current may flow more easily in some directions than in others. In these situations the simple Ohm's law (4.7) has to be replaced by a tensor relation $j_i = \sigma_{ik} E_k$, i.e. a tensor conductivity has now been substituted for the scalar σ.

The SI unit of resistivity is volt metres/ampere or **ohm metres** where the ohm is defined by

$$1\ \text{ohm} = 1\ \text{volt ampere}^{-1}$$

The unit of electrical conductivity is simply the inverse, i.e. **ohm^{-1} metre^{-1}** or **mho metre^{-1}**.

Ohm's law is an approximation to the observed behaviour of many conducting materials. It is in that sense an empirical and not a fundamental law like those we met in electrostatics. Nonetheless it is a very good approximation not only for a wide class of materials but also for a range of field strengths. Equally it is important to realize that there are many instances when Ohm's law fails. One such is the flow of electrons between the electrodes in vacuum tubes such as the vacuum diode (see §4.7). Another common example of non-ohmic behaviour occurs at semiconductor junctions.

Although it would not make sense to attempt to **derive** Ohm's law, it is possible to understand the physics behind it by using a simple model of a plasma. Suppose we have a hydrogen plasma with N electrons (each of charge $e(<0)$, mass m^-), and N protons (each of charge $-e(>0)$, mass m^+) per cubic metre and we apply a uniform (in space) and constant (in time) electric field $\mathbf{E}$. Before the field is switched on the electrons and protons will be in random motion, their speeds determined by the temperature.

The electrons will have greater thermal speeds than the protons since they are much lighter ($m^-/m^+ \simeq 1/1\,836$).

If there were no collisions between the electrons and ions then each electron would acquire a directed velocity $e\mathbf{E}t/m^-$ and each ion, $-e\mathbf{E}t/m^+$ in addition to their thermal speeds. The result would be that with particle velocities increasing indefinitely, the final state of the plasma would be two oppositely directed beams of fast particles. However, collisions between the electrons and protons produce some quite different physics. What happens is that the particles can only maintain the directed motion they acquire from the field for a limited period, namely the interval between collisions. In other words each collision wipes out any memory of the direction of motion before impact. (In a real plasma the situation is not quite so clear cut since the collisions which take place are not at all like those between, say, billiard balls in which large momentum changes occur. Collisions between charged particles – called Coulomb collisions – involve predominantly small changes in momentum so that in practice several would be needed to wipe out the memory.)

Between collisions a particle travels one **mean free path** and the velocity acquired from the field is usually quite small compared with the thermal velocity. We may think of the drift velocity arising from the **E** field as a small component superposed on a much larger random thermal component. The current in the plasma derives from this relatively slow drift of charged particles in the direction of **E**. Moreover since the protons are so much more massive than the electrons their contribution to the total current will be neglected. We are left then with an electron drift in the plasma and the force on each electron due to the field balanced by a frictional drag due to collisions. If the mean drift velocity is **v** then the momentum associated with this ordered motion will be $m^-\mathbf{v}$. The number of collisions per second will be $1/\tau$ where τ is the time between collisions. The rate of change of momentum will therefore be $m^-\mathbf{v}/\tau$ so that in a steady state,

$$\frac{m^-\mathbf{v}}{\tau} = e\mathbf{E}$$

i.e. the average drift velocity is given by

$$\mathbf{v} = \frac{e\mathbf{E}\tau}{m^-} = -\frac{|e|\,\mathbf{E}\tau}{m^-} \qquad (4.9)$$

The current density follows immediately from $\mathbf{j} = Ne\mathbf{v} = -N|e|\,\mathbf{v}$ (since we only consider the contribution from the electrons), giving

$$\mathbf{j} = \frac{Ne^2\tau}{m^-}\mathbf{E}$$

This has the form of Ohm's law, (4.7), so that we may identify the conductivity as

$$\sigma = \frac{Ne^2\tau}{m^-}$$

Although this model has been formulated in terms of a plasma, it was in fact developed originally by Drude to explain the electrical conductivity of metals. We know now of course that the dynamics of electrons in metals has to be formulated in terms of quantum mechanics. Nonetheless this old model of electrical conduction in metals still gives reasonable order of magnitude results. For instance the conductivity of copper is 6×10^7 mho m^{-1}, the density of conduction electrons $N \simeq 2 \times 10^{28}$ m^{-3}, $e = 1.6 \times 10^{-19}$ Coulomb, $m \simeq 9 \times 10^{-31}$ kg, giving $\tau \sim 10^{-13}$ sec as the time between successive collisions.

The **mobility** μ_e of an electron (or other charge carrier) is defined as its velocity per unit field so that from (4.9) we have $\mu_e = e\tau/m^-$. The drift velocity in (4.9) may then be written

$$\mathbf{v} = \mu_e \mathbf{E}$$

and the relation between the conductivity and mobility is simply $\sigma = Ne\mu_e$, i.e.

$$\sigma = \rho\mu_e$$

The dimensions of mobility are m^2 volt^{-1} sec^{-1}.

§4.3 Differential equations of the field and flow

We next summarize the equations determining the field and flow:

(i) $$\mathbf{E} = -\nabla V \tag{4.10}$$

(ii) $$\mathbf{j} = \sigma\mathbf{E} \tag{4.7}$$

(iii) $$\nabla \cdot \mathbf{j} = 0 \tag{4.6}$$

Combining all three gives a differential equation for V

(iv) $$\nabla \cdot (\sigma \nabla V) = 0 \tag{4.11}$$

For constant σ, (4.11) reduces to Laplace's equation $\nabla^2 V = 0$.

In addition we have the boundary conditions:

(v) V is continuous and at the surface of an electrode where a battery is providing charge at a definite potential,

$$V = \text{constant} \tag{4.12}$$

(vi) This last condition combined with (i) and (ii) shows that current leaves an electrode normally, and we have already seen that if the total current leaving an electrode is I, then

$$\int \mathbf{j} \cdot d\mathbf{S} = I \tag{4.13}$$

(vii) At the boundary between a conductor and an insulator or vacuum there can be no normal flow of current so that

$$\mathbf{j} \cdot \hat{\mathbf{n}} = j_n = 0 \tag{4.14}$$

Equations (i) to (vii) constitute the differential equations of steady flow.

§4.4 Heat loss

It is easy to see that there is a loss of electrical energy when a current flows, since charge is continually falling from regions of higher potential to regions of lower. We can easily calculate this loss of energy which shows itself in the generation of heat by considering what happens in a small element AB (Fig. 4.1) of a "tube of current flow" constructed in analogy with the tubes of force introduced in §2.4. In unit time a total charge $\mathbf{j} \cdot d\mathbf{S}$ flows

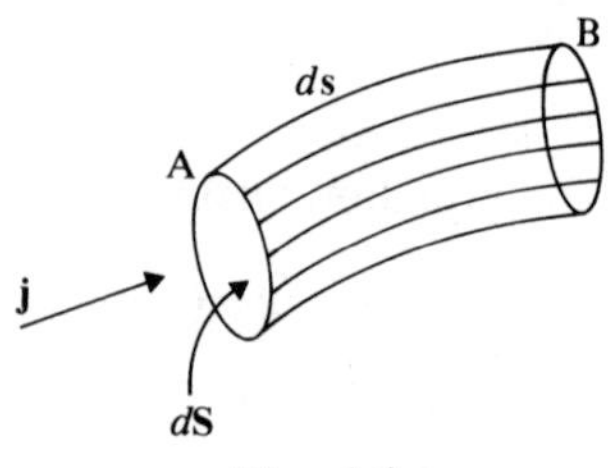

Fig. 4.1

into the volume at A and an equal charge flows out at B. The potential drop along the tube is $V_A - V_B$ and hence the rate of loss of electrical energy in this volume element is $(V_A - V_B)\mathbf{j}\cdot d\mathbf{S}$ or, since $V_A - V_B = \mathbf{E}\cdot d\mathbf{s}$ and $\mathbf{j}$ lies along $d\mathbf{s}$, this may be written $\mathbf{j}\cdot\mathbf{E}d\mathbf{r}$. The total rate of loss is therefore

$$\int \mathbf{j}\cdot\mathbf{E}\, d\mathbf{r} = \int \sigma E^2\, d\mathbf{r} \tag{4.15}$$

Now

$$\begin{aligned}\int \mathbf{j}\cdot\mathbf{E}\, d\mathbf{r} &= -\int \mathbf{j}\cdot\nabla V\, d\mathbf{r}\\ &= -\int \{\nabla\cdot(V\mathbf{j}) - V\nabla\cdot\mathbf{j}\}\, d\mathbf{r}\\ &= -\int V\mathbf{j}\cdot d\mathbf{S}\end{aligned}$$

since, by (iii), $\nabla\cdot\mathbf{j}=0$. The surface integral covers the entire surface of the conductor. At a boundary between the conductor and vacuum or an insulator, by (vii), $j_n = 0$. At each electrode, from (v) and (vi) remembering that the direction of $d\mathbf{S}$ is out of the conductor (i.e. into the electrode), we get a contribution VI where V is the electrode potential and I the current flowing out from it. The total loss of energy per unit time is $\sum VI$.

A simple interpretation of this formula is available if we realize that at an electrode of potential V from which a charge I flows in unit time, electrical energy is being provided at a rate VI. $\sum VI$ therefore represents the total rate of provision of electrical energy and hence, since the currents are steady, this must also be the rate at which electrical energy is being converted into heat energy. It is sometimes referred to as the **Joule heat loss**.

§4.5 Electromotive force

We showed in (4.15) that associated with current flow in a conductor there is an expenditure of energy at a rate $\mathbf{j}\cdot\mathbf{E}$ per unit volume. Thus steady current flow is only possible if sources of electric field are provided. A battery is just such a source of potential, the energy needed to drive a current being supplied from the chemical action between the terminals. For our purposes it is sufficient to regard a battery as a discontinuity in potential and we refer to the difference in potential between its terminals as the **electromotive force** (emf) of the battery. If we label the electric fields so provided by $\mathbf{E}'$ to distinguish them from $\mathbf{E}$, the electrostatic field which must satisfy $\nabla \times \mathbf{E} = 0$, then Ohm's law becomes

$$\mathbf{j} = \sigma(\mathbf{E} + \mathbf{E}') \tag{4.16}$$

The electromotive force, $\mathscr{E}$, is defined by

$$\mathscr{E} = \oint \mathbf{E}' \cdot d\mathbf{s} = \oint \frac{\mathbf{j} \cdot d\mathbf{s}}{\sigma} \tag{4.17}$$

since $\oint \mathbf{E} \cdot d\mathbf{s} = 0$. If we suppose that $\mathbf{j}$ is constant in the integration we might replace j by I/S so that $\mathscr{E} = I \oint ds/\sigma S$. Recalling that in (4.2) we saw that $\sigma = l/RS$, this gives $\mathscr{E} = IR$ which is the usual statement of Ohm's law, R being the resistance of the conductor.

§4.6 Networks

The resistance R depends not only on the material composing it (through σ) but also on its geometry (through l/S), and must not be confused with the resistivity η, defined by (4.8), which is solely material-dependent. Conductors characterized by resistance are themselves called resistances or resistors and denoted by -\/\/\/\/-.

Resistors may be connected to form a **network**. Suppose we have a wire AB of resistance R carrying current I. The drop in potential between its ends is RI. But if there is also a battery $\mathscr{E}$ (Fig. 4.2) the potential drop is $RI - \mathscr{E}$, since on crossing the battery in the direction from A to B there is an increase of $\mathscr{E}$ in the potential. If a set of wires are joined together then we have a network. Some or all of the wires may contain batteries.

It is an important practical problem to be able to calculate the equivalent resistance of a network but since it forms part of most

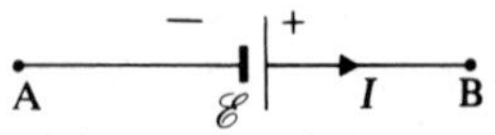

Fig. 4.2

introductory courses in electricity and is not central to the development of the theory we restrict ourselves to a statement of Kirchhoff's laws and an illustrative example. Kirchhoff's laws are two rules (rather than laws, since they contain no additional physics to what we already know) which are useful in dealing with network problems. An application of Kirchhoff's laws together with Ohm's law is sufficient to solve completely the problem of current distribution in the arms of the network. We define a **loop** to be any closed conducting path in a network of conductors, a **node** denotes any junction in the network and a **branch** to be that part of the circuit between two nodes. Then **Kirchhoff's laws** state:

I The algebraic sum of currents flowing toward a node is zero, i.e. $\sum_k i_k = 0$.

II The algebraic sum of the emfs in any loop is equal to the algebraic sum of the product of current with resistance in the same loop, i.e.

$$\sum_k \mathscr{E}_k = \sum_k i_k R_k$$

By applying these laws together with Ohm's law we can determine the i_k. The more complicated the circuit, the more tedious the algebra. In general for complicated circuits we use matrix algebra and the approach adopted is outlined using the Wheatstone bridge circuit shown in Fig. 4.3 as an example.

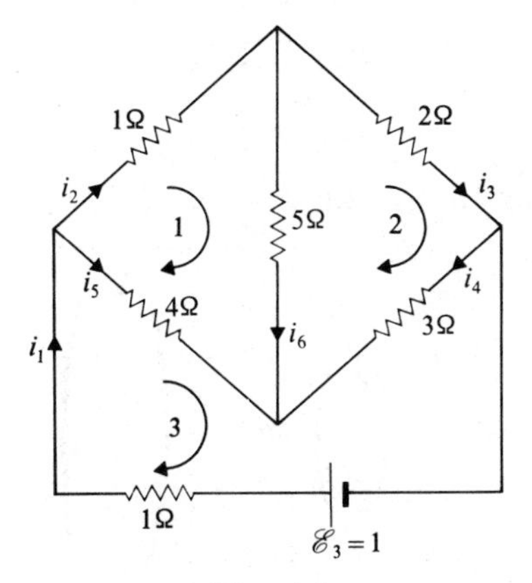

Fig. 4.3

Denote the branch currents by i_k, $k = 1 \ldots 6$. We may introduce the concept of currents in each of the 3 loops in the circuit and denote these by j_1, j_2, j_3; the loop currents are taken in a clockwise direction. It follows that the $\{j_k\}$ are related to the $\{i_k\}$ according to

$$\begin{aligned} i_1 &= j_3 & i_4 &= j_2 - j_3 \\ i_2 &= j_1 & i_5 &= -j_1 + j_3 \\ i_3 &= j_2 & i_6 &= j_1 - j_2 \end{aligned}$$

which may be expressed in matrix form as

$$\begin{pmatrix} i_1 \\ i_2 \\ i_3 \\ i_4 \\ i_5 \\ i_6 \end{pmatrix} = \begin{pmatrix} 0 & 0 & 1 \\ 1 & 0 & 0 \\ 0 & 1 & 0 \\ 0 & 1 & -1 \\ -1 & 0 & 1 \\ 1 & -1 & 0 \end{pmatrix} \begin{pmatrix} j_1 \\ j_2 \\ j_3 \end{pmatrix} \tag{4.18}$$

$$\text{i.e. } \mathbf{I} = \mathbf{MJ} \tag{4.19}$$

We further denote the voltage drop in branch k by V_k and the emfs in the loops by $\mathscr{E}_1, \mathscr{E}_2, \mathscr{E}_3$. Applying Kirchhoff's second law gives

$$\begin{aligned} \mathscr{E}_1 &= 0 = V_2 + V_6 - V_5 \\ \mathscr{E}_2 &= 0 = V_3 + V_4 - V_6 \\ \mathscr{E}_3 &= 1 = V_1 + V_5 - V_4 \end{aligned}$$

Again, expressing this result in matrix form gives

$$\mathscr{E} = \mathbf{NV} \tag{4.20}$$

where $\mathscr{E}$ and $\mathbf{V}$ are the vectors

$$\mathscr{E} = \begin{pmatrix} \mathscr{E}_1 \\ \mathscr{E}_2 \\ \mathscr{E}_3 \end{pmatrix} \text{ and } \mathbf{V} = \begin{pmatrix} V_1 \\ V_2 \\ V_3 \\ V_4 \\ V_5 \\ V_6 \end{pmatrix}$$

Moreover we can show that $\mathbf{N} = \mathbf{M}^T$. Finally we may express Ohm's law in the form

$$\mathbf{V} = \mathbf{RI} \tag{4.21}$$

where

$$\mathbf{R} = \begin{pmatrix} 1 & 0 & 0 & 0 & 0 & 0 \\ 0 & 1 & 0 & 0 & 0 & 0 \\ 0 & 0 & 2 & 0 & 0 & 0 \\ 0 & 0 & 0 & 3 & 0 & 0 \\ 0 & 0 & 0 & 0 & 4 & 0 \\ 0 & 0 & 0 & 0 & 0 & 5 \end{pmatrix} \tag{4.22}$$

Combining (4.19), (4.20) and (4.21) gives

$$\mathscr{E} = (\mathbf{M}^T\mathbf{RM})\mathbf{J} = \mathbf{PJ}$$

so that

$$\mathbf{J} = \mathbf{P}^{-1}\mathscr{E} \tag{4.23}$$

From (4.18) and (4.22) we find that

$$\mathbf{P} = \begin{pmatrix} 10 & -5 & -4 \\ -5 & 10 & -3 \\ -4 & -3 & 8 \end{pmatrix}$$

from which

$$\mathbf{P}^{-1} = \frac{adj\,\mathbf{P}}{|\mathbf{P}|} = \frac{1}{230}\begin{pmatrix} 71 & 52 & 55 \\ 52 & 64 & 50 \\ 55 & 50 & 75 \end{pmatrix}$$

$$\therefore \mathbf{J} = \mathbf{P}^{-1}\begin{pmatrix} 0 \\ 0 \\ 1 \end{pmatrix} = \begin{pmatrix} 11/46 \\ 5/23 \\ 15/46 \end{pmatrix}$$

It follows directly that $i_6 = 1/46$ which may be checked using the result in problem 24. The method outlined is quite general. In practice for complicated circuits the matrix algebra would be done using a computer.

§4.7 Breakdown of Ohm's law

In §4.2 we noted in passing that under certain conditions the flow of current in a variety of conductors could be described by Ohm's law. Equally there are many situations in which the flow of current is not governed by Ohm's law and we discuss one such in this section, i.e. current flow in a vacuum diode. We may take as a simple model for the diode the configuration in Fig 4.4. Electrons leave the cathode ($V=0$) at $x=0$ and are accelerated towards the anode ($V=V_0$) at $x=d$. Under operating conditions space charge builds up in the region between the electrodes and limits the flow of current. We shall examine the characteristics of the diode under steady state conditions.

Denoting the electric potential at a point x by $V(x)$ and the charge density by $\rho(x)$ we have from Poisson's equation

$$\frac{d^2V}{dx^2} = -\frac{\rho}{\varepsilon_0} \tag{4.24}$$

$$j = \rho v = -\frac{I}{A} \tag{4.25}$$

where A is the area of each electrode, and

$$\tfrac{1}{2}mv^2 = eV(x) \tag{4.26}$$

with $V(0)=0$ (cathode) and $V(d)=V_0$ (anode). From (4.24)–(4.26) we have

$$\frac{d^2V}{dx^2} = \frac{I}{A\varepsilon_0}\left(\frac{m}{2eV(x)}\right)^{1/2} = KV^{-1/2} \tag{4.27}$$

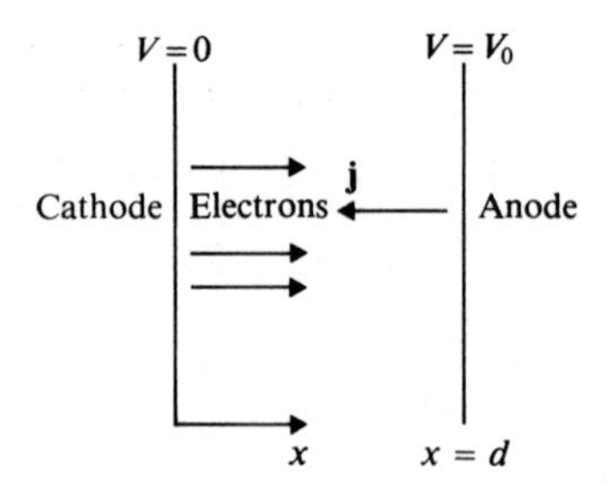

Fig. 4.4

with

$$K = \frac{I}{A\varepsilon_0}\left(\frac{m}{2e}\right)^{1/2}$$

Multiplying (4.27) by $\dfrac{dV}{dx}$ and integrating gives

$$\left(\frac{dV}{dx}\right)^2 = 4KV^{1/2} + B \tag{4.28}$$

If we suppose that the electric field is zero at the cathode then since $V(0)=0$, $B=0$ so that

$$\frac{dV}{dx} = 2K^{1/2}V^{1/4}$$

i.e.

$$\tfrac{2}{3}V^{3/4} = K^{1/2}x \tag{4.29}$$

$$\therefore\ V(x) = \left(\frac{3}{2}\right)^{4/3}\left(\frac{m}{2e}\right)^{1/3}\left(\frac{I}{A\varepsilon_0}\right)^{2/3}x^{4/3} \tag{4.30}$$

Expressing this as a current–voltage relation we have

$$I = \frac{4A\varepsilon_0}{9d^2}\left(\frac{2e}{m}\right)^{1/2}V_0^{3/2} \tag{4.31}$$

Thus we see that $I \propto V^{3/2}$ so that the behaviour of a space charge limited diode is non-ohmic. The diode characteristics are sketched in Fig. 4.5.

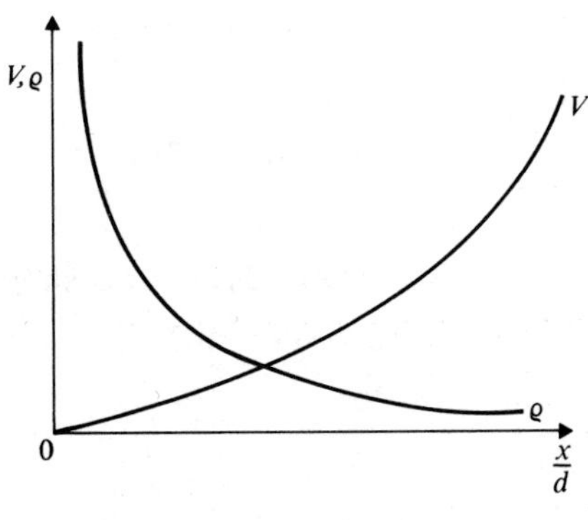

Fig. 4.5

§4.8 Examples

(A). A long line of telegraph wire $A_0A_1 \ldots A_nA_{n+1}$ is supported by n equidistant insulators at $A_1 \ldots A_n$. One end A_0 is connected to a battery, emf $\mathscr{E}$, and internal resistance R_B. The other pole of the battery and the end A_{n+1} of the telegraph wire are earthed. The resistance of each section (including A_nA_{n+1}) is R. In wet weather there is a leakage to earth at each support point, the resistance being r. Show that the current strength in A_sA_{s+1} is given by

$$\mathscr{E} \cosh(2n-2s+1)\alpha[R_B \cosh(2n+1)\alpha + \sqrt{Rr} \sinh(2n+2)\alpha]$$

where

$$2 \sinh \alpha = \sqrt{\frac{R}{r}}.$$

Applying Kirchhoff's second law at junction s:

$$\frac{V_{s-1} - V_s}{R_{s-1}} = \frac{V_s - V_{s+1}}{R_s} + \frac{V_s}{r_s}$$

and since all the R_j, r_j are equal, this gives

$$V_{s+1} - V_s\left(2 + \frac{R}{r}\right) + V_{s-1} = 0 \tag{4.32}$$

Put

$$\left(1 + \frac{R}{2r}\right) = \cosh 2\alpha \quad \text{i.e.} \quad \sinh \alpha = \frac{1}{2}\sqrt{\frac{R}{r}}$$

Then (4.32) becomes

$$V_{s+1} + V_{s-1} = 2V_s \cosh 2\alpha \tag{4.33}$$

which is satisfied by

$$V_s = A \cosh(\beta + 2s\alpha) \tag{4.34}$$

for

$$\begin{aligned} V_{s+1} + V_{s-1} &= A \cosh(\beta + [2s+2]\alpha) + A \cosh(\beta + [2s-2]\alpha) \\ &= 2A \cosh(\beta + s\alpha) \cosh 2\alpha \\ &= 2V_s \cosh 2\alpha \end{aligned}$$

Now write

$$V_s = a \cosh 2s\alpha + b \sinh 2s\alpha$$

Since $V_{n+1}=0$, this gives

$$\frac{a}{b}=-\tanh(2n+2)\alpha$$

Thus

$$V_s=\frac{a}{\sinh(2n+2)\alpha}\{\cosh 2s\alpha\sinh(2n+2)\alpha -\sinh 2s\alpha\cosh(2n+2)\alpha\}$$

$$=\frac{a\sinh(2n+2-2s)\alpha}{\sinh(2n+2)\alpha} \tag{4.35}$$

Now

$$I_{s,s+1}=\frac{V_s-V_{s+1}}{R}$$

$$=\frac{2a\cosh(2n-2s+1)\alpha\sinh\alpha}{R\sinh(2n+2)\alpha}$$

$$=a'\cosh(2n-2s+1)\alpha \tag{4.36}$$

If R_B is the internal resistance of the battery it follows that

$$\mathscr{E}-R_BI=V_0$$

i.e.

$$\mathscr{E}-R_Ba'\cosh(2n+1)\alpha=a$$

$$\therefore \mathscr{E}=R_Ba'\cosh(2n+1)\alpha+Ra'\frac{\sinh(2n+2)\alpha}{2\sinh\alpha}$$

giving finally, from (4.36)

$$I_{s,s+1}=\frac{\mathscr{E}\cosh(2n-2s+1)\alpha}{R_B\cosh(2n+1)\alpha+\sqrt{Rr}\sinh(2n+2)\alpha}$$

(B). Show that if in a closed network of conductors there are batteries of emf $\mathscr{E}_\alpha$, then the currents will distribute themselves over the network in such a way that $\sum_\alpha i_\alpha(R_\alpha i_\alpha-2\mathscr{E}_\alpha)$ is a minimum.

Consider a general variation of the network currents δi_α. Then

$$\sum_\alpha (i_\alpha+\delta i_\alpha)[R_\alpha(i_\alpha+\delta i_\alpha)-2\mathscr{E}_\alpha]$$

$$=\sum_\alpha\{i_\alpha(R_\alpha i_\alpha-2\mathscr{E}_\alpha)+2\delta i_\alpha(R_\alpha i_\alpha-\mathscr{E}_\alpha)+R_\alpha\delta i_\alpha^2\}$$

i.e. [Heat generated]$_{\text{perturbed system}}$ − [Heat generated]$_{\text{natural system}}$

$$= 2\sum_{\alpha} \delta i_{\alpha}(R_{\alpha} i_{\alpha} - \mathscr{E}_{\alpha}) + \sum_{\alpha} R_{\alpha}\delta i_{\alpha}^2$$

$$= \sum_{\alpha} R_{\alpha}\delta i_{\alpha}^2 \text{ (by Kirchhoff's second law)}$$

$$> 0$$

i.e. the natural currents are so distributed over the network that the heat generated is a minimum.

§4.9 Problems

1. Twelve equal wires of resistance R are joined at their ends to form the edges of a cube. If current enters and leaves at opposite corners of the cube, show that the resistance is $5R/6$, and if it enters and leaves at two ends of one wire the resistance is $7R/12$.

2. A set of resistances $R_1 \ldots R_n$ are joined in series. Show that the total resistance R equals $R_1 + \ldots + R_n$. But if they are joined in parallel

$$\frac{1}{R} = \frac{1}{R_1} + \frac{1}{R_2} + \ldots + \frac{1}{R_n}$$

3. Show that in steady flow the strength of a tube of flow is constant along its length.

4. Show that the lines of current flow are refracted on crossing the boundary between two media of different conductivity.

5. A conductor is in the form of a cylinder of arbitrary cross-section S bounded by two almost parallel plane sections. If t is the distance between corresponding points on the two ends, show that the resistance R is given by

$$\frac{1}{R} = \sigma \int \frac{dS}{t}$$

6. Results established previously for electrostatics have analogues in the flow of steady currents. Problem 25 in §3.10 was concerned with showing that the electrostatic energy of a charge distributed over a set of conductors was a minimum. Establish the

equivalent result for steady current flow, i.e. if given total currents flow from certain electrodes the current density in the medium distributes itself in such a way that the Joule heat loss is a minimum.

7. If, when a set of electrodes are kept at potentials $V_1, V_2, \ldots$ the currents leaving them are $I_1, I_2 \ldots$ and if when the potentials are changed to $V'_1, V'_2 \ldots$ the currents are $I'_2, I'_2 \ldots$ then $\sum_j I_j V'_j = \sum I'_j V_j$ (cf. problem 2, §3.10).

8. Use the result of example B, §4.8, applied to a closed network of conductors in which there are batteries of fixed emf to show that if any two points of a network are joined by a conductor of arbitrary resistance the total heat generated is increased. Show also that if the original network is an open one in which there are no batteries its resistance is decreased by joining any two points.

9. Use Green's reciprocal theorem (problem 7) to show that if AB and CD are two arms of a closed network containing no batteries, then the current in AB when an emf $\mathscr{E}$ is introduced in CD is the same as the current in CD when the emf is introduced in AB. If a battery in AB produces no current in CD, AB and CD are called conjugate conductors. Show that in this case a current entering the network at A and leaving at B produces no current in CD.

10. C is a closed curve in the xy plane lying entirely on the positive aide of the y-axis. This curve is now rotated through 180° about the y-axis. The volume between the curved surface so formed and the two plane ends is filled with a uniform conducting material. If when the two ends are electrodes the resistance is R, show that $\dfrac{1}{R} = \dfrac{\sigma}{\pi} \displaystyle\int \frac{dS}{x}$, where the integration is over the area embraced by the curve C.

11. A and B are opposite ends of a diameter AOB of a thin spherical shell of radius a and thickness t. Current enters and leaves by two small circular electrodes of radius c whose centres are at A and B. If I is the total current and P is a point on the shell such that the angle $POA = \theta$, show that the magnitude of the current vector at P is $I/(2\pi a t \sin\theta)$. Deduce that the resistance of the conductor is $\dfrac{1}{\pi\sigma t} \ln \cot \dfrac{c}{2a}$.

12. The space between two coaxial cylinders of radii a and b and of length t is filled with a medium of conductivity σ. Show that the resistance between the two cylinders is $\dfrac{1}{2\pi\sigma t}\ln\dfrac{b}{a}$.

13. Suppose that electrons leaving a cathode at $z=0$ with zero initial velocity are subject to a constant electric field of magnitude 1 000 volts m^{-1}. Find $\dot{z}(t)$, $z(t)$ and $\dot{z}(z)$. If 10^{14} electrons per second leave the cathode and the beam has a uniform cross section 10^{-6} m^2 determine the current density $j(z)$ and the charge density $\rho(z)$ as functions of z.

14. In §4.2 we defined the mobility of a charge carrier as its average drift velocity when subject to an electric field of unit strength. The mobilities of positive ions are important quantities in practice. The mobility of He^+ ions in pure helium at 1 atmosphere is observed to be 0.01 m^2 $volt^{-1}$ s^{-1}, whereas the mobility of electrons in helium under the same conditions is 0.10 m^2 $volt^{-1}$ s^{-1}. Calculate the values of τ_+ and τ_- corresponding to the observed mobilities.

15. In a semiconductor such as germanium, current is carried by electrons and holes. A hole may be thought of as if it had a positive charge $|e|$ and an effective mass comparable to that of the electron. The discussion in §4.2 must now be adapted to allow for the fact that the conductivity is now a function of both electron and hole densities and mobilities, i.e. $\sigma=\rho_e\mu_e+\rho_h\mu_h=N_e e\mu_e-N_h e\mu_h$. If in pure germanium the electron and hole mobilities at a temperature of 300 K are 0.36 and 0.17 m^2 $volt^{-1}$ s^{-1} respectively calculate the conductivity and determine the fraction due to electrons.

16. Show that if the conductivity of any part of a conductor is increased the resistance of the whole conductor will be decreased.

17. A square net made of wire of uniform cross-section is composed of n^2 identical square meshes. Each side of a mesh has a resistance of 1 ohm. If a current enters at one of the corners of the net and leaves from the opposite corner calculate the resistance of the net for $n=2$ and $n=3$. [Use symmetry of the circuit.]

18. Consider an electron beam of cylindrical-cross section. Suppose that the electrons move with velocity v_0 and that the current

carried by the beam is I_0. If the beam radius is a show that the electric field inside and outside the beam is given by

$$E_{in} = rI_0/2\pi\varepsilon_0 a^2 v_0 \qquad r<a$$

$$E_{out} = I_0/2\pi\varepsilon_0 r v_0 \qquad r>a$$

19. Show that the potential function $V = A\left(\frac{1}{r_1} - \frac{1}{r_2}\right)$ represents the flow of current between two small spherical electrodes in a infinite medium of uniform conductivity σ, r_1 and r_2 being the distances from the centres of the two electrodes. If the electrodes are of small radii c_1 and c_2 placed a distance a apart show that the resistance between them is approximately $R_1 + R_2$ where $R_1 = (a - c_1)/\sigma a c_1$, $R_2 = (a - c_2)/\sigma a c_2$.

This system could be used as an oceanographic probe. If the electrodes are attached to insulated cables and suspended in the sea with the circuit completed by running the insulated cables back to the ship, explain how you could measure the conductivity of sea water. Taking $c_1 = c_2 = 0.5$ m, $a = 100$ m and the length of each of the insulated cables to be 70 m estimate the conductivity of sea water from the measured resistance of 102 Ω. Assume the cables are made of copper and have a cross-section 10^{-5} m^2.

20. AB is a uniform telegraph wire. At some unknown point C of the wire there is a fault, i.e. a resistance of unknown magnitude connecting C to the earth. B is put to earth potential and an emf is applied at A. Next, A is put to earth potential and the emf is applied at B. The total resistance is measured in each case. Show how to determine the position C from these measurements.

21. In a uniform submarine cable there is a leak resistance r per unit length and the resistance of the cable is R per unit length. At distance x along the cable the potential is V and the current is I. Show that $\frac{dV}{dx} = -RI$ and $\frac{dI}{dx} = -\frac{V}{r}$. Deduce a differential equation for V, and show that if the cable is of length l the two ends being at potential V_0 and 0, then

$$V = V_0 \operatorname{cosech} \sqrt{\frac{R}{r}}\, l \sinh\left(\sqrt{\frac{R}{r}}(l-x)\right)$$

22. Under conditions of high pressure the electrons emitted by a cathode quickly reach a drift velocity determined by their mobility

and the field. In this case the current density at any point z is $j = -\rho K \dfrac{dV}{dz}$. Assuming that the field at the cathode is zero show that

$$j \simeq 10^{-9} KV^2/z^3 \text{ (amp m}^{-2}\text{)}.$$

23. Consider two resistances R_1 and R_2 connected in parallel. A current I_0 divides between them. Use the condition that $I_0 = I_1 + I_2$ together with the requirement of minimum power dissipation to show that we get the same values for the currents given by the usual circuit theory.

24. Consider the Wheatstone bridge shown in Fig. 4.3. Show that

$$i_6 = \frac{i_1(R_3R_5 - R_2R_4)}{(R_2+R_5)(R_3+R_4)+R_6(R_2+R_3+R_4+R_5)}$$

Calculate the equivalent resistance of the network.

5

Magnetostatics

§5.1 Introduction

The sciences of electricity and magnetism evolved from their beginnings until the early nineteenth century quite independently of one another. In 1820 an event occurred which changed that and sparked a period of rapid development [1]. What happened in that year was the discovery by the Danish natural philosopher, Oersted, that an electric current could exert forces precisely similar to those exerted by permanent magnets. His discovery was reported to the French Academy on 11 September 1820 and the flurry of activity following the announcement indicated that the science of electricity was ripe for development. Within a week, Ampère had shown that two parallel wires carrying currents in the same direction attract one another while, if the currents flow in opposite directions, they repel. The following month two other French scientists, Biot and Savart, reported their analysis of various experiments and formulated the law which bears their name (see §5.4).

During the next three years Ampère continued his experiments and analyses before publishing his results in a memoir later described by Maxwell as 'one of the most brilliant achievements in science'. Ampère called the science dealing with the mutual action of currents **electrodynamics** and his main conclusions may be summarized as follows:

(i) the effect of a current is reversed when its direction is reversed
(ii) the effect of a current flowing in a circuit twisted into small loops is the same as if these were removed
(iii) the force exerted by a closed circuit on an element of another circuit is at right angles to the latter
(iv) the force between two elements of circuits is unchanged when all linear dimensions are increased proportionately, the current strengths staying constant.

In addition to these data, Ampère made an assumption – incorrectly as it turned out – to obtain an expression for this force. He supposed that by analogy with the electrostatic law of interaction, the force between two elements of circuits acts along the line joining them. This derivation is presented in [1]. If one avoids making this assumption then the correct expression for the force may be derived. We shall return to this in §5.4, but to start the discussion of the magnetic effects of currents we first introduce the concept of a **magnetic field**. We have seen that Ampère's work was expressed in terms of the force experienced by one current-carrying circuit in the presence of another rather than in terms of currents and fields.

Throughout this chapter we shall deal with the magnetic effects of steady (i.e. time-independent) currents when the medium in which they occur is a vacuum. However, just as in our treatment of electrostatics in Chapters 2 and 3, there is very little difference if the medium is air. The actual difference will be discussed in Chapter 6.

§5.2 Definition of magnetic field

In Chapter 2 we defined the **electric field** strength by the ratio $\mathbf{E} = \mathbf{F}/e$ where $\mathbf{F}$ is force experienced by a test charge e initially at rest in the field. It is a natural development therefore to try to generalize the expression for electrostatic force $\mathbf{F} = e\mathbf{E}$ to include the effects of moving charges, i.e. $\mathbf{F} = e\mathbf{E} + \mathbf{F}'$. We have seen from §5.1 that magnetic effects are associated with electric currents, which in turn represent charges in motion.

The interaction of currents or charges in motion will be described in terms of a **magnetic field**, $\mathbf{B}$, which we shall define as a function of the additional force $\mathbf{F}'$ acting on a charged particle in motion in this field. It is known from very precise measurements that a test particle moving in this field experiences a force $\mathbf{F}'$ proportional to the strength of the magnetic field $\mathbf{B}$ and perpendicular to the velocity of the particle. We define $\mathbf{B}$ therefore by the relation

$$\mathbf{F}' = e\mathbf{v} \times \mathbf{B} \tag{5.1}$$

Thus the total force acting on a particle of charge e moving with velocity **v** is

$$\mathbf{F} = e[\mathbf{E} + \mathbf{v} \times \mathbf{B}] \tag{5.2}$$

We shall regard (5.2) as defining **B**. On first sight it does not appear a particularly convenient way of determining the magnetic field at any point and is a departure from the approach traditional in introductory courses (i.e. in terms of the torque on a search coil carrying a current). Naturally the various approaches are completely equivalent although the one adopted here seems preferable for a theoretical development of electrodynamics. The force **F** is in fact a key expression in classical electrodynamics and is known as the **Lorentz force**.

The units of **B** follow from (5.2); clearly the $\mathbf{v} \times \mathbf{B}$ term must have the same units as **E**, namely volt metre^{-1}. Thus the unit of **B** is volt sec. metre^{-2}. A volt sec. is known as a **weber** so that in consequence **B** is expressed in terms of webers meter^{-2}. The quantity for which the weber is the SI unit is the magnetic flux, to be defined in §5.7. The canon of the Système Internationale provides another label for the units of **B** by introducing the tesla = weber metre^{-2}. However, in practice a far more common unit for magnetic field strengths is the **gauss** where 1 gauss = 10^{-4} tesla.

A further point about **B** should be mentioned here. We have introduced **B** as the strength of the magnetic field, whereas in some books it is called the **magnetic induction**, a label that was introduced to distinguish **B** from another vector **H** (to be defined in Chapter 6). Older accounts of electrodynamics (including previous editions of this book) used **H** as the primary magnetic field and then defined **B** in terms of **H**. However, the **fundamental** magnetic field is **B** and it is unnecessary to perpetuate the other name. The relation between **B** and **H** will be discussed in Chapter 6 when we come to consider the magnetic properties of materials.

§5.3 Force on a current

Suppose now that instead of considering the force on just a single test particle we ask what happens when we have many moving charges, such as the conduction electrons in a metal or electrons in a plasma where we consider the positive carriers, the ions, to be fixed. In the latter case for a unit volume of plasma if N is the electron density we except, assuming that $\mathbf{E}=\mathbf{0}$, that

$$\mathbf{F}=Ne\mathbf{v}\times\mathbf{B}=\mathbf{j}\times\mathbf{B} \tag{5.3}$$

where $\mathbf{j}$ is the current density, introduced in (4.1). The total force acting due to the current flowing is therefore

$$\mathbf{F}_{\text{tot}}=\int_V \mathbf{j}\times\mathbf{B}\,d\mathbf{r} \tag{5.4}$$

where V is the volume of the plasma. For a plasma column – or equally for a current flowing in a wire – we may rewrite (5.4) as

$$\mathbf{F}_{\text{tot}}=\int\!\!\int(\mathbf{j}\times\mathbf{B})\,d\mathbf{s}\cdot d\mathbf{S} \tag{5.5}$$

where $d\mathbf{s}$ is an element of length of the column with its sense taken in the direction of $\mathbf{j}$ and $d\mathbf{S}$ is an element of area. Then since $\mathbf{j}$ and $d\mathbf{s}$ are parallel (5.5) may be written

$$\mathbf{F}_{\text{tot}}=\int\!\!\int(d\mathbf{s}\times\mathbf{B})(\mathbf{j}\cdot d\mathbf{S})=\int I\,d\mathbf{s}\times\mathbf{B} \tag{5.6}$$

where I is the current flowing [cf. (4.3)]. The force on an element $d\mathbf{s}$ is therefore given by

$$d\mathbf{F}=I\,d\mathbf{s}\times\mathbf{B} \tag{5.7}$$

where we have dropped the subscript on $\mathbf{F}$. If we have a current flowing in a closed loop of wire then the force on this closed circuit is simply

$$\begin{aligned}\mathbf{F}&=\oint I\,d\mathbf{s}\times\mathbf{B}\\ &=I\oint d\mathbf{s}\times\mathbf{B}\end{aligned} \tag{5.8}$$

§5.4 The Biot–Savart law

The results of Ampère's experiments on the ponderomotive forces exerted by current-carrying circuits on one another may be shown (see [1]) to lead to the following expression for the force

$$\mathbf{F}_2=\frac{\mu_0}{4\pi}I_1I_2\oint_1\oint_2\frac{d\mathbf{s}_2\times[d\mathbf{s}_1\times\mathbf{r}_{12}]}{r_{12}^3} \tag{5.9}$$

This represents the force on circuit 2 (see Fig. 5.1) due to circuit 1, hence the subscript 2 on $\mathbf{F}$; I_1 and I_2 are the currents flowing in the circuits, $d\mathbf{s}_1$, $d\mathbf{s}_2$ are respective elements of the circuits at a distance $\mathbf{r}_{12}=\mathbf{r}_2-\mathbf{r}_1$ apart. The constant factor $\mu_0/4\pi$ has been introduced to ensure compatibility between the experimental law (5.9) and the system of units (cf. Chapter 1) and is, by definition, precisely

$$\mu_0=4\pi\times10^{-7}\text{ newton amp}^{-2} \tag{5.10}$$

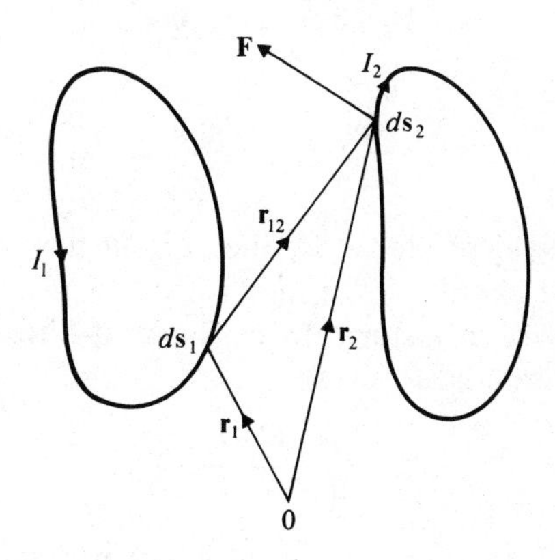

Fig. 5.1

Equation (5.9) provides the definition of the ampere. We remarked in §2.1 that the coulomb was actually defined in terms of the ampere (1 coulomb = 1 amp sec., cf. (4.1)) so that Coulomb's law itself is used to determine ε_0.

At first sight (5.9) appears not to conform to Newton's third law, on account of the asymmetry of the integrand. To demonstrate that $\mathbf{F}_2=-\mathbf{F}_1$, as it must, we expand the integrand in (5.9) to give

$$\mathbf{F}_2=\frac{\mu_0}{4\pi}I_1I_2\oint_1\oint_2\left[\frac{(d\mathbf{s}_2\cdot\mathbf{r}_{12})\,d\mathbf{s}_1}{r_{12}^3}-\frac{(d\mathbf{s}_1\cdot d\mathbf{s}_2)\mathbf{r}_{12}}{r_{12}^3}\right]$$

It is immediately apparent that the first term in the integrand is a perfect differential with respect to $d\mathbf{s}_2$, i.e.

$$\oint_2\frac{d\mathbf{s}_2\cdot\mathbf{r}_{12}}{r_{12}^3}=\oint_2\boldsymbol{\nabla}_2\left(\frac{1}{r_{12}}\right)\cdot d\mathbf{s}_2=0$$

and consequently

$$\mathbf{F}_2 = -\frac{\mu_0}{4\pi} I_1 I_2 \oint_1 \oint_2 \frac{(d\mathbf{s}_1 \cdot d\mathbf{s}_2)\mathbf{r}_{12}}{r_{12}^3} = -\mathbf{F}_1 \tag{5.11}$$

However, despite the symmetry of (5.11) it is preferable to begin our discussion of magnetostatics from (5.9). The reason is not hard to find if we recall (5.8) Clearly we may cast (5.9) immediately in this form with

$$\mathbf{F}_2 = I_2 \oint_2 d\mathbf{s}_2 \times \mathbf{B}_2 \tag{5.8a}$$

where

$$\mathbf{B}_2 = \frac{\mu_0}{4\pi} I_1 \oint_1 \frac{d\mathbf{s}_1 \times \mathbf{r}_{12}}{r_{12}^3} \tag{5.12}$$

is the magnetic field produced by the current flowing in circuit 1 at the position of circuit 2. We may regard (5.12) as a generalization of a result known historically as the **Biot–Savart law** which is expressed in differential form, i.e.

$$d\mathbf{B}_2 = \frac{\mu_0}{4\pi} \frac{I_1\, d\mathbf{s}_1 \times \mathbf{r}_{12}}{r_{12}^3} \tag{5.13}$$

The problem with this form is that it is not possible to attribute any physical meaning to it, since one cannot isolate a current element $I\,d\mathbf{s}$! It may only be interpreted as meaning that the correct $\mathbf{B}_2$ is obtained by integrating (5.13) round the complete circuit. However, the same $\mathbf{B}_2$ would be obtained if we added to (5.13) any function the integral of which round a closed path is zero.

§5.5 Field of a straight wire carrying a current

Let us consider the application of the results of §5.4 to a straight wire carrying a current I in the positive z-direction. To compute $\mathbf{B}$ at a point $P(r, \theta, 0)$ we may use (5.12)

i.e.

$$\mathbf{B}_P = \frac{\mu_0 I}{4\pi} \int_{-\infty}^{\infty} dz \frac{\hat{\mathbf{z}} \times (\mathbf{r} - \mathbf{z})}{|\mathbf{r} - \mathbf{z}|^3}$$

since $\mathbf{r}_{12} = \mathbf{r} - \mathbf{z}$. Consequently

$$\mathbf{B}_P = \frac{\mu_0 I}{4\pi} |\mathbf{r}| \hat{\boldsymbol{\theta}} \int_{-\infty}^{\infty} \frac{dz}{(r^2 + z^2)^{3/2}}$$

Setting $z = r \tan \alpha$, the integral is simply evaluated to give $2/r^2$ and so

$$\mathbf{B}_P = \frac{\mu_0 I}{2\pi r} \hat{\boldsymbol{\theta}} \tag{5.14}$$

In component form $B_r = 0$, $B_\theta = \dfrac{\mu_0 I}{2\pi r}$, $B_z = 0$, which is the usual result, familiar from elementary physics. The lines of **B** are therefore circles with centres on the wire in planes perpendicular to the current.

While it is possible to use (5.12) in highly symmetric situations like the one just considered, its use in general is rather severely circumscribed by the difficulty in evaluating the integrals. We will turn to other methods of determining **B** in §5.7.

§5.6 The laws of magnetostatics

In Chapter 2 we wrote down an expression for the electric field due to volume and surface distributions of charge, (2.4), as opposed to that for a set of discrete charges. We may take a similar step concerning the magnetic field. If instead of a current flowing in a circuit we prescribe a continuous distribution then, introducing the current density **j**, we may write the analogues to (5.8a) and (5.12) as

$$\mathbf{F} = \int \mathbf{j}(\mathbf{r}) \times \mathbf{B}(\mathbf{r})\, d\mathbf{r} \tag{5.15}$$

with

$$\mathbf{B}(\mathbf{r}) = \frac{\mu_0}{4\pi} \int \frac{\mathbf{j}(\mathbf{r}') \times (\mathbf{r} - \mathbf{r}')}{|\mathbf{r} - \mathbf{r}'|^3}\, d\mathbf{r}' \tag{5.16}$$

Now, from (5.16) form

$$\begin{aligned} \nabla \cdot \mathbf{B} &= \frac{\mu_0}{4\pi} \nabla \cdot \int_V \frac{\mathbf{j}(\mathbf{r}') \times (\mathbf{r} - \mathbf{r}')}{|\mathbf{r} - \mathbf{r}'|^3}\, d\mathbf{r}' \\ &= -\frac{\mu_0}{4\pi} \nabla \cdot \int_V \mathbf{j}(\mathbf{r}') \times \nabla \left(\frac{1}{|\mathbf{r} - \mathbf{r}'|} \right) d\mathbf{r}' \end{aligned} \tag{5.17}$$

Interchanging the order of the divergence operator and the integration and remembering that the former acts only on terms involving $\mathbf{r}$ we find, on using

$$\nabla\cdot(\mathbf{A}\times\mathbf{C})=\mathbf{C}\cdot\nabla\times\mathbf{A}-\mathbf{A}\cdot\nabla\times\mathbf{C}$$

that

$$\nabla\cdot\mathbf{B}=\frac{\mu_0}{4\pi}\int_V \mathbf{j}(\mathbf{r}')\cdot\nabla\times\nabla\left(\frac{1}{|\mathbf{r}-\mathbf{r}'|}\right)d\mathbf{r}'$$

and since

$$\nabla\times\nabla\left(\frac{1}{|\mathbf{r}-\mathbf{r}'|}\right)\equiv 0$$

then

$$\nabla\cdot\mathbf{B}=0 \tag{5.18}$$

This is one of the basic laws of magnetostatics and corresponds to the result obtained in §2.3 for the electrostatic field, $\nabla\times\mathbf{E}=0$. We may interpret (5.18) by recalling the other basic equation of vacuum electrostatics, i.e. $\nabla\cdot\mathbf{E}=\rho/\varepsilon_0$. From this it is clear that there is no analogue to electric charge in magnetostatics. In fact this will turn out to be true in general, not simply for static fields.

To complete the analogy with electrostatics let us form $\nabla\times\mathbf{B}$ starting from (5.16).

$$\begin{aligned}\nabla\times\mathbf{B}&=-\frac{\mu_0}{4\pi}\nabla\times\int_V \mathbf{j}(\mathbf{r}')\times\nabla\left(\frac{1}{|\mathbf{r}-\mathbf{r}'|}\right)d\mathbf{r}'\\&=-\frac{\mu_0}{4\pi}\int_V\left[\mathbf{j}(\mathbf{r}')\nabla^2\left(\frac{1}{|\mathbf{r}-\mathbf{r}'|}\right)-(\mathbf{j}(\mathbf{r}')\cdot\nabla)\nabla\left(\frac{1}{|\mathbf{r}-\mathbf{r}'|}\right)\right]d\mathbf{r}'\end{aligned} \tag{5.19}$$

Now

$$\nabla\left[\mathbf{j}(\mathbf{r}')\cdot\nabla\left(\frac{1}{|\mathbf{r}-\mathbf{r}'|}\right)\right]=\mathbf{j}(\mathbf{r}')\times\nabla\times\nabla\left(\frac{1}{|\mathbf{r}-\mathbf{r}'|}\right)+[\mathbf{j}(\mathbf{r}')\cdot\nabla]\nabla\left(\frac{1}{|\mathbf{r}-\mathbf{r}'|}\right)$$

the first term on the right-hand side being identically zero. Noting that

$$\nabla\left(\frac{1}{|\mathbf{r}-\mathbf{r}'|}\right)=-\nabla'\left(\frac{1}{|\mathbf{r}-\mathbf{r}'|}\right)$$

we may transform the left-hand side by writing

$$\int\left[\mathbf{j}(\mathbf{r}')\cdot\nabla'\left(\frac{1}{|\mathbf{r}-\mathbf{r}'|}\right)\right]d\mathbf{r}'=\int\left[\nabla'\cdot\left(\frac{\mathbf{j}(\mathbf{r}')}{|\mathbf{r}-\mathbf{r}'|}\right)-\frac{\nabla'\cdot\mathbf{j}(\mathbf{r}')}{\mathbf{r}-\mathbf{r}'|}\right]d\mathbf{r}'$$

The first term vanishes by using Gauss' theorem; the second is identically zero since $\nabla\cdot\mathbf{j}=0$. Thus (5.19) reduces to

$$\nabla\times\mathbf{B}=\mu_0\int_V \mathbf{j}(\mathbf{r}')\delta(\mathbf{r}-\mathbf{r}')\,d\mathbf{r}'$$

i.e.

$$\nabla\times\mathbf{B}=\mu_0\mathbf{j} \tag{5.20}$$

This second equation of magnetostatics corresponds to the differential form of Gauss' law in electrostatics, i.e. $\nabla\cdot\mathbf{E}=\rho/\varepsilon_0$. The magnetostatic counterpart of Gauss' law is readily found by using Stokes' theorem and (5.20). Then

$$\int_S \nabla\times\mathbf{B}\cdot d\mathbf{S}=\mu_0\int_S \mathbf{j}\cdot d\mathbf{S}=\oint_C \mathbf{B}\cdot d\mathbf{s}$$

where S is an open surface bounded by a closed curve C. By definition $\int_S \mathbf{j}\cdot d\mathbf{S}=I$, the current passing through C, so that

$$\oint_C \mathbf{B}\cdot d\mathbf{s}=\mu_0 I \tag{5.21}$$

which is **Ampère's law;** (5.21) may be compared with Gauss' law, (2.18). We should emphasize that Ampère's law of itself does not determine $\mathbf{B}$; we must also use $\nabla\cdot\mathbf{B}=0$.

We now have the differential equations of magnetostatics as well as electrostatics so that this is a convenient place to summarize them.

Table 5.1

Magnetostatics	Electrostatics
$\nabla\cdot\mathbf{B}=0$	$\nabla\times\mathbf{E}=0$
$\nabla\times\mathbf{B}=\mu_0\mathbf{j}$	$\nabla\cdot\mathbf{E}=\rho/\varepsilon_0$

§5.7 Magnetic field lines and flux tubes

Just as we found electric field lines and tubes of force convenient for picturing the electrostatic field in Chapters 2 and 3, their magnetic analogues are no less useful in magnetostatics. The **B** field can be mapped by lines drawn so that the tangent to the curve at any field point lies in the direction of **B** at that point. The geometry of the magnetic field lines is a matter of vital importance in many practical problems, for example the trapping of charged particles or plasma in "magnetic bottles" (see §5.11 and §8.14). For such considerations it is essential to understand the behaviour of **B** as a function of coordinates. The equation for the field lines is, by analogy with (2.13),

$$\frac{dx}{B_x} = \frac{dy}{B_y} = \frac{dz}{B_z} \tag{5.22}$$

Formally, (5.22) has two integrals

$$f(x, y, z) = \text{const}; \qquad g(x, y, z) = \text{const}$$

each of which are surfaces consisting of field lines. The intersection of any two such surfaces specifies a field line.

Magnetic tubes of force or flux tubes may be introduced in precisely the same way as in §2.4, with flux defined in the same way as §2.5, i.e. the **magnetic flux** across a surface of cross-section $d\mathbf{S}$ is defined as

$$d\Phi = \mathbf{B} \cdot d\mathbf{S}$$

and the unit is the **weber**. It follows from $\nabla \cdot \mathbf{B} = 0$ that magnetic flux is conserved inside a tube of force. A tube may branch at singular points of the field lines into two or more tubes. Using tubes of force enables us to introduce a concept which is often important in practice when considering magnetic field geometries (see §5.12). This is simply the ratio of the volume of the tube to the enclosed magnetic flux. For an element of the tube of length $d\mathbf{s}$ we have

$$\frac{d\mathbf{S} \cdot d\mathbf{s}}{\mathbf{B} \cdot d\mathbf{S}} = \frac{dS\, ds \cos\theta}{B\, dS \cos\theta} = \frac{ds}{B}$$

so that the quantity we have introduced, known as the **specific volume** of a tube of force, is simply

$$V = \int \frac{ds}{B}$$

where the integral is taken either around a closed field line or between points corresponding to the beginning and end of a tube of force.

We have already calculated the field due to a straight wire carrying a steady current in §5.5 where we found that the **B** field lines were concentric circles centred on the wires in planes perpendicular to the current. As a further example let us consider what the **B** field of a solenoid looks like.

A solenoid is a long piece of wire wound on a cylindrical surface in the form of a helix (see Fig. 5.2). Suppose that there are N turns of wire over a length L. If the length of the solenoid is much greater than its diameter then when current flows in the wire it is observed that the magnetic field outside the solenoid is much weaker than within. The fact that the **B** field is concentrated inside the coil and that $\nabla \cdot \mathbf{B} = 0$ means that, away from the ends, the field lines run parallel to the axis. The strength of the field is easily found from Ampère's law, (5.21), if we take C to be a rectangle as drawn in Fig. 5.2. Then

$$\oint_C \mathbf{B} \cdot d\mathbf{s} = BL + B_{\text{outside}} L = \mu_0 NI$$

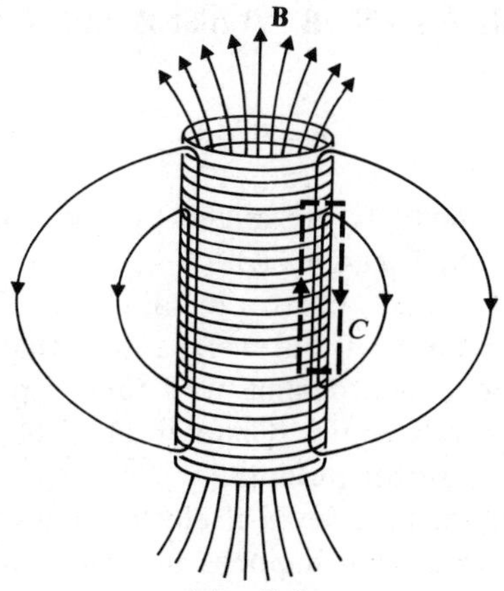

Fig. 5.2

and since $B \gg B_{\text{outside}}$,

$$B \simeq \mu_0 NI/L \tag{5.23}$$

At the ends of the solenoid the **B** lines diverge and since $\nabla \cdot \mathbf{B} = 0$ the field lines must return to the opposite end of the solenoid, joining up to give closed curves. The magnetic field is calculated in §5.13 without making the approximation which gives (5.23).

§5.8 The magnetic potentials

In §2.3 we used the fact that $\nabla \times \mathbf{E} = 0$ to express the electrostatic field in terms of the potential V as $\mathbf{E} = -\nabla V$. It is clear from Table 5.1 that an analogous situation exists in magnetostatics whenever $\mathbf{j} = 0$. Then we may write

$$\mathbf{B} = -\mu_0 \nabla V_m \tag{5.24}$$

where V_m is the **magnetic scalar potential**. Since $\nabla \cdot \mathbf{B} = 0$ it follows that the scalar potential V_m satisfies Laplace's equation

$$\nabla^2 V_m = 0$$

In §3.7 we considered the potential due to a layer of electrostatic dipoles. By analogy, the magnetic scalar potential has, mathematically, the same properties as the electric double layer.

Turning to the complete magnetostatic equations in Table 5.1 we see immediately from $\nabla \cdot \mathbf{B} = 0$ that **B** may always be expressed as

$$\mathbf{B} = \nabla \times \mathbf{A} \tag{5.25}$$

Clearly **A** is not completely prescribed by this equation since, if ψ is **any** scalar function, $\nabla \times (\mathbf{A} + \nabla\psi) = \nabla \times \mathbf{A}$. So to complete the definition we need to add an extra condition on $\nabla \cdot \mathbf{A}$. In magnetostatics we take $\nabla \cdot \mathbf{A} = 0$ but we shall see that a different choice is appropriate when dealing with time-dependent phenomena in electrodynamics (Chapter 11). The vector **A** so defined is called the **vector potential**.

In §2.6 we combined the laws of electrostatics to give Poisson's equation $\nabla^2 V = -\rho/\varepsilon_0$. What happens in magnetostatics? The answer is quickly found by substituting (5.25) into the differential version of Ampère's law, (5.20):

$$\nabla\times\nabla\times\mathbf{A}=\nabla(\nabla\cdot\mathbf{A})-\nabla^2\mathbf{A}=\mu_0\mathbf{j}$$

$$\therefore\quad \nabla^2\mathbf{A}=-\mu_0\mathbf{j}$$

This is (for each component of **A**) Poisson's equation; mathematically the component equations are identical with (2.20) and we shall turn to the solution of the equation in Chapter 7. For present purposes we may note that provided **A** vanishes at infinity and there are no surface currents (or current sheets, as they are usually called) we may by analogy with (2.8) write

$$\mathbf{A}=\frac{\mu_0}{4\pi}\int_V\frac{\mathbf{j}(\mathbf{r}')}{|r-r'|}\,d\mathbf{r}' \tag{5.26}$$

It is a straightforward exercise (problem 9) to show that **A** given by (5.26) satisfies the auxiliary condition $\nabla\cdot\mathbf{A}=0$.

Throughout this chapter in discussing the equations of magnetostatics we have referred to their electrostatic counterparts. However, having drawn this parallel, it would be misleading to suggest that the magnetic vector potential **A** is as useful in computing magnetic fields as was V in electrostatics. In §2.3 we remarked that calculating **E** is usually made easier if we first determine V and then take its gradient. However, in magnetostatics using **A** is seldom of much advantage over computing **B** directly. Its importance has to be judged in the basic role it plays in electrodynamic theory (cf. Chapter 11), and not merely as a way of finding **B**.

As an example of the use of the vector potential let us find the magnetic field due to a current flowing in a loop of radius a. Choose a system of cylindrical coordinates (r, θ, z) with the origin at the centre of the loop. The azimuthal angle θ will be measured from the plane passing through the Oz axis and the point P(r, θ, z) at which the field is to be found. It is clear that the field will be axially symmetric, i.e. it will not depend on θ. Then the components of **B** expressed in terms of the vector potential are simply

$$B_r=-\frac{\partial A_\theta}{\partial z};\qquad B_\theta=0;\qquad B_z=\frac{1}{r}\frac{\partial}{\partial r}(rA_\theta)$$

i.e. the vector potential has only one component $A_\theta(r, z)$.

If we suppose the current flowing in the loop is I then we may express (5.26) as

$$\mathbf{A}(\mathbf{r}) = \frac{\mu_0}{4\pi}\int_V \frac{\mathbf{j}(\mathbf{r}')\, d\mathbf{r}'}{|\mathbf{r}-\mathbf{r}'|} = \frac{\mu_0}{4\pi}\int_S \oint \frac{\mathbf{j}(d\mathbf{S}' \cdot d\mathbf{s}')}{|\mathbf{r}-\mathbf{r}'|}$$

i.e.

$$\mathbf{A}(\mathbf{r}) = \frac{\mu_0 I}{4\pi}\oint \frac{d\mathbf{s}'}{|\mathbf{r}-\mathbf{r}'|}.$$

From this it follows that A_θ is given by

$$A_\theta(\mathbf{r}) = \frac{\mu_0 I}{4\pi}\oint \frac{\cos\theta\, ds'}{|\mathbf{r}-\mathbf{r}'|} \tag{5.27}$$

Now $|\mathbf{r}-\mathbf{r}'| = (r^2+z^2+a^2-2ar\cos\theta)^{1/2}$ and $ds' = a\,d\theta$ so that (5.27) becomes

$$A_\theta(r, z) = \frac{\mu_0 I}{2\pi}\int_0^\pi \frac{a\cos\theta\, d\theta}{(r^2+z^2+a^2-2ar\cos\theta)^{1/2}}$$

This integral cannot be evaluated in terms of elementary functions. It may be expressed by introducing the **elliptic integrals**, $E(k)$ and $K(k)$, defined by

$$E(k) = \int_0^{\pi/2} (1-k^2\sin^2\phi)^{1/2}\, d\phi; \; K(k) = \int_0^{\pi/2} (1-k^2\sin^2\phi)^{-1/2}\, d\phi$$

in terms of which (problem 10)

$$A_\theta(r, z) = \frac{\mu_0 I}{\pi k}\sqrt{\frac{a}{r}}\left[\left(1-\frac{k^2}{2}\right)K - E\right]$$

where $k^2 = 4ar/[(a+r)^2+z^2]$.

The field near the axis of the loop ($r \ll a$) is given by the vector potential (problem 11)

$$A_\theta = \frac{\mu_0 I a^2}{4}\frac{r}{[(a+r)^2+z^2]^{3/2}}$$

§5.9 Field of a small loop: the magnetic dipole

The vector potential may also be used to advantage in determining **B** due to a small loop of current. By "small" we mean that we shall only be interested in knowing **B** at distances from the loop which are large compared to the dimensions of the loop. We shall find that any small loop is equivalent to a **magnetic dipole**; in other words it gives rise to a B field just like the $\mathbf{E}$ field found in §3.5 for an electric dipole.

Suppose that a current I is flowing in the loop; we may then replace $\mathbf{j}\,d\mathbf{r}'$ in (5.26) by $I\,d\mathbf{s}$ where $d\mathbf{s}$ is an element of the loop. Then

$$\mathbf{A}(\mathbf{r})=\frac{\mu_0 I}{4\pi}\oint\frac{d\mathbf{s}}{|\mathbf{r}-\mathbf{s}|}$$

Since we are interested in the field at large distances from the loop, i.e. $|\mathbf{r}|\gg|\mathbf{s}|$, we may expand $|\mathbf{r}-\mathbf{s}|^{-1}$ in powers of s/r. To first order

$$|\mathbf{r}-\mathbf{s}|^{-1}=\frac{1}{r}\left[1+\frac{\mathbf{s}\cdot\mathbf{r}}{r^2}+\dots\right]$$

i.e.

$$\mathbf{A}(\mathbf{r})=\frac{\mu_0 I}{4\pi}\left[\frac{1}{r^3}\oint(\mathbf{s}\cdot\mathbf{r})\,d\mathbf{s}+\dots\right] \tag{5.28}$$

since the first term in the expansion gives no contribution to $\mathbf{A}$. We may rewrite (5.28) (problem 12) in the form

$$\mathbf{A}(\mathbf{r})=\frac{\mu_0 I}{8\pi r^3}\oint(\mathbf{s}\times d\mathbf{s})\times\mathbf{r} \tag{5.29}$$

We define

$$\mathbf{m}=\frac{I}{2}\oint\mathbf{s}\times d\mathbf{s} \tag{5.30}$$

as the **magnetic moment** of the circuit so that the expression for $\mathbf{A}$ becomes

$$\mathbf{A}(\mathbf{r})=\frac{\mu_0}{4\pi}\frac{\mathbf{m}\times\mathbf{r}}{r^3} \tag{5.31}$$

Geometrically we may interpret $\mathbf{m}$ as I times the area of the loop, $\frac{1}{2}\oint\mathbf{s}\times d\mathbf{s}$.

It remains to determine $\mathbf{B}$ by calculating $\nabla\times\mathbf{A}$; from (5.31)

$$\begin{aligned}\nabla\times\mathbf{A}&=\frac{\mu_0}{4\pi}\left[\mathbf{m}\nabla\cdot\left(\frac{\mathbf{r}}{r^3}\right)-(\mathbf{m}\cdot\nabla)\frac{\mathbf{r}}{r^3}\right]\\&=-\frac{\mu_0}{4\pi}(\mathbf{m}\cdot\nabla)\frac{\mathbf{r}}{r^3}\end{aligned} \tag{5.32}$$

since $\nabla \cdot \left(\frac{\mathbf{r}}{r^3}\right) = 0$. To reduce (5.32) further note that

$$\nabla\left(\mathbf{m} \cdot \frac{\mathbf{r}}{r^3}\right) = (\mathbf{m} \cdot \nabla)\frac{\mathbf{r}}{r^3} + \mathbf{m} \cdot \nabla \times \left(\frac{\mathbf{r}}{r^3}\right)$$

The last term vanishes so that (5.32) becomes

$$\nabla \times \mathbf{A} = -\frac{\mu_0}{4\pi}\left[\frac{\mathbf{m}}{r^3} - \frac{3(\mathbf{m}\cdot\mathbf{r})\mathbf{r}}{r^5}\right]$$

i.e.

$$\mathbf{B} = \frac{\mu_0}{4\pi r^3}\left[\frac{3(\mathbf{m}\cdot\mathbf{r})\mathbf{r}}{r^2} - \mathbf{m}\right] \tag{5.33}$$

Comparing this form with (3.9) for the electric dipole field we see that the formulae are structurally identical. As a result of this **m** is often called the **magnetic dipole moment** even though there are no individual magnetic poles to correspond to the charges in the electric dipole. In particular since we have shown that the forces on small coils are of the same type as the forces on electric dipoles discussed in Chapter 3, it follows that in a field **B**, a small coil of magnetic dipole moment **m** has

$$\text{potential energy} = -\mathbf{m} \cdot \mathbf{B} \tag{5.34}$$

and experiences a

$$\text{couple} = \mathbf{m} \times \mathbf{B} \tag{5.35}$$

Likewise the mutual potential energy W between two small coils of magnetic dipole moment $\mathbf{m}_1$ and $\mathbf{m}_2$ separated by a distance **r** is simply the analogue of (3.12), namely

$$W = \frac{\mu_0}{4\pi r^3}\left[\mathbf{m}_1 \cdot \mathbf{m}_2 - 3\frac{(\mathbf{m}_1 \cdot \mathbf{r})(\mathbf{m}_2 \cdot \mathbf{r})}{r^2}\right] \tag{5.36}$$

Before leaving the magnetic dipole we might perhaps wonder why (5.33) should be identical in form to its electric counterpart when there is such a difference between the differential equations in Table 5.1. The answer to this apparent anomaly lies in the far-field approximation used in obtaining both (3.9) and (5.33) so that we have $\nabla \cdot \mathbf{B} = 0$, $\nabla \times \mathbf{B} \simeq 0$ and $\nabla \times \mathbf{E} = 0$, $\nabla \cdot \mathbf{E} \simeq 0$.

§5.10 Field at an infinite current sheet

In §2.7 we calculated the jump in **E** across a charge layer and found that

$$E_{n2} - E_{n1} = \frac{\sigma}{\varepsilon_0} \tag{5.38}$$

In this section we shall consider the magnetic analogue of (5.38) and calculate the change in **B** at a current sheet, infinite in extent and of thickness Δ. Suppose a current flows in the x- direction, with current density **j** amp. m^{-2}. The associated **B** field is shown in Fig. 5.3 as may easily be seen by picturing the conducting sheet made up of an array of straight wires carrying uniform steady current along Ox. The corresponding magnetic field for each wire in the array, has B_y and B_z components. The B_y contributions for neighbouring wires, being in opposite directions, cancel, leaving only a net B_z field. Applying Ampère's law, (5.22), to the rectangular circuit in Fig. 5.3 with one side of length z_0 in front of the sheet and the other behind it, gives

$$\oint \mathbf{B} \cdot d\mathbf{s} = B_{z1} z_0 - B_{z2} z_0 = \mu_0 j z_0 \Delta$$

i.e.

$$B_{z1} - B_{z2} = \mu_0 J_s \tag{5.39}$$

where $J_s = j\Delta$ is the **surface current density** corresponding to the (volume) current density **j**. We may think of (5.39) as the magnetic analogue of the electrostatic result, (5.38), which expresses the discontinuity in the **normal** component of the electric field at a sheet of charge in terms of the surface charge density, σ. For a current sheet we have found a corresponding discontinuity, this

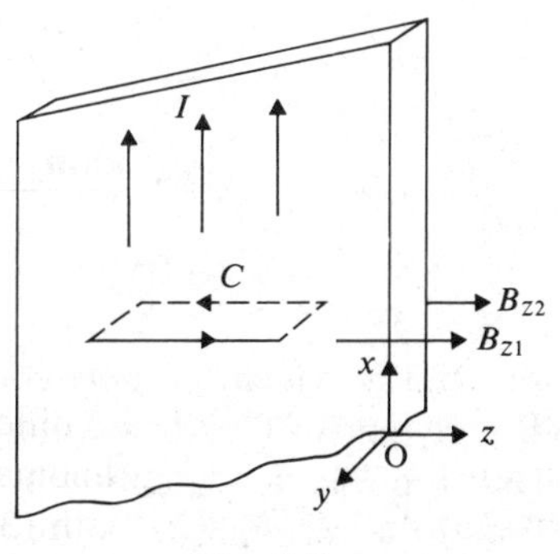

Fig. 5.3

time relating the **tangential** component of the magnetic field to the surface current density J_s. Looking back at the contrast between the electrostatic and magnetostatic fields (Table 5.1) this is precisely what we might have anticipated.

In §2.7 we remarked that provided no charges other than those in the charge sheet were involved, the arrangement is symmetrical and so $E_1 = \dfrac{\sigma}{2\varepsilon_0} = E_2$. Similarly here, if the only current source is the current sheet

$$B_{z2} = -B_{z1} = \frac{\mu_0 J_s}{2} \tag{5.40}$$

Obviously (5.39) still holds in the general case in which another current source is present; Fig. 5.4 represents the total **B** field in the neighbourhood of a current sheet when there is an additional uniform field in the z-direction.

We may draw yet another parallel if we return to §3.3 in which we calculated the mechanical force on a charged conducting surface of magnitude $\sigma^2/2\varepsilon_0$ per unit area of the surface. To determine the force per unit area on the current sheet in Fig. 5.3 we have only to evaluate

$$\begin{aligned}
\mathbf{F} &= \int_{-\Delta/2}^{\Delta/2} \mathbf{j} \times \mathbf{B}\, dy \\
&= -|\mathbf{j}|\hat{\mathbf{y}}\left[\int_{-\Delta/2}^{0} B_{z2}\, dy + \int_{0}^{\Delta/2} B_{z1}\, dy\right] \\
&= -\hat{\mathbf{y}}|\mathbf{j}|\frac{\Delta}{2}(B_{z1} + B_{z2}) \\
&= -\hat{\mathbf{y}}\frac{J_s}{2}(B_{z1} + B_{z2})
\end{aligned} \tag{5.41}$$

Using (5.39) this becomes

$$\mathbf{F} = \frac{1}{2\mu_0}(B_{z2}^2 - B_{z1}^2)\hat{\mathbf{y}} \tag{5.42}$$

Thus referring to Fig. 5.3 in which the **only** current source is the sheet itself, $B_{z1} = B_{z2}$ giving $\mathbf{F} = 0$ as one would expect. Figure 5.4 on the other hand represents a situation in which $B_{z1} > B_{z2}$ so that $(B_{z2}^2 - B_{z1}^2) < 0$ and the resultant force on the sheet is in the direction $-\hat{\mathbf{y}}$.

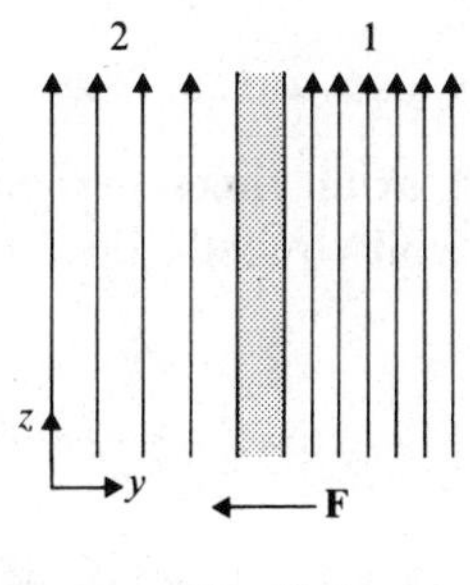

Fig. 5.4

§5.11 Motion of a charged particle in a magnetic field

In (5.1) we defined the magnetic field in terms of the force acting on a particle of charge e moving with velocity $\mathbf{v}$. The equation of motion for the particle may therefore be written

$$m\frac{d\mathbf{v}}{dt}=e\mathbf{v}\times\mathbf{B}$$

or, in terms of the displacement of the particle $\mathbf{r}(t)$,

$$m\ddot{\mathbf{r}}=e\dot{\mathbf{r}}\times\mathbf{B} \tag{5.43}$$

In the case of some magnetic fields it is possible to integrate this equation of motion. The simplest case is that in which $\mathbf{B}$ is uniform in space. Choosing the z-axis to be along the direction of the field we may write

$$\mathbf{B}=B_0\hat{\mathbf{z}}$$

so that from (5.43)

$$\ddot{\mathbf{r}}=\frac{eB_0}{m}(\dot{\mathbf{r}}\times\hat{\mathbf{z}}) \tag{5.44}$$

From this it follows by scalar multiplication with $\dot{\mathbf{r}}$ and $\hat{\mathbf{z}}$ in turn that

$$(\ddot{\mathbf{r}}\cdot\dot{\mathbf{r}})=0; \qquad (\ddot{\mathbf{r}}\cdot\hat{\mathbf{z}})=\ddot{z}=0$$

Integrating each equation once with respect to time gives

$$\dot{r}^2=\text{const.}; \qquad \dot{z}^2=\text{const.}=v_{\parallel}^2$$

where $v_{\parallel}$ denotes the component of velocity parallel to the z-axis.

Then since $\frac{m}{2}\dot{r}^2 \equiv \frac{m}{2}v^2$ is also constant (as we know it must be since the Lorentz force is at all times perpendicular to the velocity of the particle and consequently does no work on it) it follows that

$$\frac{m}{2}(v^2 - v_\parallel^2) \equiv \frac{m}{2}v_\perp^2 = \text{const.}$$

Moreover, since we have not so far used the fact that B_0 is spatially uniform these results are consequently not restricted to uniform fields.

To determine the trajectory of the particle we must write down the x and y components of (5.44):

$$\ddot{x} = \Omega\dot{y}; \qquad \ddot{y} = -\Omega\dot{x}$$

where $\Omega = |e|\, B_0/m$. Differentiating each equation once with respect to time gives

$$\dddot{x} + \Omega^2\dot{x} = 0; \qquad \dddot{y} + \Omega^2\dot{y} = 0$$

which have solutions

$$\dot{x}(t) = v_\perp \cos(\Omega t + \alpha); \qquad \dot{y}(t) = -v_\perp \sin(\Omega t + \alpha)$$

with α determined by the initial conditions. Integrating once again we find

$$\begin{aligned} x(t) &= \frac{v_\perp}{\Omega}\sin(\Omega t + \alpha) + x_0 \\ y(t) &= \frac{v_\perp}{\Omega}\cos(\Omega t + \alpha) + y_0 \\ z(t) &= v_\parallel t + z_0 \end{aligned} \tag{5.45}$$

where again (x_0, y_0, z_0) are determined by the initial conditions. The trajectory of the particle may be found from (5.45). Referred to the plane $z = v_\parallel t + z_0$ moving with the particle, the orbit is a circle with centre (x_0, y_0), i.e.

$$(x - x_0)^2 + (y - y_0)^2 = \frac{v_\perp^2}{\Omega^2}$$

The radius of this circle is $v_\perp/\Omega$ and is known as the **Larmor radius**, r_L, and the frequency of rotation, Ω, as the **Larmor frequency** or the **cyclotron frequency**. Thus to an observer at rest relative to the particle the orbit is a helix whose axis is parallel to the z-axis, i.e. to **B** (Fig. 5.5).

The motion of the particle may also be determined when a constant, uniform electric field is added (problem 20). The effect of an electric field **E**, perpendicular to **B**, is to produce a drift of the particle orthogonal to both **E** and **B** (see Fig. 5.14).

The solution of (5.43) for magnetic fields which are non-uniform i.e. for $\mathbf{B}=\mathbf{B}(\mathbf{r})$ is in general not possible analytically; these problems have to be solved numerically.

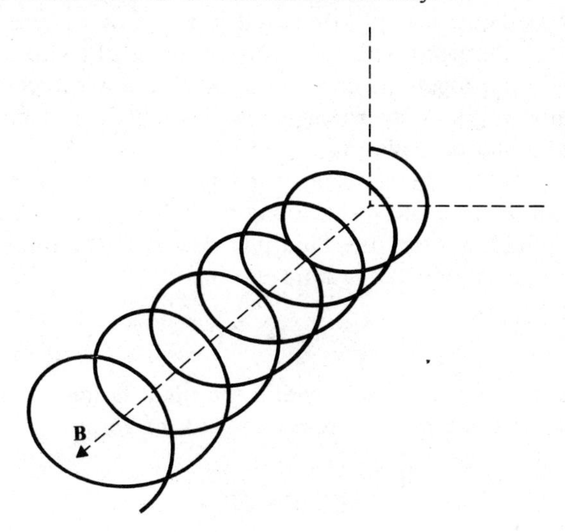

Fig. 5.5

§5.12 Hydromagnetics

In §5.10 we found the force per unit area on a current sheet. This force is normal to the surface of the sheet. Throughout this chapter we have contrasted magnetostatics with electrostatics. We now want to draw an interesting parallel with hydrostatics. If we were to regard the situation in Fig. 5.4 as one in which region 1 (where the magnetic field is higher) is a region of higher pressure than region 2 we might by analogy with hydrostatics express (5.42) as

$$\mathbf{F}=(P_2^M-P_1^M)\hat{\mathbf{y}} \tag{5.46}$$

where P^M is a **magnetic pressure** defined by

$$P^M = \frac{B^2}{2\mu_0} \tag{5.47}$$

The concept of magnetic pressure was first introduced by Faraday, who pictured the tubes of force as elastic filaments, these filaments being under tension in the direction of the field and compressed in the transverse directions. Faraday's concepts were expressed mathematically by Maxwell in terms of a stress tensor. These ideas, combined with the hydrodynamics of an electrically conducting fluid, together form the basis of a new subject **magnetohydrodynamics** or **hydromagnetics**. The conducting fluid might be a liquid metal or a plasma.

We may explore the concept of magnetic pressure more formally by considering the force acting on unit volume of such a conducting fluid. Writing the fluid pressure as P, the total force acting on unit volume of the fluid is

$$\mathbf{F} = \mathbf{j} \times \mathbf{B} - \nabla P \tag{5.48}$$

where $\mathbf{j}$ is the current density within the fluid. Using Ampère's law, $\nabla \times \mathbf{B} = \mu_0 \mathbf{j}$, we may express (5.48) in the form

$$\begin{aligned}
\mathbf{F} &= -\frac{\mathbf{B}}{\mu_0} \times (\nabla \times \mathbf{B}) - \nabla P \\
&= -\nabla\left(\frac{B^2}{2\mu_0}\right) + \frac{1}{\mu_0}(\mathbf{B} \cdot \nabla)\mathbf{B} - \nabla P \\
&= -\nabla\left(P + \frac{B^2}{2\mu_0}\right) + \frac{1}{\mu_0}(\mathbf{B} \cdot \nabla)\mathbf{B}
\end{aligned} \tag{5.49}$$

We see at once from this form that $\frac{B^2}{2\mu_0}$ may indeed be regarded as a magnetic pressure term in the way suggested by (5.47). It is important to recognize that this is only part of the effect of the $\mathbf{j} \times \mathbf{B}$ force; in general there will also be a contribution from $\frac{1}{\mu_0}(\mathbf{B} \cdot \nabla)\mathbf{B}$. This contribution vanished in the current sheet example because in that case the spatial dependence of $\mathbf{B}$ occurred at right angles to $\mathbf{B}$. In such cases we may express the condition for static equilibrium as

$$P+\frac{B^2}{2\mu_0}=\text{constant} \tag{5.50}$$

In general we may show (problems 22–24) that the stress due to a magnetic field amounts to an isotropic magnetic pressure $B^2/2\mu_0$ and a **tension** B^2/μ_0 along the magnetic field lines as illustrated in Fig. 5.6. Stretching the tube of force increases the tension, which means that the field is increased.

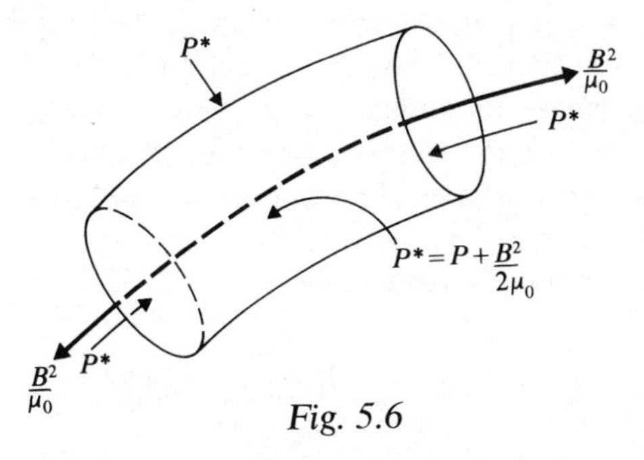

Fig. 5.6

We may follow this analogy further. Just as transverse waves can be set up using an elastic string, so by "plucking" a tube of force might it not be possible to create transverse waves propagating along the magnetic field lines? This suggestion was made by Alfvén in 1942 in work which marks the beginning of magnetohydrodynamics. The existence of the waves was subsequently confirmed and they are now known as **Alfvén waves**. Since the velocity of a transverse wave on a string is just $(T/\rho)^{1/2}$ where T is the tension in the string and ρ the density of the elastic medium it follows – if our analogy is valid – that the phase velocity of Alfvén waves v_A is just

$$v_A=\frac{B}{(\mu_0\rho)^{1/2}} \tag{5.51}$$

where ρ is now the mass density of the conducting fluid. The existence of Alfvén waves and confirmation of (5.51) was established by careful experiments in plasmas [4]. Returning to (5.48), we may express the condition for **equilibrium** of a plasma (or other conducting fluid) in a magnetic field as

$$\nabla P=\mathbf{j}\times\mathbf{B} \tag{5.52}$$

From this it follows that

$$\mathbf{B} \cdot \nabla P = 0 \tag{5.53}$$

and

$$\mathbf{j} \cdot \nabla P = 0 \tag{5.54}$$

i.e. both **B** and **j** lie on surfaces of constant pressure. If we suppose these **isobaric** surfaces are closed then (5.53) states that no magnetic field line passes through a surface and we may picture the surface as made up from a winding of **B** field lines. Similarly from (5.54) the same isobaric surface could be viewed as a winding of lines of **j**; these lines will in general intersect the **B** field lines quite arbitrarily. The cross-section in Fig. 5.7 shows a set of nested surfaces on which the pressure increases in passing from the outside towards the axis and the currents are such that $\mathbf{j} \times \mathbf{B}$ points **towards** the axis. The important implication of this is that a plasma may be contained by the $\mathbf{j} \times \mathbf{B}$ force, an arrangement known as **magnetic confinement**.

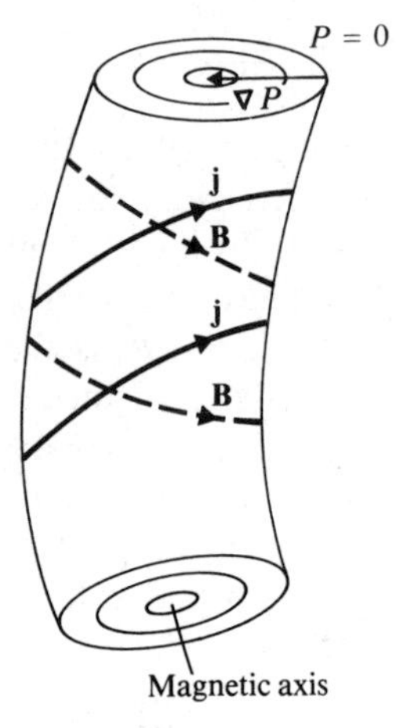

Fig. 5.7

An example of this containment is provided by the **linear pinch** which is a cylindrical column of plasma with current flowing axially. Associated with this will be an azimuthal **B** field and a $\mathbf{j} \times \mathbf{B}$ force directed towards the axis as in Fig. 5.8. Under the action of the $\mathbf{j} \times \mathbf{B}$ force the plasma is squeezed or "pinched" into a filament along the axis until an equilibrium is reached when the

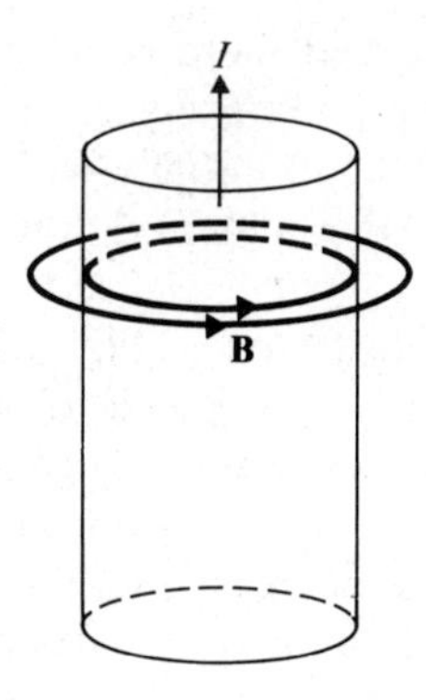

Fig. 5.8

pressure of the plasma just balances the compressing electrodynamic force. A simple analysis of the pinch effect is the subject of problems 26 and 27.

Only qualitative arguments are needed to show that the equilibrium of the linear pinch is inherently unstable. Consider Fig. 5.9, which shows an axially symmetric perturbation of the equilibrium configuration, which might be induced in an actual experiment by random fluctuations. Since for a given current the magnetic field at the surface of the plasma column is proportional to r^{-1} it follows that the magnetic pressure $B^2/2\mu_0$ is greater in the region where the perturbation squeezes the plasma into a neck and decreases where the plasma column fattens into a bulge. Since the plasma pressure P is uniform throughout (since the fluid may flow freely along the column) it follows that in the region

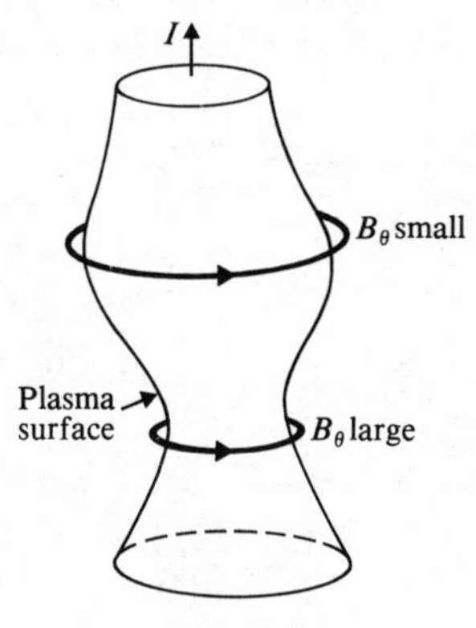

Fig. 5.9

where the plasma is squeezed there is no longer equilibrium between the magnetic and plasma pressures and the perturbation which we assumed to be small initially will grow rapidly. This results in a disruption of the column, known as a **sausage instability**.

The **kink instability**, illustrated in Fig. 5.10, is a further example of the tendency of the linear pinch to break up. In this case the perturbation grows because the magnetic pressure on the concave side of the kink is increased (**B** lines crowd together) while that on the convex side is decreased (**B** lines spread apart). It follows that the plasma column is also unstable with respect to bending.

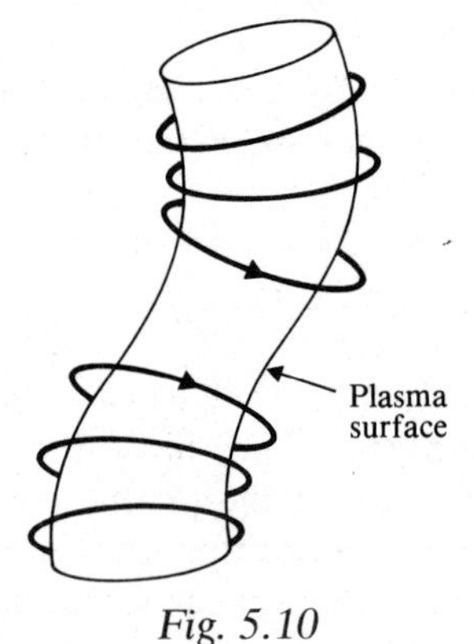

Fig. 5.10

Practical schemes for containing plasma have to be much more sophisticated than the simple linear pinch discussed in this section. One such is the axially symmetric toroidal system illustrated schematically in Fig. 5.11 and known as a **tokamak**. In this device

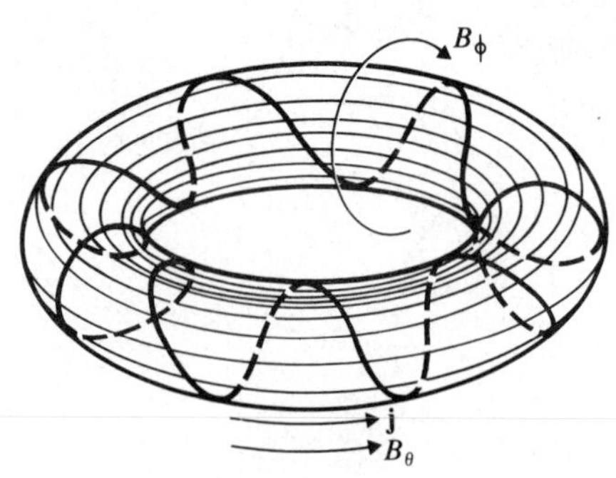

Fig. 5.11

the plasma is contained by the magnetic field of the current flowing along its axis **and** by a very powerful magnetic field **applied** parallel to the current in order to suppress the kind of instabilities to which the linear pinch was susceptible. On this configuration rest the hopes of present-day attempts to contain hot plasmas long enough for thermonuclear fusion to occur.

§5.13 Examples

(A). In §5.7 an approximate expression was found for the magnetic field within a **long** solenoid. Show that without making this approximation, the field is given by $B_z(z_0) = \frac{\mu_0 NI}{2L}(\cos\alpha + \cos\beta)$ where α and β are the angles shown in Fig. 5.12.

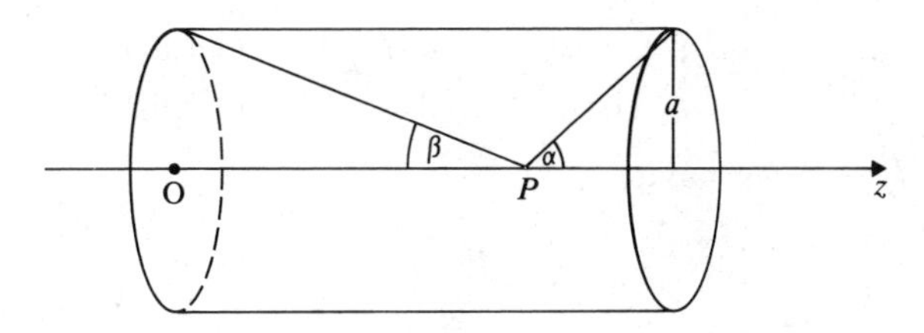

Fig. 5.12

As in §5.7, the solenoid is composed of N turns of wire wound on a cylinder of radius a and length L so that the number of turns in dz is $N dz/L$. Then, using the result of problem 3 in §5.14, it follows directly that

$$B_z(z_0) = \frac{\mu_0 NI}{L}\frac{a^2}{2}\int_0^L \frac{dz}{[(z_0 - z)^2 + a^2]^{3/2}}$$

Setting $z - z_0 = a \tan\theta$ gives

$$B_z(z_0) = \frac{\mu_0 NI}{2L}[\sin\theta_2 - \sin\theta_1]$$

where $\theta_1 = -\tan^{-1}\frac{z_0}{a}$, $\theta_2 = \tan^{-1}\frac{(L - z_0)}{a}$. In terms of α, β the required form for B_z follows immediately.

Making the approximation for a long solenoid ($L \gg a$) means that $\cos\alpha \simeq \cos\beta \simeq 1$ (provided z_0 is not near either end) and we retrieve (5.23). Would you have expected this?

(B). Determine the magnetic field due to equal and opposite currents I flowing in two parallel wires.

Taking the direction of the current in one wire to be in the positive z direction, with a second wire, as shown in Fig. 5.13, let us determine the vector potential $\mathbf{A}$ at a field point P. Suppose the wires are each of length $2L$. Clearly $A_x = 0 = A_y$ and in the midplane

$$A_z = \frac{\mu_0 I}{4\pi}\left[\int_{-L}^{L} \frac{dz}{(z^2+r_1^2)^{1/2}} - \int_{-L}^{L} \frac{dz}{(z^2+r_2^2)^{1/2}}\right]$$

Substituting $\xi = \dfrac{z}{r}$ and letting $\xi_1 = \dfrac{L}{r_1}$, $\xi_2 = \dfrac{L}{r_2}$ we find

$$A_z = \frac{\mu_0 I}{2\pi}\int_{\xi_2}^{\xi_1} \frac{d\xi}{(1+\xi^2)^{1/2}} = \frac{\mu_0 I}{2\pi}\ln\left|\frac{\xi_1+(1+\xi_1^2)^{1/2}}{\xi_2+(1+\xi_2^2)^{1/2}}\right|$$

Letting $L \to \infty$ we have

$$A_z = \frac{\mu_0 I}{2\pi}\ln\frac{r_2}{r_1}$$

so that the curves $A_z =$ constant are (in the plane perpendicular to the wires) circles with centres on the line joining the points of intersection of the wires with the plane. If the wires are a distance a apart then

$$A_z = \frac{\mu_0 I}{4\pi}\ln\left[\frac{(x-a)^2+y^2}{x^2+y^2}\right]$$

and the magnetic field follows directly by differentiating, i.e.

$$B_x = \frac{\partial A_z}{\partial y}, \qquad B_y = -\frac{\partial A_z}{\partial x}$$

Note that the equation for the magnetic field lines $\dfrac{dx}{B_x} = \dfrac{dy}{B_y}$ corresponds to $\dfrac{\partial A_z}{\partial x}dx + \dfrac{\partial A_z}{\partial y}dy = dA_z = 0$ so that the field lines are determined by the curves $A_z =$ constant sketched in Fig. 5.13.

(a)

Fig. 5.13a Currents opposed

§5.14 Problems

1. An infinite straight wire whose cross-section is a circle of radius a carries a uniform current I. Show that the magnetic field is given by

$$\mathbf{B}=\frac{\mu_0 I}{2\pi r}\hat{\boldsymbol{\theta}} \qquad r>a$$

$$=\frac{\mu_0 I r}{2\pi a^2}\hat{\boldsymbol{\theta}} \qquad r<a$$

2. Show that two wires carrying currents in the same direction attract one another. This is the reverse of the analogous problem in electrostatics illustrating the fundamental distinction between the two fields.

3. By applying the Biot–Savart law show that the magnetic field at points on the axis of a circular loop of radius a is given by

$$\mathbf{B}=\frac{\mu_0 I a^2}{2}\frac{\hat{\mathbf{z}}}{(a^2+z^2)^{3/2}}$$

where I is the current flowing in the loop.

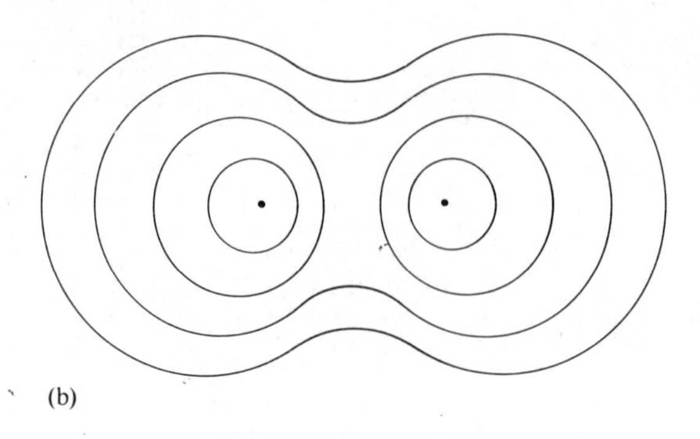

Fig. 5.13b Currents in same direction

4. Show that the magnetic field at the centre of a square coil of side $2a$ lying in the $z=0$ plane and carrying a current I is $\mathbf{B}=\sqrt{2}\mu_0 I\hat{\mathbf{z}}/\pi a$.

5. Two equal circular coils of radius a are placed opposite one another, a distance $2b$ apart. They carry the same current in the same direction. Show that at the point midway between the two centres the first three differential coefficients of the field are zero if $2b=a$. Such an arrangement which gives an approximately uniform field is known as a **Helmholtz coil**.

6. Consider a toroidal winding with N turns in which the current flowing is I. By applying Ampere's law (5.21) show that

$$\mathbf{B}=\frac{\mu_0 NI}{2\pi r}\hat{\boldsymbol{\theta}}, \qquad a<r<b$$
$$=0, \qquad r<a, \qquad r>b,$$

where a and b are respectively the inner and outer radii of the torus.

7. A current I flows in an infinite thin wire coincident with the z-axis. The return flow takes place along a parallel wire through the point $(R, 0, 0)$ where R is very large. Show that the magnetostatic potential at a point (r, θ, z) is $V_m=-\mu_0 I(\pi-\theta)/2\pi$.

8. Show that (5.26) gives an infinite value for the vector potential of an infinite straight current. But if we suppose that the current

returns along a parallel wire, then at a point P the vector potential has a magnitude $\frac{\mu_0 I}{4\pi}\ln\left(\frac{b}{a}\right)$ where a and b are the shortest distances from P to the two wires.

9. Show that $\mathbf{A}$ in (5.26) satisfies the condition $\nabla \cdot \mathbf{A} = 0$.

10. Show that in terms of the elliptic integrals $E(k)$ and $K(k)$ defined in §5.8 the vector potential for a current-carrying loop of radius a is

$$\mathbf{A}(r, z) = \frac{\mu_0 I \hat{\boldsymbol{\theta}}}{\pi k}\left(\frac{a}{r}\right)^{1/2}\left[\left(1 - \frac{k^2}{2}\right)K - E\right]$$

11. Show that the field near the axis of the loop in the previous question is determined by the vector potential

$$\mathbf{A}(r, z) = \frac{\mu_0 I a^2}{4}\hat{\boldsymbol{\theta}}\frac{r}{[(a+r)^2 + z^2]^{3/2}}.$$

12. Show from (5.28) that $\mathbf{A}(\mathbf{r}) = \frac{\mu_0 I}{8\pi r^3}\oint(\mathbf{s} \times d\mathbf{s}) \times \mathbf{r}$.

13. Verify that the vector potential of a constant field $\mathbf{B} = B\hat{\mathbf{x}}$ may be written

$$A_x = 0, \qquad A_y = -aBz, \qquad A_z = (1-a)By$$

where a is an arbitrary constant.

Similarly show that for a constant field $\mathbf{B} = B\hat{\boldsymbol{\theta}}$ in spherical polar coordinates we may write

$$A_r = 0, \qquad A_\theta = 0, \qquad A_\phi = \frac{Br}{2}\sin\theta$$

Prove that in cylindrical coordinates if $\mathbf{B} = B\hat{\mathbf{z}}$ then

$$A_r = 0, \qquad A_\theta = \frac{Br}{2}, \qquad A_z = 0.$$

14. Show that if a given magnetic field is axially symmetric and represented in cylindrical coordinates by the vector potential $(0, A_\theta(r, z), 0)$, the equation for the field lines is $rA_\theta(r, z) =$ constant.

15. Verify that the vector **A** whose components are

$$\left(\frac{\lambda yz}{(x^2+y^2)r}, -\frac{\lambda xz}{(x^2+y^2)r}, 0\right)$$

is such that

$$\nabla\times\mathbf{A}=\left(\frac{\lambda x}{r^3}, \frac{\lambda y}{r^3}, \frac{\lambda z}{r^3}\right); \qquad \nabla\cdot\mathbf{A}=0$$

Show that **A** would be the vector potential for a magnetic monopole of strength λ at the origin. Deduce from this the value of **A** for a magnetic dipole at the origin pointing along Oz.

16. Starting from (5.36) for the mutual potential energy between two small magnetic dipoles show that the couple **G** exerted on $\mathbf{m}_1$ is

$$\mathbf{G}=\frac{\mu_0}{4\pi r^3}\left[\mathbf{m}_1\times\mathbf{m}_2-\frac{3(\mathbf{m}_1\times\mathbf{r})(\mathbf{m}_2\cdot\mathbf{r})}{r^2}\right]$$

Obtain the same result by writing $\mathbf{G}=\mathbf{m}_1\times\mathbf{B}$ where **B** is the field due to the dipole of moment $\mathbf{m}_2$.

17. In the previous question why is the couple exerted on $\mathbf{m}_1$ by $\mathbf{m}_2$ not equal and opposite to that exerted on $\mathbf{m}_2$ by $\mathbf{m}_1$?

18. Find the force acting on two small current-carrying loops with magnetic moments $\mathbf{m}_1$ and $\mathbf{m}_2$.

19. An infinite conducting strip of width a carries a uniform current I. Show that

$$B_x=\frac{\mu_0 I}{4\pi a}\ln\left[\frac{\left(y-\frac{a}{2}\right)^2+x^2}{\left(y+\frac{a}{2}\right)^2+x^2}\right]$$

$$B_y=\frac{\mu_0 I}{2\pi a}\left[\tan^{-1}\left(\frac{a+2y}{2x}\right)+\tan^{-1}\left(\frac{a-2y}{2x}\right)\right]$$

20. A charged particle of charge e and mass m moves in constant uniform electric and magnetic fields, $\mathbf{E}=(0, E, 0)$, $\mathbf{B}=(0, 0, B_0)$. Show that the effect of the electric field perpendicular to the magnetic field is to cause the particle to drift with a velocity $\mathbf{v}_E=(\mathbf{E}\times\mathbf{B})/B_0^2$ (see Fig. 5.14).

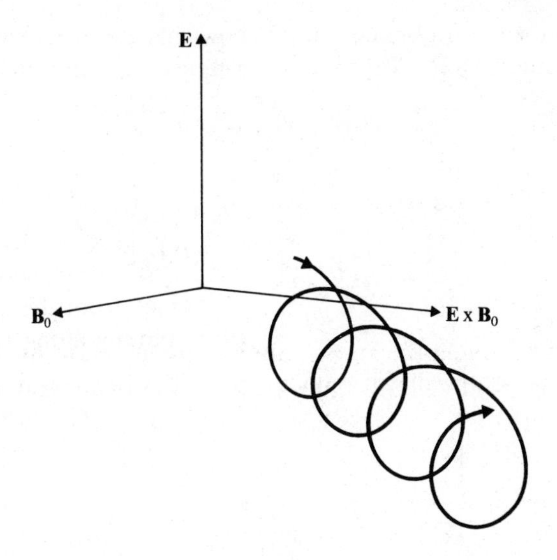

Fig. 5.14

21. Assume that each moving charge contributes to expressions (5.26), (5.16) for **A** and **B** independently of all other charges. Then a charge e moving in free space with velocity **v** has associated with it a vector potential **A** and a magnetic field **B** given by

$$\mathbf{A} = \frac{\mu_0}{4\pi}\frac{e\mathbf{v}}{r}, \qquad \mathbf{B} = \frac{\mu_0}{4\pi}\frac{e\mathbf{v}\times\mathbf{r}}{r^3}$$

Deduce that if **E** is the electrostatic field due to the charge,

$$\mathbf{B} = \mathbf{v}\times\mathbf{E}$$

[N.B. These formulae are **only** true if the velocity **v** is much less than the velocity of light.]

22. Show that the $(\mathbf{j}\times\mathbf{B})$ force may be written using Ampère's law in the form

$$(\mathbf{j}\times\mathbf{B})_i = \frac{\partial}{\partial x_k}\left(\frac{B_i B_k}{\mu_0}\right) - \frac{\partial}{\partial x_k}\left(\frac{B^2\delta_{ik}}{2\mu_0}\right)$$

23. From the form obtained in 22 show that the magnetohydrostatic condition $\mathbf{j} \times \mathbf{B} = \nabla P$ may be written in tensor form as

$$\frac{\partial \mathcal{T}_{ik}}{\partial x_k} = 0$$

where $\mathcal{T}_{ik}$ is the stress tensor given by

$$\mathcal{T}_{ik} = \left[\left(P + \frac{B^2}{2\mu_0} \right) \delta_{ik} - \frac{B_i B_k}{\mu_0} \right]$$

24. By transforming the stress tensor in the previous question to principal axes show that it reduces to the diagonal form

$$\begin{bmatrix} P + \dfrac{B^2}{2\mu_0} & 0 & 0 \\ 0 & P + \dfrac{B^2}{2\mu_0} & 0 \\ 0 & 0 & P - \dfrac{B^2}{2\mu_0} \end{bmatrix}$$

25. Calculate the phase velocity of Alfvén waves in the solar corona which consists of ionized hydrogen with approximately 10^{12} particles m^{-3}. Assume that $B = 1\,000$ gauss.

26. Consider a cylindrical tube of plasma in which an axial current flows. The plasma is pinched by the $\mathbf{j} \times \mathbf{B}$ force to form a filament along the axis of the tube. The equilibrium condition (5.52) gives

$$\frac{dP}{dr} = -\frac{B}{\mu_0 r} \frac{d}{dr} (rB) \tag{5.55}$$

Suppose that the plasma pressure vanishes at $r = r_0$ and that $P = N(r)\kappa T$, where κ is Boltzmann's constant and T is the temperature (which is assumed to be constant across the pinch). Show from (5.55) that

$$2N\kappa T = \frac{\pi}{2\mu_0} (rB)^2_{r=r_0} \tag{5.56}$$

where $N = \int_0^{r_0} n(r) \,.\, 2\pi r \, dr$.

27. Defining the total current flowing in the plasma filament by $I = \int_0^{r_0} j \,.\, 2\pi r \, dr$ show from the expression (5.56) that

$$I^2 = \frac{16\pi}{\mu_0} N\kappa T$$

This is known as **Bennett's relation** for the linear pinch.

28. Assuming that the pinch is stable and that $N = 10^{20}$ particles m^{-1}, determine from Bennett's relation the current required to produce thermonuclear temperatures (of the order of 10^8 K).

6

Electrostatics and magnetostatics of materials

§6.1 Introduction

In the development of the subject so far we considered, in Chapters 2 and 3, the electrostatic field when charges and conductors are present, while in Chapters 4 and 5 we dealt with steady currents and their associated magnetic fields. In neither case have we allowed for the presence of materials (other than the conductors themselves). In other words the field equations we have derived are those of vacuum electrodynamics. We now wish to examine the response of dielectric and magnetic media to electric and magnetic fields. We remarked in Chapter 1 that this introduces the important distinction between **microscopic** and **macroscopic** fields, a distinction which involves a very careful discussion of averaging procedures. These averages are carried out over regions which are macroscopically small but microscopically large. The procedure lies beyond the scope of the present treatment; a careful discussion is given in [4].

In discussing the electrostatics of conductors in Chapter 3 we saw that a condenser whose plates carry charges $\pm Q$ at potentials V_1 and V_2 is a reservoir of electrical energy. The amount of energy stored was shown to be $\frac{1}{2}CV^2$ where $V = V_1 - V_2$. The capacity (or capacitance) C is a function of the plate geometry, e.g. for a simple condenser constructed from parallel plates of area A a distance t apart, $C = \varepsilon_0 A/t$ provided we ignore fringe fields. In practice however such condensers are rare. Much more common are those with insulating materials or **dielectrics** between their plates. Historically it was Faraday who discovered that if the space between the plates of a condenser was filled with an insulating material such as glass or mica, its capacity was multiplied by a constant K, which depended on the material but was independent of the shape of the condenser. K is called the **dielectric constant** and values range from 1 for free space and 4 for Pyrex glass to about 80 for water. For air at normal pressure

$K = 1.0006$ which means that the results of Chapters 2 and 3 are virtually unaffected if we use air instead of a vacuum.

§6.2 Polarization

To explain this phenomenon we must consider in more detail what happens when an electric field $\mathbf{E}$ acts on a dielectric material. We drew a rough distinction in Chapter 3 between conducting materials on the one hand and insulators or dielectrics on the other. When a good conductor is subject to an electric field the conduction electrons respond very quickly to this field and a current flows. A perfect insulator on the other hand is a dielectric material with no free electrons. When a field $\mathbf{E}$ is applied, the bound electrons in the molecules of the dielectric material move in a direction opposite to $\mathbf{E}$ while the nuclei are displaced in the direction of $\mathbf{E}$. The displacements are usually very small on a molecular scale (i.e. much less than a molecular radius) since very powerful restoring forces are quickly established due to the perturbed molecular configuration. We say that the medium is **polarized** by the field. As a result of this polarization each molecule becomes a tiny dipole whose strength will depend on $\mathbf{E}$. Let us suppose that in each molecule the charges are separated by a distance $\boldsymbol{\delta}$. If there are N molecules in unit volume we can define a dipole moment/unit volume, $\mathbf{P}$ with

$$\mathbf{P} = Ne\boldsymbol{\delta} \tag{6.1}$$

If the field is not too large the strength of each microscopic dipole is proportional to $\mathbf{E}$ and we may write it $\alpha\mathbf{E}$ where α is the **polarizability**. We call $\alpha\mathbf{E}$ the induced dipole; for an isolated atom or molecule, and indeed for very many substances, it is parallel to $\mathbf{E}$. However, certain crystals and other **anisotropic** media, such as plasmas in a magnetic field, are exceptions to this rule.

There is also another type of contribution to the polarizability α. This occurs when the molecules of the dielectric are themselves permanent dipoles, such as water. In the absence of a field these permanent dipoles will be randomly oriented but, when a field is applied, a couple is established (see §3.6) which acts to align them. Consequently there will be a resultant moment in the direction of the field. Let us suppose that this effect also has been included in the value of α. The total moment induced by an electric field $\mathbf{E}$ is $N\alpha\mathbf{E}$ per unit volume†. Thus for isotropic materials, the polarization $\mathbf{P}$ is given in terms of $\mathbf{E}$ by

$$\mathbf{P} = N\alpha\mathbf{E} = \varepsilon_0\chi\mathbf{E} \tag{6.2}$$

The scalar χ is called the **electric susceptibility** and depends on the density of the material, as given by N, and on the polarizability α, both of which may vary with temperature. We should properly write χ as $\chi(E)$ but in practice over a wide range of field intensitities χ is sensibly field-independent.

Thus the effect of an electric field $\mathbf{E}$ on a dielectric is to create the polarization $\mathbf{P}$, with its associated dipole moment $\mathbf{P}d\mathbf{r}$ in each volume element $d\mathbf{r}$. In calculating the potential for such a system we may now set aside altogether any consideration of the dielectric itself, provided we imagine it replaced by the volume polarization $\mathbf{P}\ d\mathbf{r}$.

Now from (3.8) we know that the potential at any point due to a single dipole $\mathbf{p}$ is $\dfrac{1}{4\pi\varepsilon_0}\dfrac{\mathbf{p}\cdot\mathbf{r}}{r^3}$ for a dipole centred at the origin. More generally for a dipole placed at some point $\mathbf{r}'$,

$$V(\mathbf{r}) = \frac{1}{4\pi\varepsilon_0}\frac{\mathbf{p}\cdot(\mathbf{r}-\mathbf{r}')}{|\mathbf{r}-\mathbf{r}'|^3}$$

In §2.3 we noted that $(\mathbf{r}-\mathbf{r}')/|\mathbf{r}-\mathbf{r}'|^3 = -\boldsymbol{\nabla}(|\mathbf{r}-\mathbf{r}'|^{-1})$ where $\boldsymbol{\nabla}$ operates only on r. But we might equally have expressed this as $+\boldsymbol{\nabla}'(|\mathbf{r}-\mathbf{r}'|^{-1})$ where the $'$ denotes the gradient operation is carried out on the primed coordinates. This is now a more appropriate form since we want to perform operations keeping $\mathbf{r}$ fixed. So for a point dipole at $\mathbf{r}'$

$$V(\mathbf{r}) = \frac{1}{4\pi\varepsilon_0}\mathbf{p}\cdot\boldsymbol{\nabla}'\left(\frac{1}{|\mathbf{r}-\mathbf{r}'|}\right)$$

If we now picture our point dipole replaced by a volume polarization $\mathbf{P}$ so that $\mathbf{p}\to\mathbf{P}\ d\mathbf{r}$ then we can write for the potential arising from the induced polarization

$$V(\mathbf{r}) = \frac{1}{4\pi\varepsilon_0}\int \mathbf{P}(\mathbf{r}')\cdot\boldsymbol{\nabla}'\left(\frac{1}{|\mathbf{r}-\mathbf{r}'|}\right)d\mathbf{r}' \tag{6.3}$$

† There is a slight difficulty here, since dipoles in the immediate neighbourhood of any one dipole will themselves contribute to the field which polarizes the original dipole. Except when the polarization is very large this difficulty may be avoided by a suitable change in the definition of α.

Since

$$\nabla' \cdot \left(\frac{\mathbf{P}(\mathbf{r}')}{|\mathbf{r}-\mathbf{r}'|}\right) = \mathbf{P}(\mathbf{r}') \cdot \nabla'\left(\frac{1}{|\mathbf{r}-\mathbf{r}'|}\right) + \frac{\nabla' \cdot \mathbf{P}(\mathbf{r}')}{|\mathbf{r}-\mathbf{r}'|}$$

(6.3) may be written equivalently, using Gauss' theorem,

$$V(\mathbf{r}) = \frac{1}{4\pi\varepsilon_0}\left[\int \frac{\mathbf{P}(\mathbf{r}') \cdot d\mathbf{S}'}{|\mathbf{r}-\mathbf{r}'|} - \int \frac{\nabla' \cdot \mathbf{P}(\mathbf{r}')}{|\mathbf{r}-\mathbf{r}'|}\, d\mathbf{r}'\right] \tag{6.4}$$

in which the surface integral is taken over the boundary of the dielectric.

This last equation is important. It shows that when we have a polarization **P**, the resulting potential is exactly the same as we should obtain if we were to suppose that there was a volume distribution of charge $\rho_P = -\nabla \cdot \mathbf{P}$ throughout the dielectric, together with a surface distribution $\sigma_P = \mathbf{P} \cdot \hat{\mathbf{n}}$ on the boundary of the dielectric. These charges ρ_P and σ_P are referred to as **polarization charge densities** or sometimes as **bound** charge densities.

§6.3 Electric displacement D and its properties

We are now in a position to formulate the laws of the electrostatic field in the presence of dielectrics. Suppose that at any point in the medium the field is **E**, with resulting polarization **P**. Then in addition to any free charges ρ and σ that may be present, we must include contributions to the potential from the polarization or bound charges ρ_P and σ_P. Gauss' flux law now takes the form

$$\nabla \cdot \mathbf{E} = \frac{1}{\varepsilon_0}(\rho - \nabla \cdot \mathbf{P})$$

i.e.

$$\nabla \cdot (\varepsilon_0 \mathbf{E} + \mathbf{P}) = \rho$$

We define a new macroscopic field vector **D**, such that

$$\mathbf{D} = \varepsilon_0 \mathbf{E} + \mathbf{P} \tag{6.5}$$

and Gauss' law becomes

$$\nabla \cdot \mathbf{D} = \rho \tag{6.6}$$

The vector $\mathbf{D}$ was introduced by Maxwell, who called it the **electric displacement**. It plays a most important role in electrodynamics. The electrostatic field in a dielectric is seen to be in part a term $\mathbf{D}/\varepsilon_0$ with $\mathbf{D}$ related to the free charge density and in part $-\mathbf{P}/\varepsilon_0$ deriving from the polarization of the dielectric. On account of its importance in the electrodynamics of materials it is convenient to list the properties of $\mathbf{D}$.

(i) In isotropic materials for which (6.2) holds

$$\mathbf{D} = \varepsilon_0\mathbf{E} + \mathbf{P} = \varepsilon_0(1+\chi)\mathbf{E}$$

If we write

$$\varepsilon = \varepsilon_0(1+\chi) \tag{6.7}$$

then

$$\mathbf{D} = \varepsilon\mathbf{E} \tag{6.8}$$

where ε is the **permittivity** of the dielectric material. We should remember that since strictly the dielectric susceptibility is field dependent, so too is the permittivity, i.e. ε is really $\varepsilon(E)$. Moreover for anisotropic dielectrics, such as plasmas in magnetic fields, the permittivity will be a tensor ε_{ij}. Also note from (6.7) that ε has the same units as ε_0 (farad metre^{-1}). In free space $\varepsilon = \varepsilon_0$, i.e. $\mathbf{D} = \varepsilon_0\mathbf{E}$. While the electrical behaviour of materials may be characterized by χ, the susceptibility, or ε, the permittivity, it is convenient to introduce a dimensionless quantity K, the **dielectric constant**, where

$$\varepsilon = K\varepsilon_0$$

i.e.

$$K = \frac{\varepsilon}{\varepsilon_0} = 1 + \chi \tag{6.9}$$

The dimensions of the electric displacement are charge per unit area so that $\mathbf{D}$ is measured in units of **coulomb m^{-2}**.

(ii) As in the vacuum case Gauss' law may be expressed as a flux theorem in the form

$$\text{flux of } \mathbf{D} \text{ out of any surface} = \text{included free charge} \tag{6.10}$$

(iii) Since (6.10) is the counterpart of the flux theorem in §2.6 we may immediately write down several other properties of

D. For example, there are field lines of **D** given by the differential equations

$$\frac{dx}{D_x}=\frac{dy}{D_y}=\frac{dz}{D_z} \tag{6.11}$$

Moreover there are tubes of **D** and from (6.6) and (6.10) these tubes can only begin and end on free charges.

(iv) Just as in Chapter 3, $\mathbf{D}=\mathbf{0}$ in conductors and by applying the new flux theorem (6.10) as in section 3.2, we obtain the modified Coulomb's law at the surface of a conductor:

$$D_n=\sigma \tag{6.12}$$

$$E_n=\sigma/\varepsilon \tag{6.13}$$

Similarly the mechanical force on a conductor is $\sigma^2/2\varepsilon$ per unit area.

(v) Equation (6.13) enables us to complete the identification of K. For if we have two identical condensers except that one is filled with a material of uniform dielectric constant K, the other being in free space, and if they are charged so that the plates carry the same charge density σ in the two cases, then the field in the first is $1/K$ times that in the second. The potential difference between the plates is likewise different by a factor $1/K$ and consequently the capacity of the dielectric condenser is K times that of the vacuum condenser. This identifies K as defined by (6.9) with the dielectric constant introduced by Faraday.

(vi) Next suppose that we have a single charge e in a material of dielectric constant K apply the flux theorem (6.10) to a concentric sphere of radius r. By symmetry **D** has the same value at all points of the sphere and is directed radially outwards. Thus $4\pi r^2 D=e$ and so $D=e/4\pi r^2$, $E=e/4\pi\varepsilon r^2$ and $V=e/4\pi\varepsilon r$. It follows that the interaction between two charges e_1 and e_2 at positions $\mathbf{r}_1$ and $\mathbf{r}_2$ in a dielectric is governed by the force law

$$\mathbf{F}=\frac{e_1e_2}{4\pi\varepsilon}\frac{(\mathbf{r}_1-\mathbf{r}_2)}{|\mathbf{r}_1-\mathbf{r}_2|^3} \tag{6.14}$$

The inverse square law still holds but the magnitude of the force is **reduced** from its vacuum value by the factor $1/K$. An alternative way of introducing the concept of a dielectric would be to start from (6.14), parallelling the development of electrostatics in vacuum. It is one best left alone since (6.14) applies only if ε is constant and if the dielectric is isotropic and occupies all of space! Secondly it is not easy to see precisely what is meant by force in the case of solid media. For these reasons it is preferable to base the discussion of dielectrics on (6.4) and (6.5).

(vii) Since, for an isotropic medium, $\mathbf{D} = \varepsilon \mathbf{E}$ it follows that $\mathbf{D} = -\varepsilon \boldsymbol{\nabla} V$. Therefore Poisson's equation takes the form

$$\boldsymbol{\nabla} \cdot (\varepsilon \boldsymbol{\nabla} V) = -\rho \tag{6.15}$$

which simplifies for regions over which ε is constant to

$$\nabla^2 V = -\rho/\varepsilon \tag{6.16}$$

It is important to realize that unless ε is constant there is no simple integral formula for V corresponding to (2.8). In certain simple cases it may be possible to solve (6.16) directly but unless the shape of the dielectric is particularly simple we have to resort to special methods or else solve the equation numerically. We shall see in Chapter 7 that the **only** general method of solving Poisson's equation is numerical. Before any solution can be found we must know what happens at the boundary of a dielectric.

Figure 6.1 shows the boundary between a dielectric and free space. Suppose we take a unit charge round the path ABCDA. Then, since the electric field is conservative, no net work is done. In the limit in which AD is vanishingly small it follows that the work done along AB is equal to that along DC. This is conveniently expressed by saying that if E_t is the component of $\mathbf{E}$ in any direction lying in the common boundary then at an interface between two media, 1 and 2

$$E_{t1} = E_{t2} \tag{6.17}$$

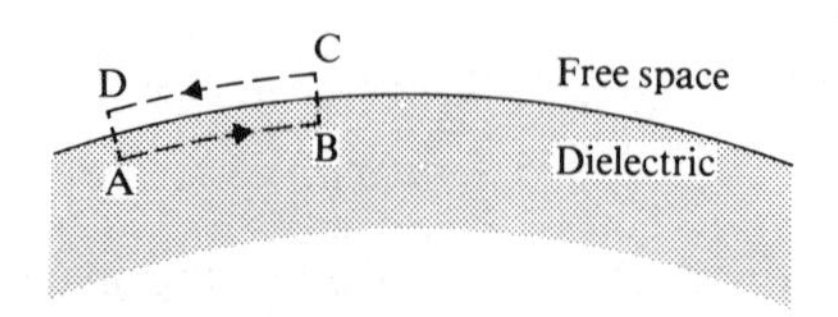

Fig. 6.1

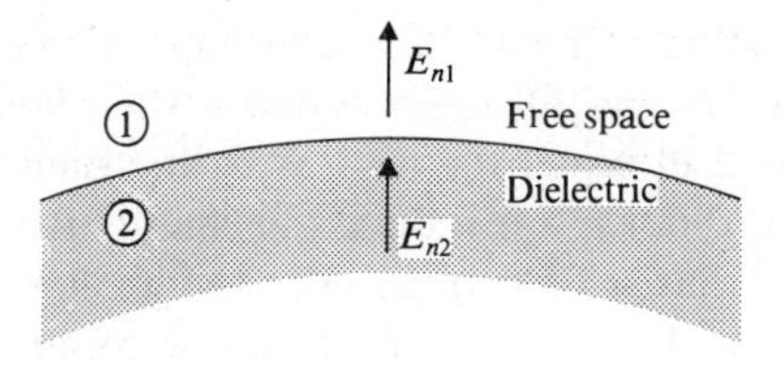

Fig. 6.2

The boundary condition for the normal component E_n may most easily be obtained by noting that when we cross a surface layer of charge σ the normal component of **E** changes by σ/ε_0. We saw in §6.2 that if the polarization is **P** there is an effective surface layer of apparent, or bound, charge equal to P_n. Thus E_n changes by P_n/ε_0, and in the notation of Fig. 6.2

$$E_{n1} - E_{n2} = P_n/\varepsilon_0$$

i.e.

$$D_{n1} = D_{n2} \tag{6.18}$$

It is left as an example to verify that (6.17), (6.18) are equally valid on passing from one dielectric to another.

§6.4 Parallel plate condenser

It is instructive to consider the particular example of a parallel plate condenser between the plates of which there is placed a slab of dielectric (Fig. 6.3). We know that in the absence of the

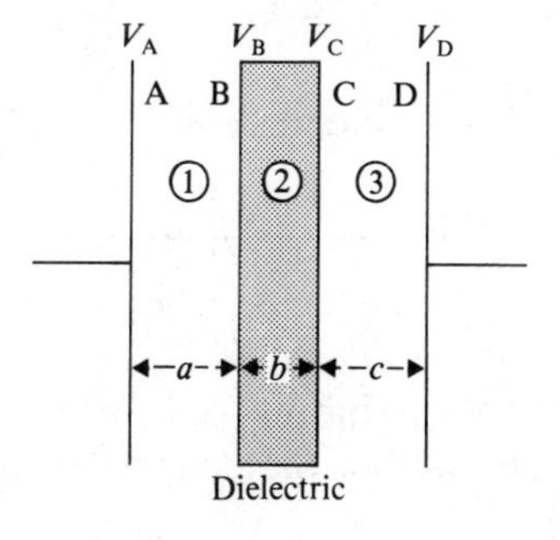

Fig. 6.3

dielectric, equal and opposite surface densities of free charge $\pm\sigma$ reside on the plates A and D while between the plates, the electric field is uniform and equal to σ/ε_0. With the dielectric inserted between the plates, dipoles are induced by the action of this field, a layer of negative charge of surface density $-\sigma_i$ appearing at B and, correspondingly, $+\sigma_i$ at C. In each of the three regions in Fig. 6.3 both **E** and **D** are constant and are directed from one plate to the other. Let the potentials at A, B, C, D be V_A, V_B, V_C, V_D respectively. Then since $\mathbf{E} = -\nabla V$ it follows that

$$E_1 = \frac{V_A - V_B}{a}, \qquad E_2 = \frac{V_B - V_C}{b}, \qquad E_3 = \frac{V_C - V_D}{c} \tag{6.19}$$

Regions 1 and 3 are free space and region 2 is filled with dielectric of permittivity ε so that

$$D_1 = \varepsilon_0 E_1 \qquad D_2 = \varepsilon E_2 \qquad D_3 = \varepsilon_0 E_3 \tag{6.20}$$

From the boundary condition (6.18) it follows that

$$D_1 = D_2 = D_3 = \beta, \quad \text{say}$$

So from (6.20)

$$E_1 = E_3 = \frac{\beta}{\varepsilon_0}, \qquad E_2 = \frac{\beta}{\varepsilon}$$

Using Coulomb's law, (6.12), we see that the charge density on plate A is β so that if A denotes the area of the plates, the condenser carries a charge βA. The total potential difference from (6.19) is

$$V_A - V_D = aE_1 + bE_2 + cE_3 = \beta\left(\frac{a}{\varepsilon_0} + \frac{b}{\varepsilon} + \frac{c}{\varepsilon_0}\right)$$

Hence the capacity of the condenser is

$$C = \frac{A\varepsilon_0}{\left(a + b\dfrac{\varepsilon_0}{\varepsilon} + c\right)}$$

Comparison with the expression in §3.8 for a condenser of thickness t in free space shows that the equivalent thickness of the condenser in Fig. 6.3 is $\left(a + \dfrac{b}{K} + c\right)$ where K is the dielectric constant.

Next suppose that A and D are attached to the terminals of a battery so that the potential difference $V_A - V_D$ is kept constant. Now let us draw the dielectric slab away from the condenser. When the slab is wholly or partly withdrawn, the capacity C is reduced, and hence the charge on the plates which is $C(V_A - V_D)$, is also reduced. This means that while the slab is being removed, a current will flow from plate A to plate D through the battery.

This condenser illustrates also the theorem on apparent charges at the end of §6.2. For inside the dielectric slab the polarization P is given by

$$\varepsilon_0 E_2 + P = D_2 = \beta$$

i.e.

$$P = \beta\left(1 - \frac{1}{K}\right)$$

Now $\mathbf{P}$ is constant so that $\nabla \cdot \mathbf{P} = 0$, and there is no apparent volume distribution of charge. But P_n at surfaces B and C has the value $\mp\beta\left(\frac{K-1}{K}\right)$ respectively. Thus according to (6.4) we may replace the dielectric slab by a layer of charge $-\beta\left(\frac{K-1}{K}\right)$ per unit area on the surface B and an equal positive layer on C and then treat the problem as though there was no dielectric present. These two layers of charge would then be responsible for changing E_1 to E_2, and E_2 to E_3 as we pass into and out of the dielectric, as is easily verified by using Gauss' flux theorem.

The field lines of **E**, **P** and **D** are sketched in Fig. 6.4.

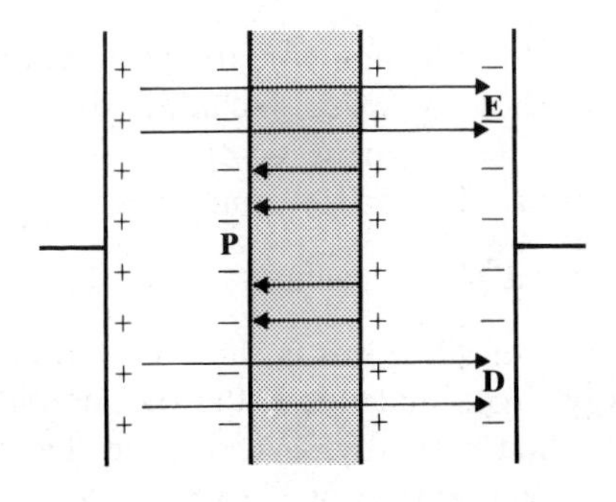

Fig. 6.4

§6.5 Energy of the electrostatic field

The expression obtained in §3.4 for the energy of the field needs to be modified when dielectrics are present. We may still start from the expression (3.6) since it is easily seen that the second proof of this result in §3.4 is unaffected by the presence of dielectrics. We must now substitute $\nabla \cdot \mathbf{D} = \rho$ and $D_n = \sigma$. Then with a parallel argument to that in §3.4 we find

$$\begin{aligned}\frac{1}{2}\int \rho V \, d\mathbf{r} &= \frac{1}{2}\int V\nabla \cdot \mathbf{D} \, d\mathbf{r} \\ &= \frac{1}{2}\int \{\nabla \cdot (V\mathbf{D}) - \mathbf{D} \cdot \nabla V\} \, d\mathbf{r} \\ &= \frac{1}{2}\int V\mathbf{D} \cdot d\mathbf{S} + \frac{1}{2}\int \mathbf{D} \cdot \mathbf{E} \, d\mathbf{r}\end{aligned}$$

The surface integral is taken outwards over the sphere at infinity – where a consideration of orders of magnitude shows that it is zero – and over the surface of each conductor, D_n being measured into the conductor. Now V is constant on each conductor, and the inward normal component, D_n, is $-\sigma$. It follows that

$$\frac{1}{2}\int \rho V \, d\mathbf{r} + \frac{1}{2}\int \sigma V \, dS = \frac{1}{2}\int \mathbf{D} \cdot \mathbf{E} \, d\mathbf{r}$$

i.e.

$$W = \frac{1}{2}\int \mathbf{D} \cdot \mathbf{E} \, d\mathbf{r} \tag{6.21}$$

Since $\mathbf{E}$ vanishes inside a conductor, the integration in (6.21) may be taken over the whole of space, including the conductors.

According to this the electrostatic energy is distributed over all space with density $\frac{1}{2}\mathbf{D} \cdot \mathbf{E}$. For isotropic dielectrics we may write this as $\frac{\varepsilon}{2}E^2$ or $D^2/2\varepsilon$. When calculating the force on any conductor using the results of §3.8, this is the appropriate expression to use for the energy. Note that this is the thermodynamic free energy and not ordinary energy, a distinction that is important if ε depends on the temperature or the pressure.

In the case of an anisotropic dielectric, e.g. a crystalline dielectric or a plasma in a magnetic field, the relation $\mathbf{D} = \varepsilon\mathbf{E}$ no longer holds and we have a situation analogous to the strain in a solid resulting from applied forces. An electric field applied to an anisotropic material along an axis of an arbitrary system of co-ordinates produces polarization components in all directions, i.e. a tensor relation holds between **P** and **E**:

$$P_i = \varepsilon_0 \chi_{ij} E_j \tag{6.22}$$

in which χ_{ij} is the susceptibility tensor. A tensor relation also holds between **D** and **E**, i.e.

$$D_i = \varepsilon_{ij} E_j \tag{6.23}$$

where ε_{ij} is the dielectric tensor. Examples involving a dielectric tensor will arise in Chapter 9.

§6.6 Minimum energy

Suppose that we are given certain conductors and each receives a given fixed total charge. We can prove that the charges distribute themselves over the surfaces of the various conductors in such a way that the energy W is a minimum. To establish this result let **E** be the true field and $\mathbf{E} + \mathbf{E}'$ the field that arises when the charges are held fixed in different positions on the conducting surfaces. In this condition the surfaces of the conductors will no longer be equipotentials, but since the total charge on each conductor is unaltered in the transformation from **E** to $\mathbf{E} + \mathbf{E}'$ we may apply Gauss' flux theorem to each conductor in the form

$$\int \varepsilon\mathbf{E} \cdot d\mathbf{S} = \text{total charge on conductor} = \int \varepsilon(\mathbf{E} + \mathbf{E}') \cdot d\mathbf{S}$$

$$\therefore \int \varepsilon\mathbf{E}' \cdot d\mathbf{S} = 0 \tag{6.24}$$

If we suppose there is no volume distribution of charge then

$$\nabla \cdot (\varepsilon\mathbf{E}) = 0 = \nabla \cdot \varepsilon(\mathbf{E} + \mathbf{E}')$$

$$\therefore \nabla \cdot \varepsilon\mathbf{E}' = 0 \tag{6.25}$$

Now consider

$$\int_V \varepsilon\mathbf{E}\cdot\mathbf{E}'\,d\mathbf{r} = -\int_V \varepsilon\mathbf{E}'\cdot\nabla V\,d\mathbf{r}$$

$$= -\int_V [\nabla\cdot(V\varepsilon\mathbf{E}') - V\nabla\cdot(\varepsilon\mathbf{E}')]\,d\mathbf{r}$$

The last term vanishes by (6.25). The other term may be transformed by Green's theorem into an integral over the sphere at infinity together with integrals over the surface of each conductor. Considerations of order of magnitude show that the contribution from the sphere at infinity is zero for any finite system of charges. The contribution from any conductor is $\int \varepsilon V\mathbf{E}'\cdot d\mathbf{S}$ where $d\mathbf{S}$ is measured outwards from the conductor. Now $V = V_S =$ constant on each S and so by (6.24)

$$\int \varepsilon\mathbf{E}\cdot\mathbf{E}'\,d\mathbf{r} = 0$$

But if $W + W'$ is the new energy

$$W' = \frac{1}{2}\int [\varepsilon(\mathbf{E}+\mathbf{E}')^2 - \varepsilon\mathbf{E}^2]\,d\mathbf{r}$$

$$= \frac{1}{2}\int \varepsilon\mathbf{E}'^2\,d\mathbf{r}$$

Since $\varepsilon > 0$, it follows that $W' > 0$, showing that in the actual configuration adopted by the charges W is a minimum.

§6.7 Stresses in the dielectric

Faraday pictured the tubes of electric displacement **D** in a state of stress, this stress giving rise to the energy $\mathbf{D}\cdot\mathbf{E}/2$ and being responsible for the apparent action-at-a-distance between charges. Let us see whether we can discover a stress system which will account for the inverse square law in an infinite dielectric of permittivity ε. We shall find, as Maxwell discovered, that it is necessary to suppose that tubes of **D** are not only in a state of tension, but that they also exert a force perpendicular to their direction on neighbouring tubes (cf. earlier discussion on magnetic tubes of force in §5.12).

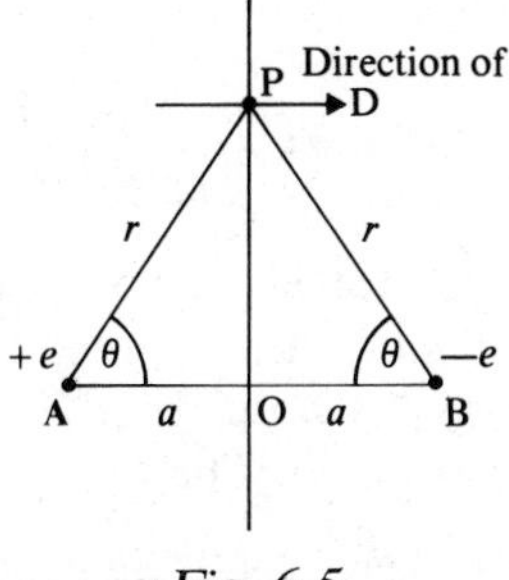

Fig. 6.5

Consider two charges $\pm e$ at A and B, a distance $2a$ apart (Fig. 6.5). According to (6.14) the force between them is $\mathbf{F} = \left(\dfrac{e^2}{16\pi\varepsilon a^2}\right)\hat{\mathbf{x}}$ where $\hat{\mathbf{x}}$ is a unit vector in the direction AB. If the tubes of **D** are in a state of stress, this is the force that they must exert across any boundary which completely separates the charges. In particular this is the force across the mid-plane PO. We choose this plane because at all points P the tubes of force cross the plane at right angles and we have only to consider the tension in the tubes: the repulsion between neighbouring tubes will make no contribution to the force **F**, for this is perpendicular to the plane PO.

Let $f(D)$ be the tension exerted across unit area of a tube where the displacement is **D**. Our problem is to calculate the form of the function f. Now the total pull across the mid-plane is $\int f(D)\, dS$, where dS is an element of area in the plane OP, and the integration extends over the whole plane. As D depends on OP and not on the azimuthal angle round AB, we can put

$$dS = 2\pi\, \mathrm{OP}\, d\,(\mathrm{OP}) = 2\pi^2 \frac{\sin\theta}{\cos^3\theta}\, d\theta$$

Thus

$$\frac{e^2}{16\pi\varepsilon a^2} = \text{total pull across plane} = \int_0^{\pi/2} f(D) 2\pi a^2 \frac{\sin\theta}{\cos^3\theta}\, d\theta$$

But the value of D at the point P is merely the sum of the values due to the charges $\pm e$ separately, so that

$$D = \frac{2e}{4\pi r^2}\cos\theta = \frac{e}{2\pi a^2}\cos^3\theta$$

Hence

$$\frac{e^2}{32\pi^2\varepsilon a^4}=\int_0^{\pi/2} f\left(\frac{e\cos^3\theta}{2\pi a^2}\right)\frac{\sin\theta}{\cos^3\theta}\,d\theta$$

This equation is to be true for all values of e. But e occurs in the form e^2 on the left-hand side; it must therefore occur in the same way on the right. This is only possible if $f(D)=\lambda D^2$, where λ is some constant to be determined. The equation then gives

$$\frac{e^2}{32\pi^2\varepsilon a^4}=\int_0^{\pi/2}\lambda\left(\frac{e\cos^3\theta}{2\pi a^2}\right)^2\frac{\sin\theta}{\cos^3\theta}\,d\theta=\frac{\lambda e^2}{16\pi^2 a^4}$$

Thus $\lambda = 1/2\varepsilon$, so that the tension $f(D)$ along a tube of force takes the form $D^2/2\varepsilon$, or $\varepsilon E^2/8\pi$, or $(\mathbf{D}\cdot\mathbf{E})/2$.

We have still to discuss the force between neighbouring tubes. To do this we replace the charge $-e$ at B in Fig. 6.5 by a charge $+e$ and consider the Coulomb repulsive force between the two charges $+e$ at A and B. We leave it as an exercise to show, by reasoning similar to that just used, that the appropriate force is obtained if each tube exerts on neighbouring tubes a repulsion $\varepsilon E^2/2$ per unit area.

These arguments have shown that in the particular case of two equal charges we need to assume a pressure and a tension along tubes of force, each of amount $\varepsilon E^2/2$. As these forces only involve the local values of ε and $\mathbf{E}$, it is reasonable to suppose that the same stress system would apply in other cases, and that the precise origin of the field is immaterial. It can indeed be proved that these stresses do give the correct mechanical forces whatever the charge distribution. But we shall content ourselves here with noticing that at the surface of a charged conductor where the tubes of force leave normally, we should expect them to exert on the conductor a mechanical force equal to $f(D)$, i.e. $\varepsilon E^2/2$ per unit area. This agrees exactly with the value previously deduced from Gauss' law in §6.3, i.e. $\sigma^2/2\varepsilon$.

From the extrapolation of the conclusions reached by considering the simple configuration of charges in Fig. 6.5 we may, by analogy with §5.12, suppose that the stress due to an electric field in a dielectric medium amounts to an isotropic pressure $\varepsilon E^2/2$ and a tension εE^2 along the lines of $\mathbf{D}$. By analogy we may introduce a stress tensor for the electric field T_{ik} given by

$$T_{ik}=\varepsilon[E_iE_k-\tfrac{1}{2}E^2\delta_{ik}]$$

In a **uniform** fluid dielectric which is in equilibrium in an electric field the electric stresses are of course balanced by the hydrostatic pressure, P. The total stress tensor may be written

$$\mathcal{T}_{ik} = -\left(P + \frac{\varepsilon E^2}{2}\right)\delta_{ik} + \varepsilon E_i E_k \tag{6.26}$$

We may apply this result to obtain the force at a dielectric-vacuum boundary. Denoting the normal and tangential components of the field by E_n, E_t in the dielectric and by primed quantities in the vacuum, we have, from (6.26)

$$-P + \frac{\varepsilon}{2}(E_n^2 - E_t^2) = \frac{\varepsilon_0}{2}(E_n'^2 - E_t'^2)$$

Applying the boundary conditions (6.17), (6.18) gives $E_t = E_t'$, $\varepsilon E_n = \varepsilon_0 E_n'$ so that

$$P = \frac{(\varepsilon_0 - \varepsilon)}{2}\left[E_t^2 + \frac{\varepsilon}{\varepsilon_0}E_n^2\right]$$

Thus there is a mechanical force on the dielectric directed from the dielectric into the vacuum (since $\varepsilon > \varepsilon_0$).

We have supposed that the dielectric fluid is incompressible. It is possible to show, though we shall not do so here, that if ε is a function of the density and temperature, i.e. $\varepsilon = \varepsilon(\rho, T)$ there are other forces exerted in the dielectric by an electric field. One of their effects is to make the dielectric expand or contract, the phenomenon being known as **electrostriction**.

§6.8 Cavities in a solid dielectric

We have hitherto defined **E** at any given point as the force on a unit charge placed at that point. This is quite satisfactory if the medium is a fluid since the charge can move and hence the force on it may be determined. But if the dielectric is solid, it is rather difficult to see what is meant by the definition of **E**, since to measure it we have to excavate a small cavity in the medium and imagine our charge to be placed inside; it is then no longer in the dielectric medium! We are therefore led to consider the field inside such a cavity. There are two types of cavity that are particularly important; we may refer to them as needle-shaped and

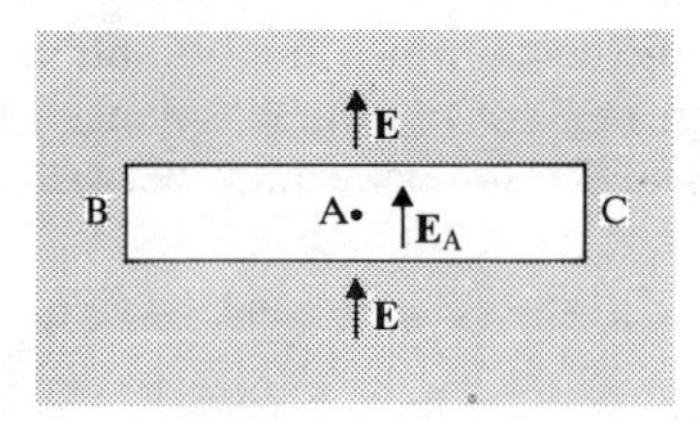

Fig. 6.6 Disc-shaped cavity; measures *D*

disc-shaped. In Fig. 6.6 we have supposed a cavity to be made having the shape of a thin disc whose plane is perpendicular to the direction of **E**; in Fig. 6.7 the cavity is needle-shaped, pointing in the same direction as **E**. Both cavities are small; all round them the dielectric permittivity is ε, but inside them it is ε_0.

Now at points inside the dielectric, but fairly near the cavities, the electric field may be regarded as the sum of two parts. One part is the undisturbed field, which was there before the cavity was made, and the other is the perturbing effect of the cavity. But if all the dimensions of the cavity are small, this latter contribution is small (cf. §§7.7, 7.8), so that the field outside the cavity is effectively the same as if there were no cavity at all. Let us call this **E**.

In Fig. 6.6 it follows from symmetry that in the central part of the cavity, the lines of force go straight across the gap; near the sides they will bend a little due to edge effects. But if the radius of the cavity is much greater than the thickness, the electric field near A may be regarded as uniform, and is related to **E** by

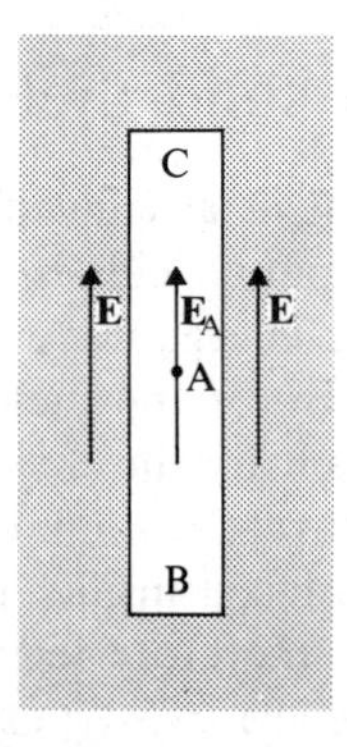

Fig. 6.7 Needle-shaped cavity; measures *E*

means of the continuity conditions (6.17), (6.18). Thus the field $\mathbf{E}_A$ in the central part of the cavity is directed across the cavity parallel to **E**, and from the fact that D_n is continuous it follows that

$$\varepsilon_0 \mathbf{E}_A = \varepsilon \mathbf{E} = \mathbf{D}$$

so that the force on a unit charge placed in the cavity of the shape of Fig. 6.6 will measure the displacement **D** in the solid and not the field **E**. Similarly in Fig. 6.7, the field at A will be parallel to **E** since the edge effects from B and C will be insignificant. So the boundary condition that the surface component of **E** is continuous tells us that

$$\mathbf{E}_A = \mathbf{E}$$

and the force on a unit charge in a needle-shaped cavity will indeed measure the internal field **E** in the solid. Cavities of different shapes will measure different combinations of **D** and **E**. We shall show in Chapter 7, for example (§7.8), that in a small spherical cavity the field is $3\mathbf{D}/(\varepsilon_0 + 2\varepsilon)$.

§6.9 Magnetic media

Thus far we have dealt with the electrical properties of materials in a simple classical and **macroscopic** way. We must now turn to the development of a macroscopic description of magnetic media in which the material in characterized by a few parameters. In Chapter 5 when we discussed the magnetic fields associated with charges in motion we supposed that the charges were moving in a vacuum. The question we must now answer is how will **B** be affected when matter is present?

In Chapter 1 we described Ampère's suggestion that each atom contained within itself minute electric currents – a hypothesis which has been amply confirmed by atomic physics. Consider now what happens in the presence of a magnetic field **B**. In many materials the minute electric currents associated with the orbital motion and spin of the electrons average to zero. When such atoms are put in a magnetic field then, on account of the changing magnetic field, tiny electric currents are generated by induction (cf. Chapter 8) in the cloud of electrons surrounding the nucleus. By Lenz's law (§8.1) these currents flow in a direction such that the magnetic fields associated with them **oppose** the inducing field **B**. The induced magnetic moments are therefore oriented in a direction opposite to **B** and this is the origin of **diamagnetism**.

In other materials the atoms do have a resultant magnetic moment on account of the fact that the currents from the orbital motions and spins of the electrons do not average to zero. While electron spins tend to pair and cancel one another there are many atoms in which the pairing is incomplete. When introduced into a magnetic field the magnetic moments of such atoms tend to align with the field. In these magnetic materials the induced magnetism enhances the external field and this behaviour is the origin of **paramagnetism**. Paramagnetism is a temperature-dependent phenomenon. Since thermal motion always tends to counter attempts by structures to arrange themselves in order, it is generally true that paramagnetic effects are more pronounced at low temperatures.

Our picture so far is one in which **all** atoms (whether or not their resultant magnetic moment vanishes), when placed in an external magnetic field, are diamagnetic while, in **some** cases, this behaviour is overridden by a paramagnetism due to a net magnetic moment. While this seems perfectly reasonable, a closer examination shows that it is completely incorrect from a classical point of view. When we talk about "net" effects we are in fact glossing over matters which need very careful consideration [4]. The correct procedure involves calculating an average of the magnetic moment for an assembly of atoms in thermal equilibrium. If we do this **classically** we find that the average magnetic moment is zero in all cases. Thus any respectable theory of magnetic media must be based on a quantum-mechanical and not a classical formulation. In those situations in which the electrons do obey classical mechanics – as in a plasma – then the classical arguments are valid. Plasmas exhibit diamagnetism as may easily be seen from the discussion in §5.11. We saw there that the net effect of many electrons gyrating in their Larmor orbits was to produce a current and so induce a magnetic field, the magnitude of which relative to the applied field **B** is given by

$$\frac{B_{\text{ind}}}{B} \sim \frac{\mu_0 j r_{\text{L}}}{B} \sim \frac{\mu_0 n m v_\perp^2}{B^2} \sim \frac{\mu_0 n W_\perp}{B^2}$$

where n is the particle density. Since the induced field opposes the applied field it is clear that a plasma is diamagnetic.

Although formal classical arguments are inadequate to explain the behaviour of magnetic media we shall use them nonetheless, in that as far as diamagnetic and paramagnetic behaviour is concerned, they give us some idea of what is going on. There is however another kind of magnetism, **ferromagnetism**, which can be understood **only** in terms of quantum mechanics. The basic facts are well known. Ferromagnetic materials – iron, cobalt, nickel together with certain alloys – when brought near a magnet, are attracted very strongly. Unlike paramagnetic materials they do not readily shed their magnetic properties when the temperature is increased. For example, nickel has to be heated to over 630 K and iron to about 1 040 K before their ferromagnetic properties are reduced to weakly paramagnetic. This property was discovered by Curie and the critical temperature at which this change of behaviour occurs is known as the **Curie temperature**.

This strange behaviour has its origin in the interactions between unpaired electron spins and so is inherently quantum mechanical. Roughly speaking what happens in, for example, iron is that there are five unpaired electrons. These electrons interact in such a way that their spins align in the same direction. Furthermore this alignment within one atom of iron is strong enough to influence its neighbours so that the atoms behave **collectively** rather than as individuals. This collective behaviour of ferromagnetic crystals was first observed by Bitter. By allowing very fine iron powder to settle on the surface of a crystal Bitter found that irregular **domains** were formed. Domains are typically $10^{-12}\,\mathrm{m}^3$ to $10^{-16}\,\mathrm{m}^3$ in volume, containing on average 10^{15} atoms. Within the domains there is almost complete alignment of the electron spins of these atoms but the domains themselves are randomly oriented so that there is no overall magnetization of the specimen. However, when a ferromagnetic substance is placed in an external magnetic field the domains rotate collectively and line up with the field.

§6.10 Magnetization current

We saw in §5.9 that a current flowing in a small loop was equivalent to a magnetic dipole of moment $\mathbf{m}$. Assuming a distribution of magnetic dipoles over a volume element $d\mathbf{r}'$ –

corresponding to the atomic currents discussed in the previous section – we may introduce a quantity **M**, the **magnetization**, defined by

$$d\mathbf{m} = \mathbf{M}\,d\mathbf{r}' \tag{6.27}$$

Then from (5.31),

$$\begin{aligned}\mathbf{A}(\mathbf{r}) &= \frac{\mu_0}{4\pi}\int \mathbf{M}(\mathbf{r}')\times\nabla'\left(\frac{1}{r}\right)d\mathbf{r}' \\ &= \frac{\mu_0}{4\pi}\left[\int \frac{\nabla'\times\mathbf{M}}{r}\,d\mathbf{r}' - \int \nabla'\times\left(\frac{\mathbf{M}}{r}\right)d\mathbf{r}'\right]\end{aligned} \tag{6.28}$$

We now use the fact that for any vector **a**

$$\int \nabla\times\mathbf{a}\,d\mathbf{r}' = -\int \mathbf{a}\times d\mathbf{S}' \tag{6.29}$$

To show this consider any constant vector **t**; then

$$\begin{aligned}\mathbf{t}\cdot\int \mathbf{a}\times d\mathbf{S}' &= \int (\mathbf{t}\times\mathbf{a})\cdot d\mathbf{S}' \\ &= \int \nabla'\cdot(\mathbf{t}\times\mathbf{a})\,d\mathbf{r}' \\ &= -\mathbf{t}\cdot\int \nabla\times\mathbf{a}\,d\mathbf{r}'\end{aligned}$$

which, since **t** is an arbitrary vector, gives (6.29). Using (6.29) in the second integral in (6.28) gives

$$\mathbf{A}(\mathbf{r}) = \frac{\mu_0}{4\pi}\left[\int \frac{\nabla'\times\mathbf{M}}{r}\,d\mathbf{r}' + \int \frac{\mathbf{M}\times d\mathbf{S}'}{r}\right] \tag{6.30}$$

Provided the surface integral may be taken over a surface outside the region in which current flows, it will be zero. From (5.26) we know that

$$\mathbf{A} = \frac{\mu_0}{4\pi}\int \frac{\mathbf{j}(\mathbf{r}')}{r}\,d\mathbf{r}'$$

and since (6.30) – with the above proviso – reads

$$\mathbf{A} = \frac{\mu_0}{4\pi}\int \frac{\nabla'\times\mathbf{M}}{r}\,d\mathbf{r}'$$

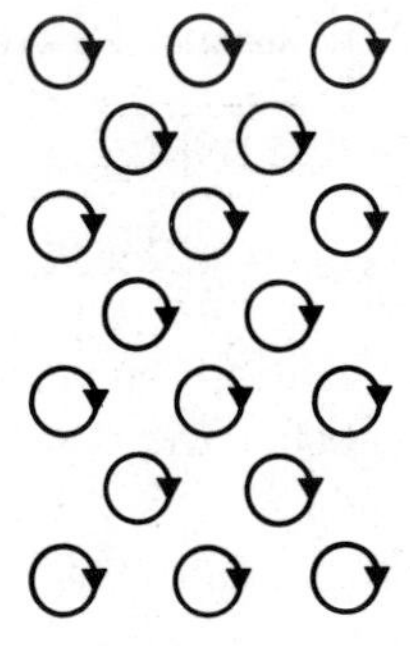

Fig. 6.8

it follows that $\nabla\times\mathbf{M}$ may be identified with a current density. We may write

$$\nabla\times\mathbf{M}=\mathbf{j}_M \tag{6.31}$$

where $\mathbf{j}_M$ is the **magnetization current density**. The subscript M is added to distinguish it from $\mathbf{j}$, the current density due to free or conduction electrons. That due to $\mathbf{M}$, the resultant magnetic moment per unit volume, originates from the Amperian currents. Unlike the conduction current, these neither transport charge nor produce ohmic heating.

Physically we may interpret (6.31) by the following model. From Fig. 6.8 we see that when the magnetization is uniform the microscopic currents tend to cancel so that the net current vanishes. However, for the non-uniform $\mathbf{M}$ in Fig. 6.9 clearly there is net excess of charge moving down at A over that in the upwards direction at B. Using this model it is straightforward (problem 28) to show that (6.31) holds.

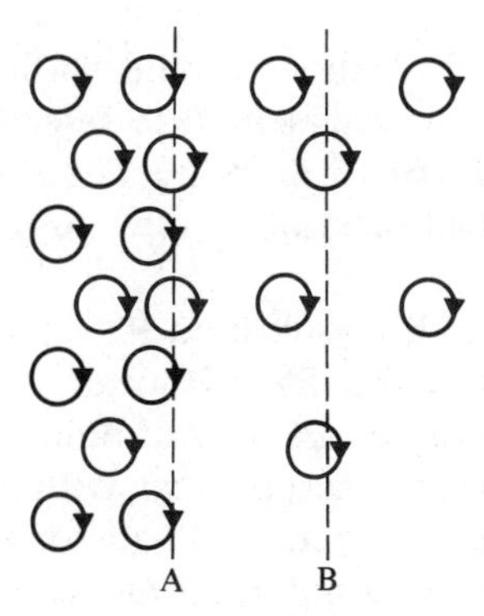

Fig. 6.9

§6.11 Equations of the magnetic field

We may now write the total current density in magnetic media as

$$\mathbf{j}_{\text{total}} = \mathbf{j} + \nabla \times \mathbf{M} \tag{6.32}$$

where $\mathbf{j}$ is the free (or conduction) current density. The new current density $\mathbf{j}_{\text{total}}$ clearly satisfies the continuity equation (4.6), since $\nabla \cdot \nabla \times \mathbf{M} \equiv 0$.

From Ampère's law, (5.21), we have

$$\begin{aligned} \nabla \times \mathbf{B} &= \mu_0 \mathbf{j}_{\text{total}} \\ &= \mu_0(\mathbf{j} + \nabla \times \mathbf{M}) \end{aligned}$$

i.e.

$$\nabla \times \left(\frac{\mathbf{B}}{\mu_0} - \mathbf{M} \right) = \mathbf{j}$$

If we introduce a new vector $\mathbf{H}$ defined by

$$\mathbf{H} = \frac{\mathbf{B}}{\mu_0} - \mathbf{M} \tag{6.33}$$

Ampère's law now reads

$$\nabla \times \mathbf{H} = \mathbf{j} \tag{6.34}$$

or, in integral form,

$$\oint \mathbf{H} \cdot d\mathbf{s} = I \tag{6.35}$$

where $\mathbf{j}$ is the true conduction current density and $I = \int \mathbf{j} \cdot d\mathbf{S}$. The quantity $\mathbf{H}$ is called the **magnetic field intensity** and is measured in units of **ampere metres**$^{-1}$, as is $\mathbf{M}$. Despite the label just attached to $\mathbf{H}$ we shall most often refer to it as the magnetic field $\mathbf{H}$.

Earlier editions of this book used $\mathbf{H}$ as the fundamental vector of the magnetic field and defined $\mathbf{B}$ in terms of $\mathbf{H}$ and the magnetization, although the desirability of doing things the other way round was recognized in an appendix. With our definition we see that the field $\mathbf{B}$ is the magnetic analogue of the electric field $\mathbf{E}$ while $\mathbf{H}$ is the magnetic counterpart of the electric displacement $\mathbf{D}$ as in Table 6.1

Table 6.1

Electrostatics	Magnetostatics
$\int \mathbf{E} \cdot d\mathbf{S} = \frac{Q}{\varepsilon_0}$	$\oint \mathbf{B} \cdot d\mathbf{s} = \mu_o I$
$\int \mathbf{D} \cdot d\mathbf{S} = Q$	$\oint \mathbf{H} \cdot d\mathbf{s} = I$

In free space $\mathbf{M} = \mathbf{0}$ and (6.33) reduces to

$$\mu_0 \mathbf{H} = \mathbf{B}$$

§6.12 Permanent magnets

A permanent magnet is to be regarded as an agglomeration of small magnetic particles and described in terms of the magnetization **M**. In this case, with no conduction currents,

$$\nabla \times \mathbf{H} = 0 \tag{6.36}$$

though of course

$$\nabla \times \mathbf{B} = \mu_0 (\nabla \times \mathbf{M}) \neq 0$$

Moreover from $\nabla \cdot \mathbf{B} = 0$ it follows that

$$\nabla \cdot \mathbf{H} = -\nabla \cdot \mathbf{M} \tag{6.37}$$

In this case the field **H** is like the electrostatic field **E**. It is a conservative field and may be derived from a scalar potential V_m, i.e.

$$\mathbf{H} = -\nabla V_m \tag{6.38}$$

We may extend the analogy further by introducing a magnetic source density $\rho_m = -\nabla \cdot \mathbf{M}$, so that from (6.37)

$$\nabla \cdot \mathbf{H} = \rho_m \tag{6.39}$$

The magnetic "charges" or **magnetic monopoles** – though they do not exist as far as **classical** electrodynamics is concerned – may be thought of as the analogues of electric charges. In particular they are the sources of the scalar potential V_m. By analogy with electrostatics we may write

$$V_m = \frac{1}{4\pi} \left[\int \frac{\mathbf{M} \cdot d\mathbf{S}'}{r} - \int \frac{\nabla \cdot \mathbf{M}}{r} d\mathbf{r}' \right] \tag{6.40}$$

where $-\nabla \cdot \mathbf{M}$ is the volume density distribution of magnetic monopoles and M_n the surface density of monopoles.

At points far enough away from the magnets that $1/r$ may be regarded as effectively the same for all the magnetic particles

$$
\begin{aligned}
V_m &= \frac{1}{4\pi}\int \mathbf{M}(\mathbf{r}')\cdot\nabla'\left(\frac{1}{r}\right) d\mathbf{r}' \\
&= -\frac{1}{4\pi}\int \mathbf{M}(\mathbf{r}')\cdot\nabla\left(\frac{1}{r}\right) d\mathbf{r}' \\
&= -\frac{1}{4\pi}\nabla\left(\frac{1}{r}\right)\cdot\int \mathbf{M}(\mathbf{r}')\, d\mathbf{r}' \\
&= -\frac{1}{4\pi}\left(\mathcal{M}\cdot\nabla\left(\frac{1}{r}\right)\right) \qquad (6.41)
\end{aligned}
$$

where $\mathcal{M} = \int \mathbf{M}(\mathbf{r}')\, d\mathbf{r}'$ is the total magnetic moment. At large distances the system behaves like a single magnetic particle of moment $\mathcal{M}$.

The field of a permanent magnet may also be described in terms of a vector potential $\mathbf{A}$ by (6.30):

$$
\mathbf{A}(\mathbf{r}) = \frac{\mu_0}{4\pi}\left[\int \frac{\nabla'\times\mathbf{M}}{r}\, d\mathbf{r}' + \int \frac{\mathbf{M}\times d\mathbf{S}'}{r}\right]
$$

in which the surface integral is now evaluated over the surface of the magnet itself.

§6.13 A uniform bar magnet

As an example consider a bar magnet of uniform circular cross-section (Fig. 6.10). The ends are perpendicular to the length of the rod, a distance l apart, and we suppose that the magnetization $\mathbf{M}$ is uniform and directed along the bar.

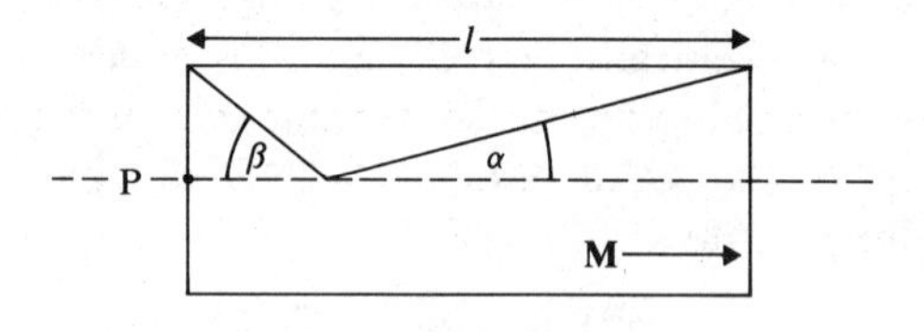

Fig. 6.10

The potential due to this magnet may be found from (6.40). Since $\nabla \cdot \mathbf{M} = 0$ the magnetic field $\mathbf{H}$ is the same as if we had a layer of positive magnetic monopoles M at one end and an equal layer of negative monopoles at the other. At large distances therefore the magnet behaves like a single magnetic dipole of moment $\pi a^2 lM$, a being the radius of either end. For closer distances it behaves as if poles $\pm\pi a^2 M$ existed at the two ends. Very close to either pole we must take account of the distribution over the whole circular end.

The magnetic field $\mathbf{B}$ can be expressed in terms of the amperian surface current density j_S. In §5.13 we saw that the magnetic field within a solenoid of length l and N turns could be expressed as

$$B = \frac{\mu_0 NI}{2l}(\cos\alpha + \cos\beta)$$

where I is the current in the winding. If we think of the $\mathbf{B}$ field in this case as that due to a solenoid wound on the surface of the magnet then, since $j_S l = NI$,

$$B = \frac{\mu_0 j_S}{2}(\cos\alpha + \cos\beta) \tag{6.42}$$

Moreover $M = \mu_0 j_S$ so that

$$\begin{aligned} H &= (B - M)/\mu_0 \\ &= \frac{j_S}{2}(\cos\alpha + \cos\beta - 2) \end{aligned} \tag{6.43}$$

We see from (6.43) that the fields B and H are in opposite directions within the magnet. At the point P

$$B_P = \frac{\mu_0 j_S}{2}\cos\alpha$$

so that on either side of the end of the magnet $B_{P-} = B_{P+}$. For P just within the magnet, (6.43) tells us that

$$H_{P-} = \frac{j_S}{2}(\cos\alpha - 2)$$

while for P just outside,

$$H_{P+} = \frac{j_S}{2}\cos\alpha$$

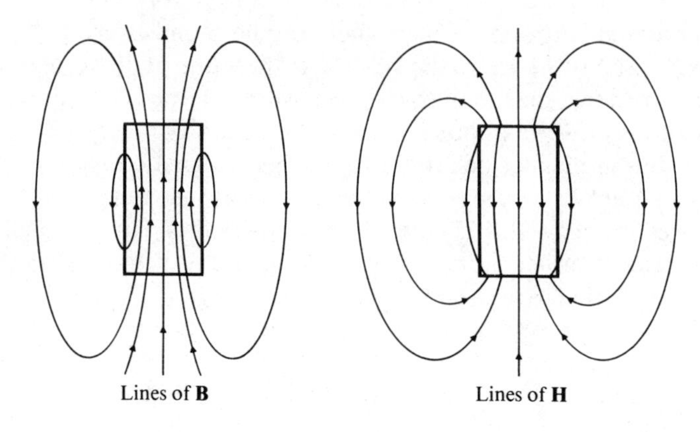

Fig. 6.11

i.e.

$$H_{P+} - H_{P-} = j_S = \frac{M}{\mu_0} \tag{6.44}$$

while

$$B_{P+} - B_{P-} = 0 \tag{6.45}$$

The field lines for both **B** and **H** are drawn in Fig. 6.11.

§6.14 Magnetic permeability

If we except ferromagnetic substances, many materials exhibit linear magnetic behaviour, i.e. the magnetization **M** is proportional to **B** and therefore to **H**. Ferromagnetic materials on the other hand are strongly nonlinear. Thus for diamagnetic and paramagnetic materials we may write

$$\mathbf{M} = \chi_m \mathbf{H} \tag{6.46}$$

so that

$$\begin{aligned} \mathbf{B} &= \mu_0(1+\chi_m)\mathbf{H} \\ &= \mu\mathbf{H} \end{aligned} \tag{6.47}$$

In (6.46) χ_m is the **magnetic susceptibility** and in (6.47) μ is the **permeability**. The ratio μ/μ_0 is the relative permeability of the material:

$$K_m = \frac{\mu}{\mu_0} = 1 + \chi_m \tag{6.48}$$

Recapitulating the equivalent results for dielectric materials from §6.3 we find

$$\mathbf{D} = \varepsilon_0(1+\chi)\mathbf{E} = \varepsilon\mathbf{E}$$

with the relative permittivity or dielectric constant given by

$$K = \frac{\varepsilon}{\varepsilon_0} = 1 + \chi \tag{6.9}$$

For diamagnetic materials $\chi_m < 0$, $\mu < \mu_0$ and $K_m < 1$, while for paramagnetic substances $\chi_m > 0$, $\mu > \mu_0$, $K_m > 1$. For ferromagnetics μ is very large.

As with dielectrics there are magnetic materials, ferrites, which are anisotropic. For such materials (6.47) is replaced by the tensor relation

$$B_i = \mu_{ij} E_j \tag{6.49}$$

where μ_{ij} is the permeability tensor.

§6.15 Boundary conditions

In analogy to the boundary conditions (6.17) and (6.18) on **E** and **D** we must discuss the continuity conditions satisfied by **B** and **H** at the boundary between two media in which the permeability is different (Fig. 6.12). Since the equation $\nabla \cdot \mathbf{B} = 0$ is unaltered by the presence of magnetic materials the normal component of **B** must satisfy the condition

$$B_{n1} = B_{n2} \tag{6.50}$$

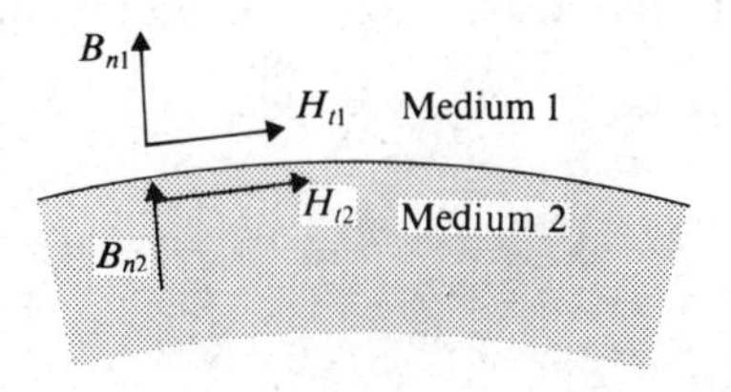

Fig. 6.12

The condition on **H** is found from (6.35) by integrating round an elemental circuit as in §5.10, i.e.

$$\oint \mathbf{H} \cdot d\mathbf{s} = H_{t1}\Delta s - H_{t2}\Delta s = I$$

If we define a surface current I_S by $I_S \Delta s = I$ this gives

$$H_{t1} - H_{t2} = I_S \tag{6.51}$$

i.e. there is a discontinuity in the tangential component of the magnetic field **H** equal to the surface current in amperes per metre at the boundary between the two media.

We ought also to include boundary conditions for **A** at a change of medium. If $\hat{\mathbf{n}}$ is a unit vector directed along the normal from medium 1 to medium 2 these are

$$\mathbf{A}_1 = \mathbf{A}_2 \tag{6.52}$$

$$\frac{\partial \mathbf{A}_1}{\partial n} - \frac{\partial \mathbf{A}_2}{\partial n} = (\chi_{m1}\mathbf{H}_1 - \chi_{m2}\mathbf{H}_2) \times \hat{\mathbf{n}} \tag{6.53}$$

It is possible to verify the sufficiency of these conditions by deducing (6.50), (6.51) from them.

§6.16 Cavities

The boundary conditions (6.50), (6.51) allow us to discuss the field inside a small cavity scooped out of the magnetic medium. Consideration of this problem is important because of the indefiniteness which attaches to our definition and measurement of **B** and **H** at points inside a medium. Our best plan is to suppose part of the medium around some point P to be removed and then to determine **B** and **H** at P inside the cavity. The argument is very similar to that already used in §6.8. In fact, since the boundary conditions for **B** and **H** in this case are exactly similar to those for **D** and **E**, the analysis of §6.8 can be applied to this

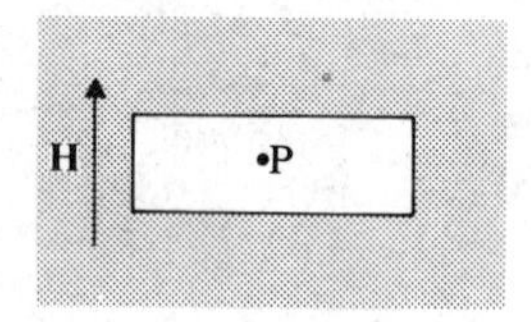

Fig. 6.13

problem just as it stands. Thus, with a disc-shaped cavity, as in Fig. 6.13, which may be compared with Fig. 6.6, the magnetic field at P inside the hollow is the same as the **B** field outside, and with a needle-shaped cavity pointing in the direction of **H**, as in Fig. 6.14, which compares with Fig. 6.7, the magnetic field at P inside the cavity is given by $\mathbf{B}_{\text{cavity}} = \mathbf{B}_{\text{outside}} - \mu_0 \mathbf{M}$.

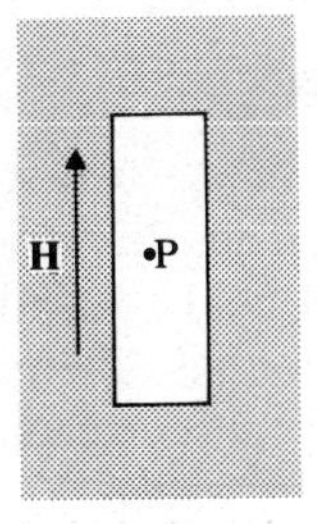

Fig. 6.14

§6.17 Problems

1. Show from a consideration of the boundary conditions (6.18) that at a change of dielectric medium field lines are refracted according to the law $K_1 \cot \theta_1 = K_2 \cot \theta_2$, where θ_1 and θ_2 are the angles between the directions of the field and the common normal to the boundary.

2. S_1 and S_2 are two equipotential surfaces completely surrounding an electrostatic system on which the net charge is Q; their potentials are V_1 and V_2 and there is no dielectric between them. Prove that if S_1 and S_2 are taken to be the plates of a condenser and the medium between them is of uniform dielectric permittivity ε, the capacity is $4\pi\varepsilon Q/(V_1 - V_2)$.

3. A spherical condenser consists of two concentric spheres of radii a and d. Concentric with these and lying between them is a spherical shell of dielectric of permittivity ε bounded by the spheres $r = b$, $r = c$. Show that if $a < b < c < d$, the capacity C of the condenser is given by

$$\frac{4\pi\varepsilon_0}{C} = \frac{1}{a} - \frac{1}{d} + \frac{(\varepsilon_0 - \varepsilon)}{\varepsilon}\left(\frac{1}{b} - \frac{1}{c}\right)$$

4. Show that the polarization vector between $r=b$ and $r=c$ in the previous question is $\dfrac{\varepsilon-\varepsilon_0}{4\pi\varepsilon}\times\dfrac{Q}{r^2}$, directed radially, Q being the charge on either plate of the condenser. Show that the equivalent bound charges are $\rho=0$ together with two surface layers

$$\sigma_b=-\frac{\varepsilon-\varepsilon_0}{4\pi\varepsilon b^2}Q \quad \text{and} \quad \sigma_c=\frac{\varepsilon-\varepsilon_0}{4\pi\varepsilon}\frac{Q}{c^2}$$

at the boundary of the dielectric.

5. A condenser is formed of the two spheres $r=a$, $r=b(b>a)$ with uniform dielectric K. The dielectric strength of the medium (i.e. the greatest permitted field strength before it conducts) is E_0. Show that the greatest potential difference between the two plates, so that the field nowhere exceeds the critical value, is $E_0a(b-a)/b$.

6. If in the previous question the potential difference is gradually increased beyond the critical value so that charge can flow into part of the dielectric, show that the condenser does not break down completely until the voltage is increased to $E_0(b-a)$. Investigate the distribution of charges when this latter condition is reached.

7. The slab of dielectric in Fig. 6.3 is partly removed from between the plates so that of the total area A an area x is covered with dielectric and $A-x$ with free space. Assuming that the lines of force go straight from one plate to the other, show that the capacity C is given by

$$C=\frac{A\varepsilon_0}{(a+b+c)}+\frac{bx(\varepsilon-\varepsilon_0)}{(a+b+c)\left(a+\dfrac{b\varepsilon_0}{\varepsilon}+c\right)}$$

Calculate the energy and show that if the charge on the plates is kept constant the dielectric is drawn into the condenser.

8. There is a point charge at the origin in an infinite dielectric medium K. Verify that the stress system of §6.7 is such that there is no resultant force on the volume confined between the spherical surfaces $r=a$, $r=b$ and a diametral plane through the charge.

9. A conducting sphere of radius a in an infinite dielectric medium receives a charge Q. It is now divided in two by a

diametral plane and the halves are slightly separated. Show by a consideration of the stresses between tubes of force that the one part exerts on the other a force $Q^2/32\pi\varepsilon a^2$. Verify that this is exactly equivalent to the vector addition of forces $\sigma^2/2\varepsilon$ per unit area of surface, where σ is the surface density of charge.

10. A condenser consists of the conducting spheres $r=a$, $r=b$. The dielectric permittivity is independent of r but varies with direction. Write down the differential equation for the potential, and show that it is satisfied by the expression $V=A+B/r$, where A and B are constants. Show that B is related to the charge Q on the inner sphere $r=a$ by the equation $Q=B\int \varepsilon\, d\omega$, $d\omega$ being an element of solid angle. Hence show that the capacity of the condenser is

$$\frac{ab}{(b-a)}\int \varepsilon\, d\omega$$

11. Show that in an electrostatic problem in which the potential depends solely upon the radial distance r, the differential equation for V is

$$\frac{1}{r^2}\frac{d}{dr}\left(r^2\varepsilon\frac{dV}{dr}\right)=-\rho$$

A charge e is placed at the origin in a medium in which the dielectric constant is $1+a/r$. Show that the potential is

$$V=\frac{e}{4\pi\varepsilon_0}\ln\frac{r+a}{r}$$

12. A small particle of polarizability α is placed in an inhomogeneous field $\mathbf{E}$. Show that the force on it is $\frac{1}{2}\alpha\boldsymbol{\nabla}E^2$. (Note that this implies that a small dielectric particle moves towards places of higher field strength.)

13. Show that the mechanical force on a conductor is $\sigma^2/2\varepsilon$ per unit area (cf. §6.3).

14. Show that the energy expended in polarizing a dielectric is $(\varepsilon-\varepsilon_0)E^2/2$ per unit volume. Deduce that the energy of the field, regarded as the sum of the energy $\dfrac{\varepsilon_0 E^2}{2}$ in free space and $\frac{1}{2}(\varepsilon-\varepsilon_0)E^2$ in the dielectric, is $\displaystyle\int\frac{\mathbf{D}\cdot\mathbf{E}}{2}\,d\mathbf{r}$.

15. V and V' are the potential functions due to two different distributions of charge on a given series of m conductors. There are no other charges present. Show that if $\mathbf{a}$ is the vector $KV'\nabla V - KV\nabla V'$, then $\nabla \cdot \mathbf{a} = 0$. Next, using the divergence theorem $\int \mathbf{a} \cdot d\mathbf{S} = \int \nabla \cdot \mathbf{a}\, d\mathbf{r}$, deduce **Green's Reciprocal Theorem** (cf. §3.10, problem 2), that if the conductors carry charges Q_j, Q_j' and their potentials are V_j, V_j', then

$$\sum Q_j V_j' = \sum Q_j' V_j$$

16. $S_1 \ldots S_m$ are m conductors carrying charges $Q_1 \ldots Q_m$ at potentials $V_1 \ldots V_m$. A new conductor S_{m+1} is introduced carrying a charge Q_{m+1}, and the new potentials are found to be $V_1', V_2' \ldots V_{m+1}'$. The two fields are $\mathbf{E}$ and $\mathbf{E}'$. Show that

$$\int \varepsilon \mathbf{E} \cdot \mathbf{E}'\, d\mathbf{r} = \int \varepsilon \mathbf{E}' \cdot \mathbf{E}'\, d\mathbf{r} - Q_{m+1} V_{m+1}'$$

the integration being over all space outside the $m+1$ conductors. Deduce that if the new conductor is either uncharged or earthed, the electrical energy is diminished (cf. §3.10, problem 26).

17. The volume charge distribution ρ in a given electrostatic system, and the total charge on each conductor, are kept unaltered, but the dielectric permittivity changed from ε to $\varepsilon + \varepsilon'$, neither of which is necessarily constant throughout all space. The value of the field changes from $\mathbf{E}$ to $\mathbf{E} + \mathbf{E}'$. $\mathbf{E}'$ and ε' are small quantities. Show that

$$\nabla \cdot (\varepsilon \mathbf{E}') = -\nabla \cdot (\varepsilon' \mathbf{E}), \quad \text{and} \quad \int \varepsilon' \mathbf{E} \cdot d\mathbf{S} = -\int \varepsilon \mathbf{E}' \cdot d\mathbf{S}$$

over each conductor. Further, writing $\mathbf{E} = -\nabla V$, and using Green's theorem, prove that

$$\int \varepsilon \mathbf{E} \cdot \mathbf{E}' d\mathbf{r} = -\int \varepsilon' \mathbf{E} \cdot \mathbf{E}\, d\mathbf{r}$$

and deduce that the change in energy is $-\frac{1}{2} \int \varepsilon' \mathbf{E}^2\, d\mathbf{r}$. (Notice that this proves **Thomson's theorem** that an increase in the dielectric constant without alteration of charges decreases the energy of the field.)

18. Show that the field inside the cavities of Figs. 6.6 and 6.7 may be obtained by the use of the polarization charges discussed in §6.2.

19. Show that lines of **H** are refracted at a change of medium, and that if the angles made with the normal are θ_1 and θ_2, then

$$\mu_1 \cot \theta_1 = \mu_2 \cot \theta_2$$

20. A torus, or anchor ring, is obtained by rotating a circle about an axis in its own plane distant f from the centre. The volume so formed is filled with soft iron of permeability μ. A total of N turns of wire are wound closely and evenly round the ring, and the wire carries a current I. Show that at points inside the ring the magnetic field B is $\mu NI/2\pi r$, where r is the distance from the axis of rotation. A small air-gap is now made in the iron by cutting away a thin sector bounded by two planes through the central axis which make a small angle α with each other. Show that the total induction in the air-gap is thus reduced by a factor $\left[1+\left(\frac{\mu}{\mu_0}-1\right)\frac{\alpha}{2\pi}\right]^{-1}$. This is one way of creating a strong magnetic field by means of a current, if α is kept small.

21. Deduce the formula $(\mathbf{m}\cdot\text{grad})\,\mathbf{H}$ for the force on a magnetic particle in a field **H** by regarding the particle as a pair of equal and opposite magnetic poles a small distance apart.

22. One sphere completely surrounds another. The space between is uniformly magnetized. Show that there is no field inside the cavity. Write down an expression for the magnetostatic potential at any point.

23. At a point on the axis inside a bar magnet there is a small spherical hole. Show that the field at the centre of this hole is $-\frac{2}{3}M+\frac{M}{2}(\cos\alpha+\cos\beta)$ where α and β are the same angles as in §6.13. If the magnet is very long deduce that the field is $\frac{1}{3}M$ in a direction opposite to the magnetization.

24. Show that for a uniformly magnetized solid $\mathbf{A}=\frac{\mu_0}{4\pi}\mathbf{M}\times\int\frac{d\mathbf{S}'}{r}$.

Next show that for a sphere of radius a and centre O, $\int\frac{d\mathbf{S}'}{r}$ is a vector in the direction from O to the point P from which r is measured. If $\mathbf{OP}=\mathbf{R}$, prove that

$$\int\frac{d\mathbf{S}'}{r}=\frac{\pi\mathbf{R}}{R^3}\int_{|a-R|}^{a+R}(a^2+R^2-r^2)\,dr$$

Deduce that the magnetic vector potential for a uniformly magnetized sphere is given by

$$\text{outside the sphere } \mathbf{A} = \frac{\mu_0 a^3}{3} \frac{\mathbf{M} \times \mathbf{R}}{R^3}$$

$$\text{inside the sphere } \mathbf{A} = \frac{\mu_0}{3} \mathbf{M} \times \mathbf{R}$$

Interpret your result.

25. Show that the magnetostatic potential V_m for a uniformly magnetized solid is the same, at outside points, as the potential due to a magnetic shell closely surrounding the body and of strength Mz, where z is measured in the same direction as the magnetization $\mathbf{M}$.

26. A cylindrical bar magnet of length l is uniformly magnetized with magnetization $\mathbf{M}$. Show that the vector potential given by (6.30) is precisely the same as we should find for a current lM flowing uniformly round the curved surface of the magnet.

27. Show that the boundary conditions (6.52), (6.53) for $\mathbf{A}$ lead to the more familiar conditions (6.50), (6.51) for $\mathbf{B}$ and $\mathbf{H}$.

28. By considering two adjacent volume elements in a magnetic material show that the net current at their interface in the direction Oz is $-\dfrac{\partial M_x}{\partial y} \Delta x \, \Delta y$. Repeat the argument for adjoining elements along Ox and hence deduce (6.31).

7

Potential theory and applications

§7.1 Mathematical equivalence of all potential problems

The solution of a problem in electrostatics may be regarded as known when we have determined the electrostatic potential at all points, in other words when we have solved Poisson's equation

$$\nabla^2 V = -\rho/\varepsilon_0 \qquad (7.1)$$

subject to the boundary conditions appropriate to some particular configuration of electrodes and applied potentials. The same is true for the distribution of current and for magnetostatics. We have seen, for example, in Chapter 5 that the magnetic vector potential $\mathbf{A}$ also satisfies Poisson's equation. Mathematically therefore the differential equations which determine the potential are structurally identical; the differences occur mainly in the boundary conditions. For this reason the mathematics used to solve one problem may often be used immediately to solve another. In this chapter we shall consider a series of problems, closely related, to which we may give the name of potential problems, since they all consist in solving the potential equation under various given conditions. They will be solved by a variety of methods. The only general way of solving potential problems is by numerical methods and is outlined in §7.17.

Let us first consider the equations that must be satisfied by a potential function. For convenience we shall denote the potential by ϕ, without stating whether the resulting function applies to an electrostatic, magnetostatic or current-flow problem. For regions which are charge free, $\nabla^2 V = 0$, while for those which are current free, $\nabla^2 \mathbf{A} = 0$ so that in many cases we may consider Laplace's equation

$$\nabla^2 \phi = 0 \qquad (7.2)$$

rather than Poisson's equation.

The boundary conditions vary somewhat according to the type of problem. Thus

$$\phi = \text{constant} \tag{7.3}$$

is the condition on any conductor in electrostatics or electrode in current flow; and $\int \frac{\partial \phi}{\partial n} dS$ is related to the total charge on the conductor or to the total current from an electrode. With a finite system of charges, currents or magnets

$$\phi \to 0 \text{ at infinity} \tag{7.4}$$

However, if there is a uniform field E_0 at infinity in the z direction this latter condition must be replaced by

$$\phi = -E_0 z + \phi' \tag{7.4a}$$

where ϕ' is finite at infinity.

Moreover there can be no singularities in ϕ except at isolated charges, double layers, electrodes or magnets. Near an isolated charge e for example, $\phi = \frac{e}{4\pi\varepsilon r} + \phi'$ where ϕ' is finite, r being measured from the charge, so that $\phi - \frac{e}{4\pi\varepsilon r}$ is finite. Similarly at a dipole $\mathbf{m}$ in a vacuum, $\phi - \frac{\mathbf{m} \cdot \mathbf{r}}{r^3}$ is finite.

At an interface between two media other types of boundary conditions are introduced. Thus at the boundary between an insulator and a conductor carrying a current

$$\frac{\partial \phi}{\partial n} = 0 \tag{7.5}$$

If we have two media, 1 and 2, in contact, we shall generally have two distinct analytic expressions ϕ_1 and ϕ_2 valid in the two regions with boundary conditions at all common points of 1 and 2,

$$\phi_1 = \phi_2, \qquad \varepsilon_1 \frac{\partial \phi_1}{\partial n} = \varepsilon_2 \frac{\partial \phi_2}{\partial n} \tag{7.6}$$

or their equivalent.

§7.2 Boundary value problems for static fields

We saw in Chapter 2 that the solution to Poisson's equation is straightforward in regions in which ρ is specified everywhere, and which are unbounded. The problem in practice is that we do not usually know $\rho(\mathbf{r})$ in advance and in most problems we have to determine the potential in a bounded region. A problem in electrostatics for example may involve a number of conductors (electrodes) with either V or the total charge on each prescribed but the surface charge distribution σ will **not** be known in general and may not be obtained until a solution to the problem has been found.

To see how to deal with boundary conditions let us first cast Poisson's equation into integral form by using Green's theorem

$$\int_V [\phi\nabla^2\psi - \psi\nabla^2\phi]\, d\mathbf{r} = \int_S \left[\phi\frac{\partial\psi}{\partial n} - \psi\frac{\partial\phi}{\partial n}\right] dS \tag{7.7}$$

with $\psi = 1/4\pi\varepsilon_0|\mathbf{r}-\mathbf{r}'|$. Then since

$$\nabla^2\left(\frac{1}{|\mathbf{r}-\mathbf{r}'|}\right) = -4\pi\delta(\mathbf{r}-\mathbf{r}') \tag{7.8}$$

it follows from (7.7) that (for $\mathbf{r}$ within V)

$$\phi(\mathbf{r}) = \frac{1}{4\pi\varepsilon_0}\int_V \frac{\rho(\mathbf{r}')\, d\mathbf{r}'}{|\mathbf{r}-\mathbf{r}'|} + \frac{1}{4\pi}\int_S \left\{\frac{1}{|\mathbf{r}-\mathbf{r}'|}\frac{\partial\phi}{\partial n'} - \phi\frac{\partial}{\partial n'}\left(\frac{1}{|\mathbf{r}-\mathbf{r}'|}\right)\right\} dS' \tag{7.9}$$

Note that (7.9) is not in any sense a solution to the original partial differential equation but simply this equation rewritten as an integral equation. Poisson's equation is an elliptic equation and as such is overspecified if both ϕ and $\dfrac{\partial\phi}{\partial n}$ (Cauchy conditions) are prescribed on the boundary of the region V. Appropriate boundary conditions for elliptic equations are either Dirichlet (in which ϕ is specified on the boundary) or Neumann (in which $\partial\phi/\partial n$ is specified). The solution to Poisson's equation over a finite volume V with either Dirichlet or Neumann boundary conditions on S may be found from (7.9) by an approach using Green functions. Green functions satisfy Poisson's equation in which the source term is a unit charge at $\mathbf{r}'$, i.e.

$$\nabla^2 G(\mathbf{r}, \mathbf{r}') = -\frac{\delta(\mathbf{r}-\mathbf{r}')}{\varepsilon_0} \tag{7.10}$$

where

$$G(\mathbf{r}, \mathbf{r}') = \frac{1}{4\pi\varepsilon_0|\mathbf{r}-\mathbf{r}'|} + H(\mathbf{r}, \mathbf{r}') \tag{7.11}$$

with $H(\mathbf{r}, \mathbf{r}')$ harmonic (i.e. satisfying Laplace's equation) in V. With this function we may now choose H to eliminate one or other of the surface integrals in (7.9). Physically we may picture $H(\mathbf{r}, \mathbf{r}')$ as the potential due to a charge distribution **external** to V, so chosen as to satisfy Dirichlet **or** Neumann conditions on S when combined with the potential of a unit charge at the source point $\mathbf{r}'$. For example, for Dirichlet boundary conditions, $G(\mathbf{r}, \mathbf{r}') = 0$, $\mathbf{r}' \in S$, so that

$$\phi(\mathbf{r}) = \int_V G(\mathbf{r}, \mathbf{r}')\rho(\mathbf{r}')\, d\mathbf{r}' - \varepsilon_0 \int_S \phi(\mathbf{r}') \frac{\partial G}{\partial n'}\, dS' \tag{7.12}$$

When S is earthed so that $\phi_S(\mathbf{r}') = 0$ this gives

$$\phi(\mathbf{r}) = \int_V G(\mathbf{r}, \mathbf{r}')\rho\,(\mathbf{r}')\, d\mathbf{r}' \tag{7.13}$$

which is simply an expression of the superposition principle applied to an ensemble of point sources within V. Again if $\rho(\mathbf{r}') = 0$ and $\phi = \phi_S$ on S, then

$$\phi(\mathbf{r}) = -\varepsilon_0 \int_S \phi_S(\mathbf{r}') \frac{\partial G}{\partial n'}\, dS' \tag{7.14}$$

for the potential within a region whose boundary is maintained at a potential ϕ_S. Physically, the normal derivative of the Green function represents the surface charge density induced on the bounding surface S by placing unit charge at $\mathbf{r}'$ if S were an earthed conductor.

§7.3 Uniqueness theorem

In the previous section we looked at the appropriate boundary conditions for Poisson's equation to have a well-behaved solution within a region V bounded by S. We saw that specifying either the potential ϕ on a closed S or $\dfrac{\partial \phi}{\partial n}$ (which corresponds to a given surface charge density) defined a unique problem. We now want to establish that there is only one potential function satisfying all the requisite conditions. The following proof applies to electrostatics but is trivially adapted for other static conditions (steady current flow or magnetostatics).

Suppose we have an electrostatic system consisting of one set of conductors, each carrying a given total charge; another set of conductors each kept at a given potential; and a given volume distribution of charge, in the presence of given dielectrics. This is the most general type of electrostatic problem. We shall show that if ϕ is real, with its first and second derivatives continuous in a region V and on its boundary S, and if $\nabla^2\phi = \rho/\varepsilon$ with either $\phi = \text{constant}$ or $\partial\phi/\partial n = \text{constant}$ on S then it is unique.

For suppose there are two potential functions ϕ_1 and ϕ_2 which satisfy all the required conditions. Put $\phi = \phi_1 - \phi_2$ so that $\nabla^2\phi = 0$ within V and $\phi = \text{constant}$ on the surface of all conductors and in particular $\phi = 0$ on any conductor whose potential is prescribed.

Consider $\int_S \phi\varepsilon\boldsymbol{\nabla}\phi \cdot d\mathbf{S}$; by Gauss' theorem

$$\int_S \phi\varepsilon\boldsymbol{\nabla}\phi \cdot d\mathbf{S} = \int_V \boldsymbol{\nabla} \cdot (\varepsilon\phi\boldsymbol{\nabla}\phi)\, d\mathbf{r}$$

$$= \int_V [\varepsilon\boldsymbol{\nabla}\phi \cdot \boldsymbol{\nabla}\phi + \phi\boldsymbol{\nabla} \cdot (\varepsilon\boldsymbol{\nabla}\phi)]\, d\mathbf{r} = \int_V \varepsilon|\boldsymbol{\nabla}\phi|^2\, d\mathbf{r}$$

since $\boldsymbol{\nabla} \cdot (\varepsilon\boldsymbol{\nabla}\phi) = 0$ (Laplace's equation for dielectrics). The surface integral is taken over the sphere at infinity and over all the conducting surfaces. On each conductor whose potential is fixed $\phi = 0$ so that the integral vanishes. Over any of the conductors on which the total charge is fixed we know that $\displaystyle\int \frac{\partial\phi}{\partial n}\, dS$ must vanish so that again there is no contribution to $\int \phi\varepsilon\boldsymbol{\nabla}\phi \cdot d\mathbf{S}$. Also of course at infinity ϕ vanishes at least to order r^{-1} so that the integral over the sphere at infinity is also zero. Thus we are left with

$$\int_V \varepsilon|\boldsymbol{\nabla}\phi|^2\, d\mathbf{r} = 0$$

i.e.

$$\phi \equiv \phi_1 - \phi_2 = \text{constant}$$

Over some of the surfaces and at infinity $\phi_1 = \phi_2$ and so $\phi = 0$, i.e. the potential function is unique.

§7.4 Images in a plane

An important application of the uniqueness theorem is found in the **method of images** which may be illustrated most simply by an example. Consider (Fig. 7.1) two charges $\pm e$ at A and B in free space. The field lines cross from A to B as shown and the potential at any point is $\phi = \frac{e}{4\pi\varepsilon_0}\left(\frac{1}{r_1} - \frac{1}{r_2}\right)$ where r_1 and r_2 are the distances from A and B respectively. Clearly along the mid-plane XY where $r_1 = r_2$ the potential is zero. In fact we may say that

$$\phi = \frac{e}{4\pi\varepsilon_0}\left(\frac{1}{r_1} - \frac{1}{r_2}\right) \tag{7.15}$$

is a potential function which gives potential zero all over the infinite plane XY, tends to infinity like e/r at A, and tends to zero at large distances. But these are precisely the conditions that must be satisfied by the potential when a point charge e is placed at A outside an infinite conducting plane XY at zero potential (Fig. 7.2). So by the principle of uniqueness we have now solved this latter problem. It follows, therefore, that on the right-hand side of XY the potential is given by (7.15).

We may describe the situation by saying that if a charge e is at A outside the infinite conducting plane XY which is kept at zero

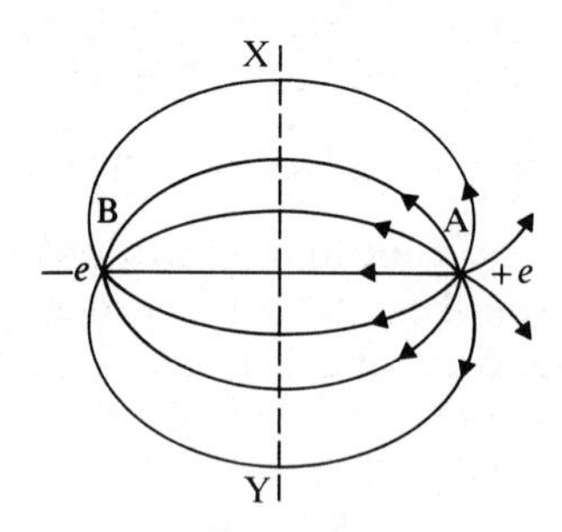

Fig. 7.1

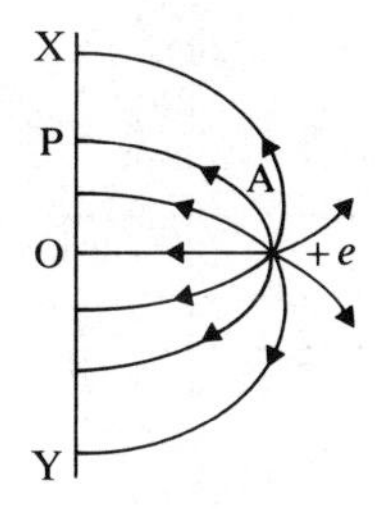

Fig. 7.2

potential, the potential on the right of XY is just the same as if the plane were removed and replaced by a charge $-e$ at B. We refer to this as the image charge of A in the plane.

The potential (7.15) is, of course, only valid on the right of XY, and the lines of force which start from A end at XY, as shown in Fig. 7.2. Unless there are some other charges present, the potential is zero everywhere to the left of XY. It is important to realize that the charge $-e$ at B has no more real existence than an optical image; all that we can say is that on the right of XY the system behaves as if we replaced the infinite plane by a charge $-e$ at B.

If we know the potential we can easily calculate the field and the density of charge at any point of XY. Thus at P in Fig. 7.2,

$$\sigma(y) = -\varepsilon_0 \frac{\partial \phi}{\partial n}$$

and on substituting the appropriate value of ϕ it is soon verified that

$$\sigma(y) = \frac{-ae}{2\pi(a^2 + y^2)^{3/2}} \tag{7.16}$$

where $a = \text{AO}$ is the shortest distance from A to the plane. The induced charge on XY is practically all concentrated near O and the density falls off as the inverse cube of AP.

Also, the force acting on the charge e at A is simply the force that would be exerted by the image charge $-e$ at B. This is an attraction towards the plane of magnitude $e^2/16\pi\varepsilon_0 a^2$.

The method of images is readily extended. Thus, let OX and OY (Fig. 7.3) be two semi-infinite perpendicular planes at zero potential, and let a charge $+e$ be placed at A. One may easily

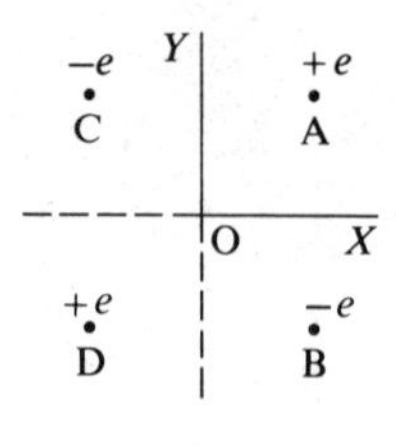

Fig. 7.3

verify that the image system consists of the original charge $+e$ at A, together with $-e$, $-e$ and $+e$ at B, C, D respectively. In the first quadrant XOY the potential at a point P is,

$$\phi = \frac{e}{4\pi\varepsilon_0}\left(\frac{1}{\mathrm{AP}} - \frac{1}{\mathrm{BP}} - \frac{1}{\mathrm{CP}} + \frac{1}{\mathrm{DP}}\right) \tag{7.17}$$

A similar method may be used if the angle XOY is of the form π/n, where n is an integer. The case in which n is not an integer cannot be solved by the method of images (problem 6).

It is evident that an image system similar to that of Figs. 7.1–7.3 would exist if the point charge at A was replaced by a line charge e parallel to the planes. Thus, corresponding to the example of Fig. 7.2, we should have a line charge e parallel to an infinite conducting plane at zero potential. The potential on the right of XY would be the same as that due to two line charges $\pm e$ at A and B, so that

$$\phi = -\frac{e}{2\pi\varepsilon_0}\ln\frac{r_1}{r_2} \tag{7.18}$$

To summarize, we have made use of the method of images to obtain a Green function for a system with plane boundaries. By making use of (7.14) we may solve the problem of determining the potential at any point P above the plane boundaries on which a potential function ϕ_S is prescribed.

§7.5 Images with spheres and cylinders

Another important image system is found as follows. Consider (Fig. 7.4) two charges $+e$ and $-e'$ at A and B. The potential due

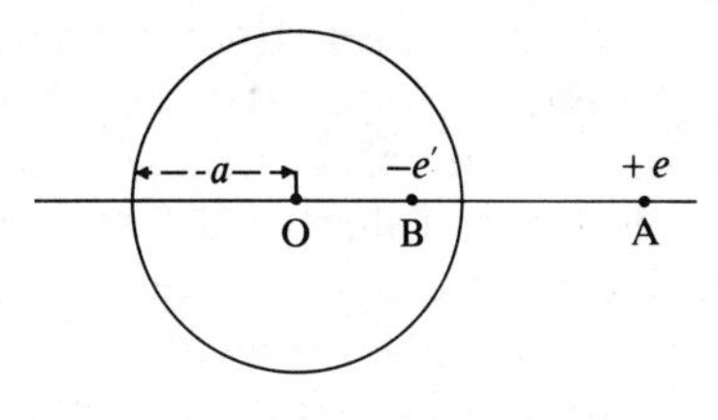

Fig. 7.4

to these by themselves is

$$\phi = \frac{1}{4\pi\varepsilon_0}\left(\frac{e}{r_1} - \frac{e'}{r_2}\right)$$

where r_1 and r_2 denote distances from A and B respectively. So $\phi = 0$ on the surface

$$r_1/r_2 = e/e' \tag{7.19}$$

This surface, which is spherical, is shown in Fig. 7.4 with its centre at some point O on the line AB. The relation between e, e' and the various distances may be put in a more useful form by letting OA $= f$, and the radius of the sphere be a. We leave it as an exercise to verify that (7.19) is equivalent to the following statement – a charge e at a point A distance f from the centre of a sphere of radius a, together with a charge $-ea/f$ $(= e')$ at a point B on the line OA such that OB $= a^2/f$, give zero potential on the sphere. A and B are obviously inverse points with respect to the sphere.

It follows, just as in §7.4, that if a charge $+e$ is placed at A outside a conducting sphere connected to earth, then the potential outside the sphere is the same as if we replaced the sphere by a charge $-ea/f$ at the inverse point B. Thus, if r_1 and r_2 have the same meaning as in (7.19), the potential outside the sphere is

$$\phi = \frac{e}{4\pi\varepsilon_0}\left(\frac{1}{r_1} - \frac{a}{fr_2}\right) \tag{7.20}$$

Inside the sphere the potential is zero. The charge $-ea/f$ at B is evidently the image charge of e at A.

Once we know the potential all other quantities are soon found. In particular, since tubes of force can only begin and end on charges, Fig. 7.4 shows us that the net number of tubes that arrive on the sphere's surface is simply the same as the number that would end on the image charge at B in the absence of the sphere. Thus the sphere carries a net negative charge $-ea/f$.

If, instead of being at zero potential the sphere is given a total charge Q, the complete potential will be the superposition of (7.20) and the potential due to an isolated sphere carrying charge $Q+ea/f$. In that case, measuring r from the centre O:

$$\phi = \frac{1}{4\pi\varepsilon_0}\left[\frac{e}{r_1} - \frac{ea/f}{r_2} + \frac{(Q+ea/f)}{r}\right] \tag{7.21}$$

In particular, if a charge e is placed at a distance f from the centre of an uncharged conducting sphere, so that $Q=0$ in the above equation; the potential at outside points is the same as that due to the original charge, together with a charge $-ea/f$ at the inverse point and a charge $+ea/f$ at the centre of the sphere. (See problem 23). The latter two charges are now the images of the first charge.

A similar system of images can be calculated for the two-dimensional case of a line charge parallel to a conducting cylinder.

By combining two or more sets of images, we can similarly discuss more complicated problems. Thus, since a dipole may be regarded as two large almost coincident charges, its image in a plane is a similar dipole and in a sphere it is a dipole at the inverse point, whose magnitude and direction are soon determined, together with a charge at the inverse point (problem 11). Other examples of images will be found in the questions at the end of the chapter. The method of images is often a very quick and useful way of short-circuiting a more elaborate solution of the potential equation.

§7.6 Other methods available for potential problems

We have seen how the Green function method may be used to solve potential problems. Often alternative methods may prove more convenient particularly if the boundaries correspond to coordinate surfaces in orthogonal coordinate systems. We usually expand ϕ as a series of harmonic functions with coefficients that must be determined from the boundary conditions.

In spherical polar coordinates r, θ, ψ the simplest harmonic function is just r^{-1} in which r may be measured from any arbitrary point (or equivalently as $1/|\mathbf{r}-\mathbf{r}_0|$ where $\mathbf{r}_0$ is an arbitrary constant vector). Other simple harmonic functions are $r\cos\theta$ and $r^{-2}\cos\theta$. Since $r\cos\theta = z$, these latter are particularly suited to problems connected with a uniform field. These three functions are particular cases of the spherical harmonics $r^n P_n(\cos\theta)$ and $r^{-(n+1)} P_n(\cos\theta)$, where $P_n(\cos\theta)$ is the Legendre polynomial of degree n in which n is a positive integer. Since $P_0 = 1$, $P_1 = \cos\theta$ we recognise at once the source of functions used earlier. The functions $r^n P_n(\cos\theta)$ remain finite when $r = 0$, but not when $r \to \infty$; the opposite is true for $r^{-(n+1)} P_n(\cos\theta)$. It is this fact which usually decides which of the two types will be required in an given problem. Each has cylindrical symmetry about the polar axis $\theta = 0$. If, however, our problem does not have this degree of symmetry we must turn to the associated Legendre polynomials, and use functions such as $r^n P_n^m(\cos\theta)\cos m\psi$.

The corresponding functions in two-dimensional polar coordinates r, θ are

$$\ln r, \qquad \ln|r - r_0|, \qquad r^n \cos n\theta, \qquad r^n \sin n\theta$$

With these latter it may be necessary to restrict n to be an integer, positive or negative, if we want to keep ϕ single-valued over the whole range of θ. In cylindrical polar coordinates r, θ, z the standard harmonic functions are $J_m(nr)\cos m\theta\, e^{\pm nz}$, and $J_m(nr)\sin m\theta\, e^{\pm nz}$, where $J_m(nr)$ is the Bessel function of order m and variable nr.

There are, of course, many other harmonic functions besides those mentioned above. But as we shall see in the following sections these are sufficient to deal with a great many important problems in electrostatics, current flow and magnetostatics.

§7.7 Spherical conductor in a uniform field

Consider the case of a conducting sphere of radius a (Fig. 7.5) placed at zero potential (i.e. connected to earth) in a uniform electric field E_0 parallel to the z-axis. If we take spherical polar coordinates with the axis along Oz, we have to find a potential ϕ

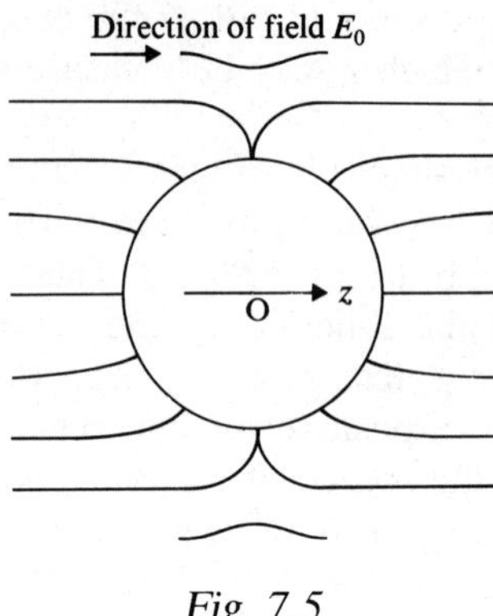

Fig. 7.5

which satisfies the following conditions

(i) $\nabla^2\phi = 0$,

(ii) $\phi = 0$ when $r = a$, for all values of θ,

(iii) $\phi + E_0 r \cos\theta$ is finite at $r =$ infinity.

The form of (iii) suggests that we try the functions referred to in §7.6 which involve $\cos\theta$ times a function of r. We shall see the justification for this later. But for the present let us see if we can get a solution

$$\phi = -E_0 r \cos\theta + \frac{A\cos\theta}{r^2} \tag{7.22}$$

in which A is an arbitrary constant to be determined. This choice automatically satisfies (i) and (iii). It also satisfies (ii) if

$$-E_0 a + \frac{A}{a^2} = 0, \quad \text{i.e.} \quad A = E_0 a^3$$

Thus the solution of this problem is simply

$$\phi = -E_0 r \cos\theta + \frac{E_0 a^3 \cos\theta}{r^2} \tag{7.23}$$

The electric field lines corresponding to this potential are shown in Fig. 7.5. We may think of the term $E_0 a^3 \cos\theta/r^2$ as representing the disturbing effect of the sphere. This disturbing effect is the same as if a dipole of moment $4\pi\varepsilon_0 E_0 a^3$ was placed at O pointing in the direction of the field E_0. (This is sometimes called the image dipole.)

It is obviously true that (7.23) provides us with a complete and unique solution. We should perhaps ask why there are no more terms? To answer this question let us suppose that there were other terms. As there is cylindrical symmetry about $\theta = 0$, the additional terms must be of the general form $r^{-(n+1)}P_n(\cos\theta)$. If we included terms $r^n P_n(\cos\theta)$, then $\phi + E_0 r\cos\theta$ would not be finite at infinity. So let us try

$$\phi = -E_0 r\cos\theta + \frac{A_1}{r^2}\cos\theta + \frac{A_2}{r^3}P_2(\cos\theta) + \ldots \tag{7.24}$$

This satisfies (i) and (iii). It satisfies (ii) if, for all values of θ,

$$0 = \left(-E_0 a + \frac{A_1}{a^2}\right)\cos\theta + \frac{A_2}{a^3}P_2(\cos\theta) + \frac{A_3}{a^4}P_3(\cos\theta) + \ldots$$

The Legendre polynomials are linearly independent. So each coefficient must separately vanish. This means that $A_1 = E_0 a^3$, $A_2 = A_3 = \ldots = 0$. Hence the extra terms vanish, and (7.23) is indeed the full solution.

Knowing ϕ we can soon get the induced charge density σ on the surface of the sphere. For

$$\frac{\sigma}{\varepsilon_0} = -\left(\frac{\partial\phi}{\partial r}\right)_{r=a} = \left[E_0\left(1 + \frac{2a^3}{r^3}\right)\cos\theta\right]_{r=a} = 3E_0\cos\theta$$

Hence

$$\sigma = 3\varepsilon_0 E_0\cos\theta \tag{7.25}$$

This shows that the induced charge is negative on the left half of the sphere, and positive on the right half, so that the total charge on the sphere is zero.

If the sphere receives a charge Q it will no longer be at zero potential. But we can determine the resultant potential by the principle of superposition. In fact,

$$\phi = -E_0 r\cos\theta + \frac{E_0 a^3}{r^2}\cos\theta + \frac{Q}{4\pi\varepsilon_0 r} \tag{7.26}$$

The potential of the sphere ($r = a$) is now $Q/4\pi\varepsilon_0 a$, and all other quantities can easily be found. If Q lies between $\pm 12\pi a^2\varepsilon_0 E_0$ the surface of the sphere will be divided into a region of negative charge and another region of positive charge.

§7.8 Dielectric sphere in a uniform field

Let us replace the conducting sphere of §7.7 by a dielectric sphere of uniform dielectric constant K_2 (Fig. 7.6) placed in a medium of dielectric constant K_1. There are now two separate regions of space, inside and outside the sphere. Let us call the two potentials ϕ_1 and ϕ_2. We must determine ϕ_1 and ϕ_2 so that

(i) $\nabla^2\phi_1 = 0$, $\nabla^2\phi_2 = 0$,
(ii) $\phi_1 + E_0 r \cos\theta$ is finite at infinity,
(iii) ϕ_2 finite in $r \le a$,
(iv) $\phi_1 = \phi_2$ on $r = a$, for all θ,
(v) $K_1 \dfrac{\partial\phi_1}{\partial r} = K_2 \dfrac{\partial\phi_2}{\partial r}$ on $r = a$, for all θ. This is the condition (6.18) of §6.3 that D_n, or εE_n, must be continuous at the dielectric interface. The results for the conducting sphere suggest that we try

$$\phi_1 = -E_0 r \cos\theta + \frac{A}{r^2}\cos\theta \tag{7.27}$$

$$\phi_2 = Br\cos\theta \tag{7.28}$$

This satisfies (i), (ii) and (iii) whatever the values of A and B. It also satisfies (iv) if $-E_0 a + \dfrac{A}{a^2} = Ba$, and (v) if

$$K_1\left(-E_0 - \frac{2A}{a^3}\right) = K_2 B$$

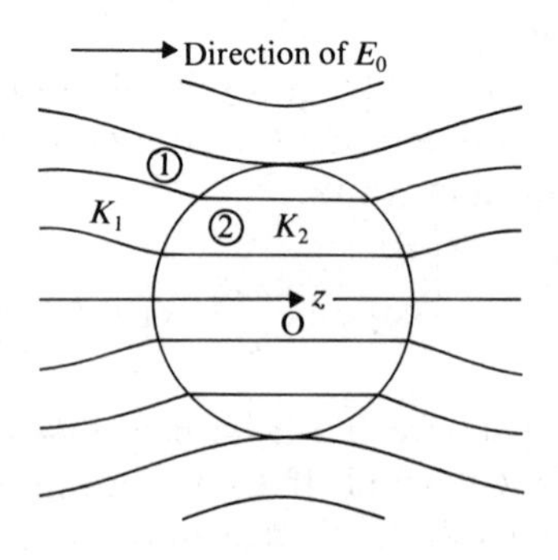

Fig. 7.6

These are two simultaneous equations for A and B, giving

$$A = \frac{K_2 - K_1}{K_2 + 2K_1} E_0 a^3, \qquad B = \frac{-3K_1}{K_2 + 2K_1} E_0 \tag{7.29}$$

Thus (7.27)–(7.29) solve our problem completely. Taking $K_1 = 1$ we have the case of a dielectric sphere in a vacuum. Taking $K_2 = 1$ we have the case of a spherical hole in a dielectric K_1. This is interesting as it adds something more to the discussion of cavities in §6.7. The fact that ϕ_2 may be written in the form $\phi_2 = Bz$ shows that the field inside the sphere is quite uniform and equal to $-B$, i.e. $\dfrac{3K_1}{K_2 + 2K_1} E_0$. The disturbing effect of the sphere is shown by the dipole term $A \cos\theta/r^2$ outside $r = a$. In Fig. 7.6 we show the lines of electric displacement D when $K_2 > K$. This shows that the sphere tends to concentrate the lines. Just the opposite holds if $K_2 < K_1$.

§7.9 Small magnet in a spherical hole

Our next example is taken from magnetostatics. Suppose (Fig. 7.7) that a small magnet of moment m lies at O point along the z-axis, at the centre of a spherical hollow of radius a surrounded by a medium of uniform magnetic permeability μ. Measuring from O as origin we have to find two potential functions ϕ_1 and ϕ_2, so that

(i) $\nabla^2 \phi_1 = 0$, $\nabla^2 \phi_2 = 0$,
(ii) $\phi_2 \to 0$ at infinity,
(iii) $\phi_1 - m \cos\theta/r^2$ is finite at $r = 0$,
(iv) $\phi_1 = \phi_2$ at $r = a$, for all θ,
(v) $\mu_0 \dfrac{\partial \phi_1}{\partial r} = \mu \dfrac{\partial \phi_2}{\partial r}$ at $r = a$, for all θ. This is the condition (6.50) of §6.15, that B_n is continuous at the boundary.

In view of (iii) it is natural to look for solutions involving $r \cos\theta$ and $\cos\theta/r^2$. We leave it as an exercise to verify that if we write

$$\phi_1 = \frac{m \cos\theta}{r^2} + Ar \cos\theta$$

$$\phi_2 = \frac{B \cos\theta}{r^2}$$

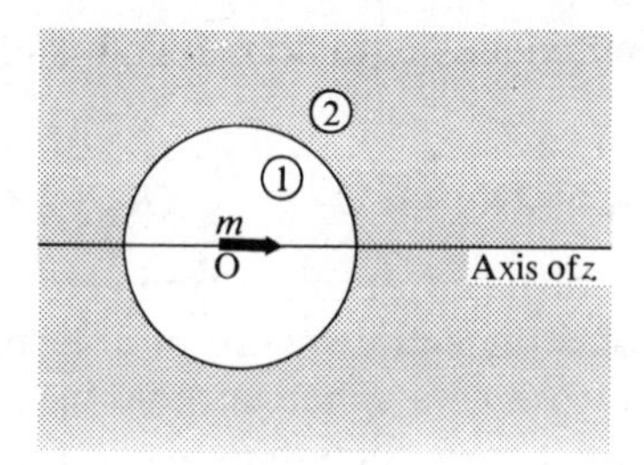

Fig. 7.7

our conditions are all satisfied by

$$A=\frac{2(\mu_0-\mu)}{\mu_0+2\mu}\frac{m}{a^3},\qquad B=\frac{3\mu_0 m}{\mu_0+2\mu} \tag{7.30}$$

§7.10 Point charge outside a dielectric sphere

In this problem we wish to find the potential function due to a point charge e placed at B (Fig. 7.8) a distance b from the centre of an uncharged dielectric sphere of radius a. The notation is shown in the figure. We have to find two potential functions ϕ_1 and ϕ_2 so that

(i) $\nabla^2\phi_1=0$, $\nabla^2\phi_2=0$,

(ii) $\phi_2-\dfrac{e}{4\pi\varepsilon_0 R}$ is finite at $R=0$, R being measured from B,

(iii) $\phi_2\to 0$ at infinity. More accurately, since the total charge is e, we shall have $\phi_2=\dfrac{e}{4\pi\varepsilon_0 r}+$higher powers of $\dfrac{1}{r}$, at infinity,

(iv) ϕ_1 is finite if $r\le a$,

(v) $\phi_1=\phi_2$ on $r=a$, for all θ,

(vi) $K\dfrac{\partial\phi_1}{\partial r}=\dfrac{\partial\phi_2}{\partial r}$ on $r=a$, for all θ. This is the continuity condition for D_n at the change of medium.

This problem is one in which we shall need a complete expansion in Legendre polynomials. So let us write

$$\phi_1=A_0+A_1 rP_1(\cos\theta)+A_2r^2P_2(\cos\theta)+\dots$$

$$\phi_2=\frac{e}{4\pi\varepsilon_0 R}+C+\frac{B_0}{r}+\frac{B_1P_1(\cos\theta)}{r^2}+\frac{B_2P_2(\cos\theta)}{r^3}+\dots \tag{7.31}$$

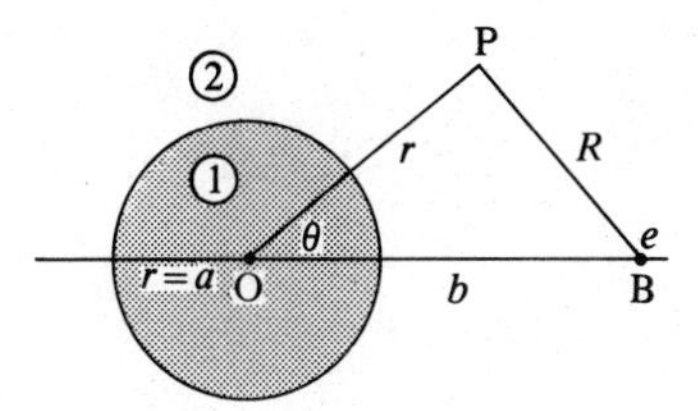

Fig. 7.8

We must find values of A_0, $A_1, \ldots B_0, \ldots C$ so that the conditions (i)–(vi) are satisfied. Now (7.31) automatically satisfies (i), (ii) and (iv). It also satisfies (iii) if $C=0$ and $B_0=0$. To deal with (v) and (vi) we have to use the fact that if $r<b$,

$$\frac{1}{R}=\frac{1}{b}+\frac{r}{b^2}P_1(\cos\theta)+\frac{r^2}{b^3}P_2(\cos\theta)+\ldots \tag{7.32}$$

Conditions (v) and (vi) now give us two equations in θ, which must be satisfied for all values of θ. Since the P_n functions are linearly independent, we may equate coefficients of each $P_n(\cos\theta)$ on the two sides of each identity. Then (v) gives

$$A_0=\frac{e}{4\pi\varepsilon_0 b},\qquad A_n a^n=\frac{B_n}{a^{n+1}}+\frac{ea^n}{4\pi\varepsilon_0 b^{n+1}},\qquad (n=1,2,\ldots)$$

and (vi) gives

$$nKA_n a^{n-1}=-\frac{(n+1)B_n}{a^{n+2}}+\frac{nea^{n-1}}{4\pi\varepsilon_0 b^{n+1}},\qquad (n=1,2,\ldots)$$

In this way each coefficient A_n and B_n is obtained, and the potential is known.

If we want to find the force on the charge at B, we shall have to calculate the field there by differentiating ϕ_2. But the first term in ϕ_2 must be omitted since this is simply the direct field of the charge and no charged particle can move itself by means of its own direct field. The effective field is therefore

$$-\frac{\partial}{\partial r}\left(\phi_2-\frac{e}{4\pi\varepsilon_0 R}\right)$$

measured at $\theta=0$, $r=b$. Now at $\theta=0$, each $P_n(\cos\theta)=1$, so that the effective field is

$$\frac{2B_1}{b^3}+\frac{3B_2}{b^4}+\dots$$

and the force on the charge at B is a force $\sum \frac{neB_{n-1}}{4\pi\varepsilon_0 b^{n+1}}$ in the direction OB.

§7.11 A two-dimensional problem in magnetostatics

Problems in two dimensions are generally best done by the methods of §7.14. But we can briefly illustrate how the methods so far discussed may also be applied, if we wish. Consider an infinitely long circular cylinder uniformly magnetized in a direction perpendicular to its axis (Fig. 7.9). Let us use cylindrical polar coordinates in which the origin is at O, the z-axis is taken to be the axis of the cylinder and the $\theta=0$ direction (or x-axis) is parallel to the permanent magnetization M. We have to solve the equations

(i) $\nabla^2\phi_1=0$, $\nabla^2\phi_2=0$,
(ii) $\phi_2\to 0$ at infinity,
(iii) ϕ_1 is finite for $r\leq a$,
(iv) $\phi_1=\phi_2$ on $r=a$,
(v) $-\frac{\partial\phi_1}{\partial r}+M\cos\theta=-\frac{\partial\phi_2}{\partial r}$ on $r=a$. This is the condition that B_n is continuous at $r=a$.

Let us expand ϕ_1 and ϕ_2 in series form.

$$\phi_1=\sum A_n r^n \cos n\theta, \qquad n=1, 2, \dots$$

$$\phi_2=\sum \frac{B_n}{r^n}\cos n\theta, \qquad n=1, 2, \dots$$

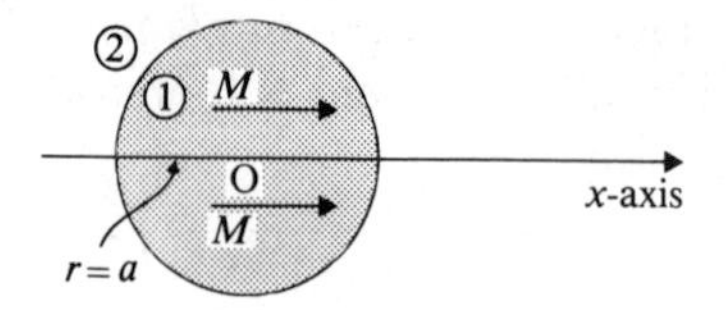

Fig. 7.9

This automatically satisfies (i), (ii) and (iii). Further, (iv) and (v) are satisfied if and only if

$$A_1 = \frac{M}{2}, \qquad B_1 = \frac{Ma^2}{2}, \qquad A_n = B_n = 0 \quad \text{when} \quad n \geq 2.$$

In this way we have obtained the complete solution.

§7.12 Another two-dimensional problem

A rather more difficult two-dimensional problem is shown in Fig. 7.10. An infinitely long line-charge of strength e at A is parallel to an infinitely long dielectric cylinder of radius a, and $\mathrm{OA} = f$. We wish to find the potential function at all points.

The potential due to a line charge e at A, without any disturbing effects due to the dielectric, is

$$\phi = -\frac{e}{2\pi\varepsilon_0} \ln R + \text{constant},$$

where R is measured from the line charge A as in the figure. We have therefore to find two potential functions ϕ_1 and ϕ_2 valid in the regions 1 and 2, such that

(i) $\nabla^2\phi_1 = 0$, $\nabla^2\phi_2 = 0$,

(ii) ϕ_1 is finite in $r \leq a$,

(iii) $\phi_2 + \dfrac{e}{2\pi\varepsilon_0} \ln R$ is finite for $r \geq a$,

(iv) $\phi_1 = \phi_2$ on $r = a$,

(v) $K\dfrac{\partial\phi_1}{\partial r} = \dfrac{\partial\phi_2}{\partial r}$ on $r = a$.

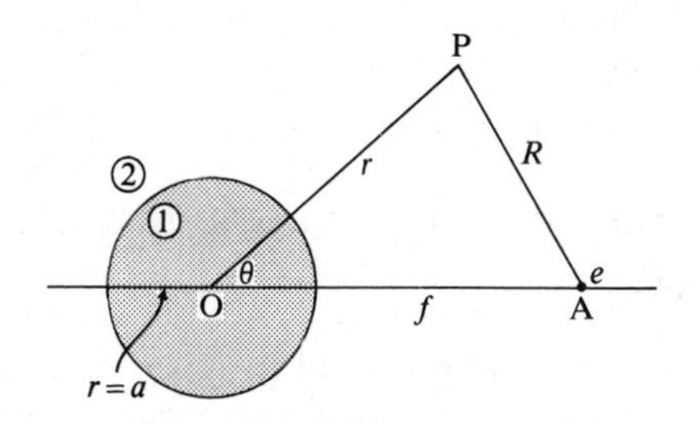

Fig. 7.10

We therefore put

$$\phi_1 = \sum_{n=0}^{\infty} A_n r^n \cos n\theta$$

$$\phi_2 = -\frac{e}{2\pi\varepsilon_0} \ln R + \sum_{n=0}^{\infty} \frac{B_n}{r^n} \cos n\theta \qquad (7.33)$$

This automatically satisfies (i), (ii) and (iii). To satisfy (iv) and (v) we use the fact that if $r<f$,

$$\log R = \log f - \frac{r}{f} \cos \theta - \frac{r^2}{2f^2} \cos 2\theta - \ldots \qquad (7.34)$$

This is most easily proved in complex coordinates by putting $z = x+iy$, expanding $\log(f-z)$ in powers of z/f, and then taking the real part. Thus condition (iv) gives, on comparing coefficients of $\cos n\theta$:

$$A_0 = -\frac{e}{2\pi\varepsilon_0} \ln f + B_0, \qquad A_n a^n = \frac{ea^n}{2\pi\varepsilon_0 n f^n} + \frac{B_n}{a^n}$$

In the same way condition (v) gives:

$$nKA_n a^{n-1} = \frac{ea^{n-1}}{2\pi\varepsilon_0 f^n} - \frac{nB_n}{a^{n+1}} \cdot (n = 1, 2, \ldots)$$

These two equations determine the whole solution, apart from an arbitrary constant. Such a constant represents an arbitrary zero of potential (in two-dimensional problems it is seldom possible to put the potential zero at infinity).

When solving such problems as these, in two or three dimensions, it is advisable to follow the same procedure as in §§7.8–7.12, i.e. to begin by making a list of all the conditions that the potential must satisfy. It is impossible then to leave out any of the conditions by mistake, and the whole solution can be set out quite systematically.

§7.13 Problems with axial symmetry – a useful device

When a three-dimensional potential problem has axial symmetry, it is generally best to use spherical polar coordinates r, θ, ψ, and to take the axis of symmetry as the polar axis $\theta = 0$. In such a case the potential is independent of ψ, and we can write

$$\phi = \sum_n \left(A_n r^n + \frac{B_n}{r^{n+1}}\right) P_n(\cos\theta) \tag{7.35}$$

where A_n and B_n are certain constants to be determined. Along the axis of symmetry $\theta = 0$ so that $P_n(\cos\theta) = 1$, and $r = z$; thus ϕ reduces to

$$\phi_{\text{axis}} = \sum_n \left(A_n z^n + \frac{B_n}{z^{n+1}}\right) \tag{7.36}$$

Now it may happen that we are able, by other means, to calculate the potential along the axis of symmetry. In that case, by comparing our result with (7.36) we determine at once the coefficients A_n and B_n. This means that the complete potential function (7.35) is now known, for points off the axis as well as for those on it.

The simplest illustration of this useful device is found in calculating the potential due to a circular wire carrying a current I. We have already determined in §5.8 the vector potential A_θ at points close to the axis of symmetry. It is straightforward to show that

$$\phi_{\text{axis}} = \frac{I}{2}\{1 - z/\sqrt{(a^2 + z^2)}\}$$

If we write this in the form required by (7.36) we need two separate expansions:

$$z < a : \phi_{\text{axis}} = \frac{I}{2}\left(1 - \frac{z}{a} + \frac{z^3}{2a^3} - \dots\right)$$

$$z > a : \phi_{\text{axis}} = \frac{I}{2}\left(\frac{a^2}{2z^2} - \frac{3a^4}{8z^4} + \dots\right)$$

It follows that the potential ϕ is given at all points by an expression of the form (7.35) where, if

$$r < a, \qquad B_n = 0, \qquad A_0 = \frac{I}{2}, \qquad A_2 = A_4 = \dots = 0,$$

$$A_1 = \frac{I}{2a}, \qquad A_{2n+1} = \frac{I}{2}(-1)^{n+1} \frac{1.3\dots(2n-1)}{2.4\dots 2n} \frac{1}{a^{2n+1}}$$

and if

$$r > a, \qquad A_n = 0, \qquad B_0 = 0, \qquad B_2 = B_4 = \ldots = 0,$$

$$B_1 = I\frac{a^2}{4}, \qquad B_{2n+1} = \frac{I}{2}(-1)^n \frac{1.3\ldots(2n+1)}{2.4\ldots(2n+2)} a^{2n+2}$$

From this it is a straightforward matter to calculate the field at all points.

§7.14 Two-dimensional problems, conjugate functions

It often happens that the potential problem which we have to solve involves only two variables, x, y. We have already met this situation in the flow of current in a thin plate (§4.9) and in the magnetic field round a long straight wire (§5.5). We now turn to a discussion of such cases, using the powerful method of complex variables.

Our fundamental problem is to seek solutions of Laplace's equation, and then to make them satisfy the relevant boundary conditions. In the two-dimensional case that we are considering, Laplace's equation is

$$\nabla^2\phi \equiv \frac{\partial^2\phi}{\partial x^2} + \frac{\partial^2\phi}{\partial y^2} = 0 \tag{7.37}$$

In order to solve this equation we put $z = x + iy$, where $i = \sqrt{-1}$.

Now consider any function of z, which we write

$$w = f(z) = f(x + iy) = \phi(x, y) + i\psi(x, y) \tag{7.38}$$

so that ϕ and ψ are the real and imaginary parts of $f(z)$. Clearly by partial differentiation of $w = f(x + iy)$

$$\frac{\partial^2 w}{\partial x^2} + \frac{\partial^2 w}{\partial y^2} = 0$$

Separating the real and imaginary parts

$$\nabla^2\phi = 0, \qquad \nabla^2\psi = 0 \tag{7.39}$$

Hence ϕ and ψ, as defined by (7.38), are each possible potential functions satisfying (7.37). Their close relationship in (7.38) suggests that we call them conjugate functions. By partial differentiation of (7.38) we have

$$\frac{\partial w}{\partial y} = if'(z) = i\frac{\partial w}{\partial x}$$

Equating the real and imaginary parts, we obtain the **Cauchy relations:**

$$\begin{aligned} \frac{\partial \phi}{\partial x} &= \frac{\partial \psi}{\partial y} = \text{real part of } f'(z) \\ \frac{\partial \psi}{\partial x} &= -\frac{\partial \phi}{\partial y} = \text{imaginary part of } f'(z) \end{aligned} \tag{7.40}$$

These two statements may be summed up in the one statement that if s_1 and s_2 are perpendicular directions related in the anti-clockwise fashion of Fig. 7.11, then

$$\frac{\partial \phi}{\partial s_1} = \frac{\partial \psi}{\partial s_2} \tag{7.41}$$

In particular, if s_1 is taken along a contour $\phi = \text{constant}$, $\partial\phi/\partial s_1 = 0$, so that $\partial\psi/\partial s_2 = 0$, showing that ψ is constant along the s_2 direction. Now for different values of the constant, $\phi(x, y) =$ constant represents a whole family of curves. Thus $\phi(x, y) = \text{constant}$, and $\psi(x, y) = \text{constant}$, are two families of curves which cut orthogonally.

It is usual to regard ϕ as the potential function, and then ψ is called the stream function, though we could of course interchange ϕ and ψ. The contours $\phi = \text{constant}$ are equipotentials, so that $\psi = \text{constant}$ must be the equation of the field lines. In fluid dynamics $\psi = \text{constant}$ represents the lines flow, and this accounts for ψ being called the stream function.

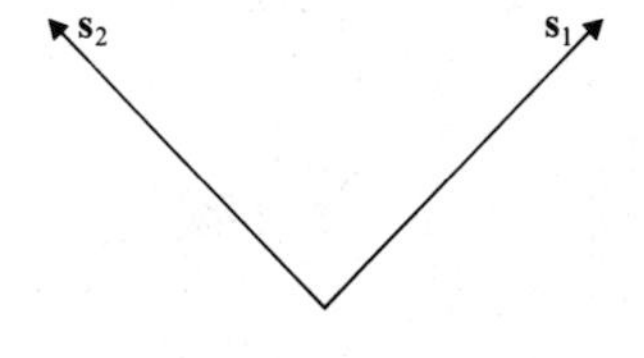

Fig. 7.11

An example will illustrate these ideas. Let us take $f(z)=z^2$, so that (7.38) becomes

$$w=z^2=x^2-y^2+2ixy$$

i.e.

$$\phi(x, y)=x^2-y^2, \qquad \psi(x, y)=2xy$$

We see that $\phi=0$ on the two perpendicular lines $y=\pm x$; the other equipotentials are the rectangular hyperbolae $x^2-y^2=K$ (constant) shown in Fig. 7.12. The field lines are the orthogonal hyperbolae $xy=$ constant, shown dotted in the diagram. This potential function would solve the problem of a condenser formed by two conductors, one of which is formed by the planes $y=\pm x$, and the other is any member (e.g. PQRS) of the family of hyperbolae $x^2-y^2=K$.

It is not difficult to calculate the charge distribution in such a problem. For if σ is the density of charge at a point Q, $\dfrac{\sigma}{\varepsilon_0}=-\dfrac{\partial\phi}{\partial n}$ where $\partial/\partial n$ denotes differentiation along the normal from right to left. According to (7.41) this is

$$\frac{\sigma}{\varepsilon_0}=+\frac{\partial\psi}{\partial s}, \tag{7.42}$$

where $\partial/\partial s$ denotes differentiation along the conductor from R to Q. Hence the total charge between Q and R per unit thickness

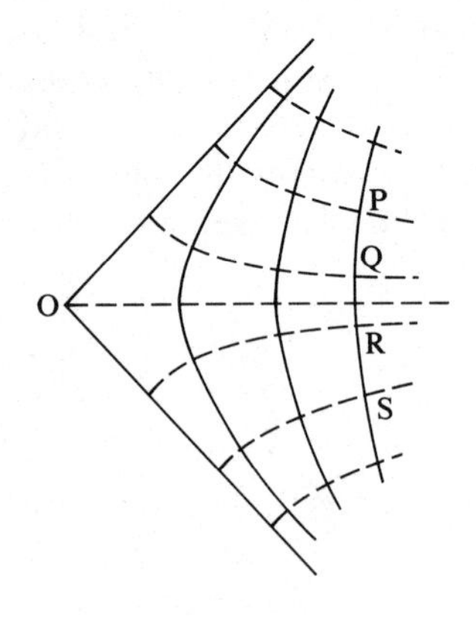

Fig. 7.12

perpendicular to the xy plane is q, where

$$q = \int \sigma \, ds = \varepsilon_0(\psi_Q - \psi_R) \tag{7.43}$$

This gives us a ready means of calculating capacities.

We can also calculate the field **E**, since, from (7.40)

$$\begin{aligned} E_x &= -\partial\phi/\partial x = -\text{real part of } f'(z), \\ E_y &= -\partial\phi/\partial y = \text{imaginary part of } f'(z) \end{aligned} \tag{7.44}$$

From this it follows that

$$E = |f'(z)| = \left|\frac{dw}{dz}\right| \tag{7.45}$$

and the direction that E makes with the x-axis is

$$-\arg f'(z) = \pi - \arg \frac{dw}{dz} \tag{7.46}$$

These last two equations suggest that it is often easier to work directly in terms of w rather than in terms of ϕ and ψ. This is indeed the case, as can be seen from the following special cases.

(a) Line charge *e* at the origin. We have already seen that the potential ϕ for a line charge e at the origin is $\phi = -\dfrac{e}{2\pi e_0} \ln r$.

This is obviously the real part of the function $-\dfrac{e}{2\pi\varepsilon_0} \ln z$, so that

$$\phi = -\frac{e}{2\pi\varepsilon_0} \ln r, \qquad \psi = -\frac{e}{2\pi\varepsilon_0}\theta, \qquad w = -\frac{e}{2\pi\varepsilon_0} \ln z \tag{7.47}$$

If the line charge is at the point z_0, then by a simple change of origin

$$w = -\frac{e}{2\pi\varepsilon_0} \ln (z - z_0) \tag{7.48}$$

(b) Line electrode of strength *I*. For a line electrode in a medium of conductivity σ from which a current I per unit thickness is flowing, it follows from (7.48) that

$$w = -\frac{I}{2\pi\sigma} \ln (z - z_0) \tag{7.49}$$

(c) Line doublet *m* at the origin. If the doublet is in the x direction, then by regarding it as the superposition of two large adjacent line charges, it can be seen that

$$\phi = \frac{2mx}{r^2}, \qquad \psi = -\frac{2my}{r^2}, \qquad w = \frac{2m}{z} \tag{7.50}$$

If the doublet is in the y direction

$$\phi = \frac{2my}{r^2}, \qquad \psi = \frac{2mx}{r^2}, \qquad w = \frac{2im}{x} \tag{7.51}$$

If the doublet makes an angle α with the x-axis,

$$w = \frac{2me^{i\alpha}}{z} \tag{7.52}$$

and if it is at the point $z = z_0$, then

$$w = \frac{2me^{i\alpha}}{z - z_0} \tag{7.53}$$

(d) Magnetic field of uniform current *I* in a straight wire. We see from §5.5 that if the current flows through the origin perpendicular to the xy plane, then the magnetostatic potential is $-\frac{I\theta}{2\pi}$. This is the real part of $\frac{iI}{2\pi} \ln z$, so that

$$\phi = -\frac{I\theta}{2\pi}, \qquad \psi = \frac{I}{2\pi} \ln r, \qquad w = \frac{iI}{2\pi} \ln z \tag{7.54}$$

If the current flows through the point z_0,

$$w = \frac{iI}{2\pi} \ln (z - z_0) \tag{7.55}$$

(e) Uniform field E_0. With a uniform field E_0 in the direction Ox

$$\phi = -E_0 x, \qquad \psi = -E_0 y, \qquad w = -E_0 z \tag{7.56}$$

With a field E_0 in the direction Oy,

$$\phi = -E_0 y, \qquad \psi = +E_0 x, \qquad w = iE_0 z \tag{7.57}$$

and if the field makes an angle α with the x-axis,

$$w = -E_0 e^{-i\alpha} z \tag{7.58}$$

By suitable combinations of (7.47)–(7.58) it is possible to solve many problems of this nature. We give one example.

A line source e is at the point z_0 above the infinite plane $y = 0$, which is kept at zero potential. By the method of images we know that the potential is the same as that due to a line charge e at z_0 and a line charge $-e$ at the image point. This is the complex conjugate $\bar{z}_0$. Hence from (7.18) and (7.48)

$$\begin{aligned} w &= -\frac{e}{2\pi\varepsilon_0} \ln(z - z_0) + \frac{e}{2\pi\varepsilon_0} \ln(z - \bar{z}_0) \\ &= -\frac{e}{2\pi\varepsilon_0} \ln \frac{z - z_0}{z - \bar{z}_0} \end{aligned} \tag{7.59}$$

From this it is easy to calculate the field and charge distribution.

§7.15 Conformal representation

A complex variable approach can sometimes be used very effectively to transform one problem into another, which can then be solved. To do this we consider the transformation

$$\zeta = f(z), \tag{7.60}$$

which maps the z-plane on to the ζ-plane, i.e. a point $z = x + iy$ in the z-plane is transformed into a point $\zeta = \xi + i\eta$ in the ζ-plane. Since $d\zeta = f'(z)dz$, it follows that

$$|d\zeta| = |f'(z)||dz|, \qquad \arg d\zeta = \arg f'(z) + \arg dz$$

Thus in the immediate neighbourhood of $z = z_0$, where $\zeta = \zeta_0 = f(z_0)$ all distances in the ζ-plane are $|f'(z_0)|$ times as large as the corresponding distances in the z-plane, and small arcs are turned through an angle $\arg f'(z_0)$. Any small element of area in the z-plane becomes an element of area in the ζ-plane having the same shape as before but whose dimensions are each $|f'(z_0)|$ as great, and which is turned through an angle $\arg f'(z_0)$. For this reason the process is described as **conformal representation.**

The importance of the transformation $\zeta = f(z)$ is that if we have any function $\phi(x, y)$ which in terms of ξ and η may be written $\Phi(\xi, \eta)$ then

$$\frac{\partial^2\Phi}{\partial\xi^2}+\frac{\partial^2\Phi}{\partial\eta^2}=M^2\left(\frac{\partial^2\phi}{\partial x^2}+\frac{\partial^2\phi}{\partial y^2}\right) \tag{7.61}$$

where M, which is called the modulus of the transformation, is defined by

$$M=1\bigg/\left|\frac{d\zeta}{dz}\right|=\left|\frac{dz}{d\zeta}\right| \tag{7.62}$$

Thus if ϕ is a potential function in xy space, so that $\nabla^2\phi=0$, then $\dfrac{\partial^2\Phi}{\partial\xi^2}+\dfrac{\partial^2\Phi}{\partial\eta^2}=0$, so that Φ is a potential function in $\xi\eta$ space. In other words, potential functions transform into potential functions. Also, any curve or boundary along which ϕ is constant becomes a new curve or boundary along which Φ is constant.

This type of transformation is particularly useful because, if there is a line charge e at a point z_0, then in the transformed problem there is an equal line charge e at the related point ζ_0 where $\zeta_0=f(z_0)$. For in the neighbourhood of z_0 and ζ_0,

$$\begin{aligned}\Phi(\xi, \eta)=\phi(x, y)&=-\frac{e}{2\pi\varepsilon_0}\ln|z-z_0|+\text{terms finite at } z=z_0\\ &=-\frac{e}{2\pi\varepsilon_0}\ln\left|\frac{dz}{d\zeta}(\zeta-\zeta_0)\right|+\text{terms finite at } \zeta=\zeta_0\\ &=-\frac{e}{2\pi\varepsilon_0}\ln|\zeta-\zeta_0|+\text{terms finite at } \zeta=\zeta_0\end{aligned}$$

This result indicates that the potential corresponds to a line charge e in the ζ-plane at the transformed point ζ_0.

We can combine together the results of this paragraph by saying that the potential function for given boundaries and charges in the z-plane is precisely equivalent to the potential function for the transformed boundaries and charges in the ζ-plane. If we can solve this problem, then by transforming back to the z-plane we shall be able to solve the original problem. The only difficulties that may arise are connected with the zeros of M and $1/M$. If points z_0 occur in the region of interest for which $M=0$ or $M=$ infinity, it will be necessary to investigate more closely the nature of the transformation near these points. It is often possible to avoid such singularities and we shall not discuss them further.

In more complicated problems it is sometimes an advantage to apply two or more conformal transformations successively. In this way we can simplify the problem stage by stage until it can finally be solved.

§7.16 Examples

Consider first (Fig. 7.13a) the problem of a line charge e at a point P where $z = z_0$, placed parallel to two infinite planes OX,OR. The planes, which are kept at zero potential, make an angle π/n with each other. Our analysis applies equally well for all values of n, so that it completes the discussion of this problem that could be given by the method of images, as in §7.4, where it was necessary to suppose that n was an integer.

Take O as origin and apply the transformation $\zeta = z^n$, using dashes to denote the transformed points. This transformation maps the space between OX and OR into the whole space above the axis R'O$'X'$ in the ζ-plane (Fig. 7.13b). The line charge e at P transforms to an equal line charge at P$'$ given by $\zeta_0 = z_0^n$. The new problem is therefore that of a line charge e placed above an infinite plane at zero potential. This has already been solved in (7.18) by the method of images, and

$$\Phi(\xi, \eta) = -\frac{e}{2\pi e_0} \ln |\zeta - \zeta_0| + \frac{e}{2\pi\varepsilon_0} \ln |\zeta - \bar{\zeta}_0|$$

Putting this back in terms of the z-plane, we see that the potential required for our original problem is

$$\phi(x, y) = -\frac{e}{2\pi\varepsilon_0} \ln |z^n - z_0^n| + \frac{e}{2\pi\varepsilon_0} \ln |z^n - \bar{z}_0^n|$$

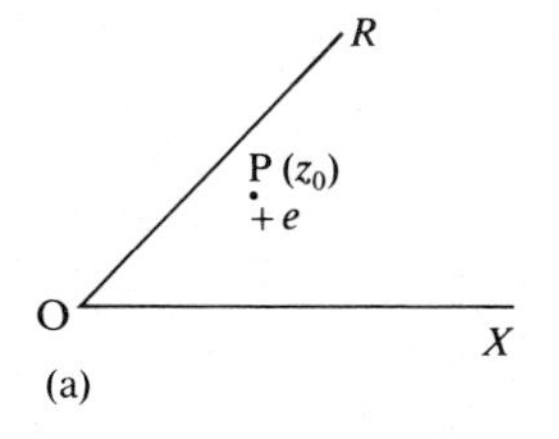

Fig. 7.13a z-plane

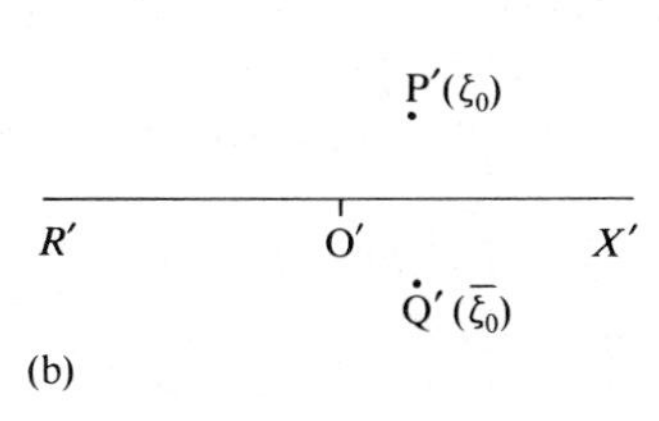

Fig. 7.13b ζ-plane

In terms of the w-function (§7.14),

$$w = -\frac{e}{2\pi\varepsilon_0}\ln\frac{z^n - z_0^n}{z^n - \bar{z}_0^n} \tag{7.63}$$

If we set $e=1$ this expression gives the Green function for the **general** wedge-shaped region; particular cases were discussed by the method of images in §7.4.

Our second example is the current distribution in a thin sheet (see Fig. 7.14a) of infinite length bounded by the lines AB, CD for which $y = \pm\frac{\pi}{2}$, when a current I per unit thickness of sheet enters at P($x=a$, $y=0$) and leaves at Q($x=-a$, $y=0$). In the z-plane we have to find a potential function ϕ such that

$$\phi \to \text{infinity like } -2I\ln|z-a| \text{ at } z=a,$$

$$\phi \to \text{infinity like } -2I\ln|z+a| \text{ at } z=-a,$$

$$\frac{\partial\phi}{\partial n}=0, \text{ on } y=\pm\frac{\pi}{2} \text{ for all } x.$$

The transformation $\zeta = e^z$ is suited to this problem: for it gives

$$\xi = e^x\cos y,\ \eta = e^x \sin y \tag{7.64}$$

and it is straightforward to verify that in passing from the z-plane to the ζ-plane (Fig. 7.14b)

(i) the line AB becomes A′B′, in which $\xi = 0$, $-\infty < \eta \le 0$,
(ii) the line CD becomes C′D′, in which $\xi = 0$, $0 \le \eta < +\infty$,
(iii) P and Q become P′ and Q′ where $\zeta_{P'} = e^a$, $\zeta_{Q'} = e^{-a}$,
(iv) the region between AB and CD becomes the half-plane $\xi > 0$.

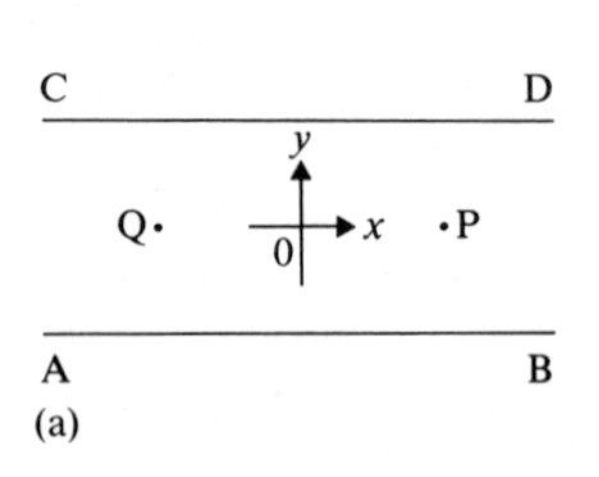

Fig. 7.14a z-plane

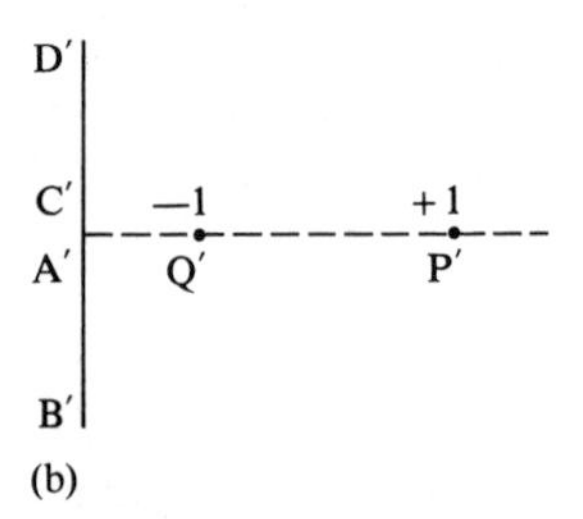

Fig. 7.14b ζ-plane

Thus the corresponding problem that we must solve in the ζ-plane is that of a current I entering at P′ and leaving at Q′, with $\partial\Phi/\partial n = 0$ on the contour B′D′ where $\xi = 0$. We solve this by the method of images. The potential function in the relevant part of the ζ-plane is the same as that due to $+I$ at $\zeta = -e^{a}$, and $-I$ at $\zeta = -e^{-a}$; these last two points are the mirror images of P′ and Q′ in the line $\xi = 0$. So the potential function (see (7.49)) is

$$w = -\frac{I}{2\pi\sigma}\{\ln(\zeta - e^{a}) + \ln(\zeta + e^{a}) - \ln(\zeta - e^{-a}) - \ln(\zeta + e^{-a})\}$$

$$= -\frac{I}{2\pi\sigma}\ln\left(\frac{\zeta^2 - e^{2a}}{\zeta^2 - e^{-2a}}\right)$$

In terms of the z-plane co-ordinates, the solution of our original problem is given by

$$w = -\frac{I}{2\pi\sigma}\ln\left(\frac{e^{2z} - e^{2a}}{e^{2z} - e^{-2a}}\right) \tag{7.65}$$

In this way the flow of current is completely determined. Since in the neighbourhood of the two points $z = \pm a$, the equipotentials are small circles, this potential function also solves the problem when the electrodes at P and Q are small circles instead of mathematical points.

§7.17 Finite-difference solutions of static field problems

We have now examined a number of analytic methods for the solution of Laplace's equation and we have seen in §7.2 how, by the method of Green functions, we may in principle construct a solution to Poisson's equation. In practice, finding a Green function or solving Laplace's equation is only possible for systems with some degree of symmetry. We now turn to the **only** truly general method of solving static field problems, namely the finite-difference solutions of the Laplace and Poisson equations. In this the original partial differential equation is replaced by a finite-difference approximation in which the difference quotients are substituted for the actual derivatives. The problem space is subdivided into a grid of orthogonal lines having a finite number of intersections or **mesh-points**. We then write the difference equation corresponding to each mesh-point and solve these equations simultaneously to obtain values for the function–which are approximations for the exact values–at each mesh-point.

As an example of the procedure consider a one-dimensional Poisson's equation

$$\frac{d^2\phi}{dx^2} = x \tag{7.66}$$

subject to the boundary conditions $\phi(0)=0$, $\phi(1)=1$. A finite difference mesh is set up over the region $0 \le x \le 1$; in Fig. 7.15 x_1, x_N are the boundary points and $\{x_j\}$, $j = 2, 3 \ldots N-1$, the interior mesh-points. The continuous variables x and ϕ are replaced by discrete variables x_i, ϕ_i i.e.

$$x \to x_i; \qquad \phi \to \phi_i = \phi(x_i) \tag{7.67}$$

In a numerical solution we seek the function values ϕ_i at the mesh points x_i; intermediate values of ϕ are then found by an interpolation procedure. Choosing a uniform mesh size h so that $x_{j+1} = x_j + h$, $\forall j$ and using Taylor's theorem to express function derivatives in terms of function values we have, assuming that the functions possess derivatives up to order 4

$$\phi(x_{i+1}) = \phi(x_i + h) = \phi(x_i) + h\phi'(x_i) + \frac{h^2}{2}\phi''(x_i)$$

$$+\frac{h^3}{6}\phi'''(x_i) + \frac{h^4}{24}\phi^{iv}(\xi) \qquad x_i \le \xi \le x_i + h$$

$$\phi(x_{i-1}) = \phi_i(x_i - h) = \phi(x_i) - h\phi'(x_i) + \frac{h^2}{2}\phi''(x_i)$$

$$-\frac{h^3}{6}\phi'''(x_i) + \frac{h^4}{24}\phi^{iv}(\eta) \qquad x_i - h \le \eta \le x_i$$

Adding these gives

$$\phi(x_i + h) + \phi(x_i - h) = 2\phi(x_i) + h^2\phi''(x_i) + O(h^4)$$

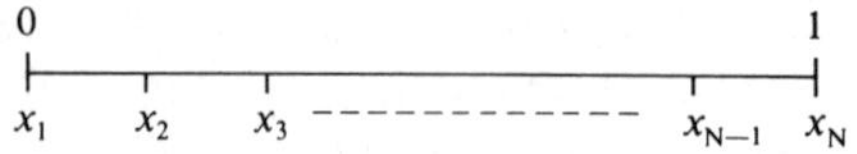

Fig. 7.15

i.e.

$$\phi''(x_i)=\frac{\phi(x_{i+1})+\phi(x_{i-1})-2\phi(x_i)}{h^2} \tag{7.68}$$

Similarly, by subtracting, we find

$$\phi'(x_i)=\frac{\phi(x_{i+1})-\phi(x_{i-1})}{2h}+O(h^2) \tag{7.69}$$

When the potential is a function of two variables x, y, i.e. $\phi(x, y)$ then with mesh spacing h along Ox and k along Oy we have (cf. Fig. 7.16)

$$\frac{\partial\phi}{\partial x}(x_i, y_j)=\frac{\phi(x_{i+1}, y_j)-\phi(x_{i-1}, y_j)}{2h} \tag{7.70}$$

$$\frac{\partial\phi}{\partial y}(x_i, y_j)=\frac{\phi(x_i, y_{j+1})-\phi(x_i, y_{j-1})}{2k} \tag{7.71}$$

$$\frac{\partial^2\phi(x_i, y_j)}{\partial x^2}=\frac{\phi(x_{i+1}, y_j)+\phi(x_{i-1}, y_j)-2\phi(x_i, y_j)}{h^2} \tag{7.72}$$

$$\frac{\partial^2\phi(x_i, y_j)}{\partial y^2}=\frac{\phi(x_i, y_{j+1})+\phi(x_i, y_{j-1})-2\phi(x_i, y_j)}{k^2} \tag{7.73}$$

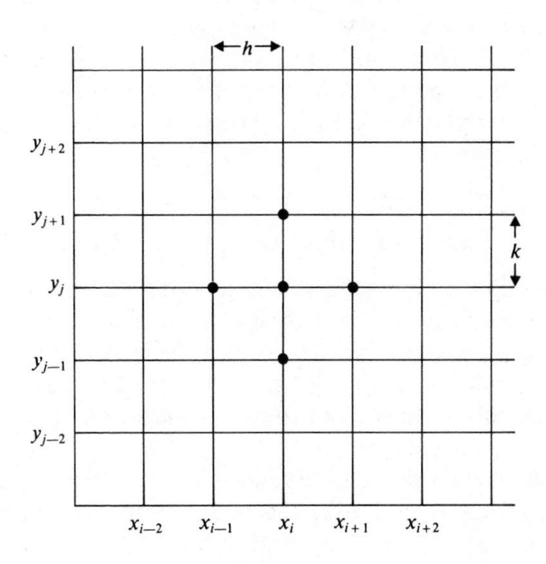

Fig. 7.16

Again we require that fourth derivatives with respect to both variables exist.

Hence approximating Poisson's equation

$$\frac{\partial^2 \phi}{\partial x^2}+\frac{\partial^2 \phi}{\partial y^2}=S(x, y)$$

using equal grid spacings $h=k$ in the x, y directions we get, to $O(h^2)$

$$\begin{aligned}\phi(x_{i-1}, y_j)+\phi(x_{i+1}, y_j)+\phi(x_i, y_{j+1})+\phi(x_i, y_{j-1})\\ -4\phi(x_i, y_j)-h^2S(x_i, y_j)=0\end{aligned} \quad (7.74)$$

§7.18 Numerical solution of Poisson's equation

If Poisson's equation is to be solved within a rectangle with boundary values prescribed then (7.74) must be satisfied at all the interior mesh points. If there are $N-1$ interior mesh points in the x direction and $M-1$ such points in the y direction we have a total of $(N-1)(M-1)$ interior mesh points. Requiring (7.74) to be satisfied at these points produces $(N-1)(M-1)$ linear equations for the $(N-1)(M-1)$ unknowns.

These may easily be solved by the Gauss–Seidel iterative method in which the $(p+1)$th iterate for the variable at (x_i, y_j) is found from the pth iterate at neighbouring points by rewriting (7.74) in the form

$$\begin{aligned}\phi^{(p+1)}(x_i, y_j)=\tfrac{1}{4}[\phi^{(p)}(x_{i-1}, y_j)+\phi^{(p)}(x_{i+1}, y_j)\\ +\phi^{(p)}(x_i, y_{j-1})+\phi^{(p)}(x_i, y_{j+1})-h^2S(x_i, y_j)]\end{aligned} \quad (7.75)$$

To start the interaction we give the function values $\phi(x_i, y_j)$ at the internal mesh points the **average** of the prescribed boundary values. For each sweep over the region the maximum change in $|\phi^{(p+1)}(x_i, y_j)-\phi^{(p)}(x_i, y_j)|$ should be determined and the iteration ended when this is reduced to some satisfactory level i.e. the iteration has converged.

It may be shown that convergence of the method is assured if the strategy for sweeping over the mesh points has certain properties. A strategy with the required properties is obtained if we sweep along grid lines parallel to either coordinate axis always

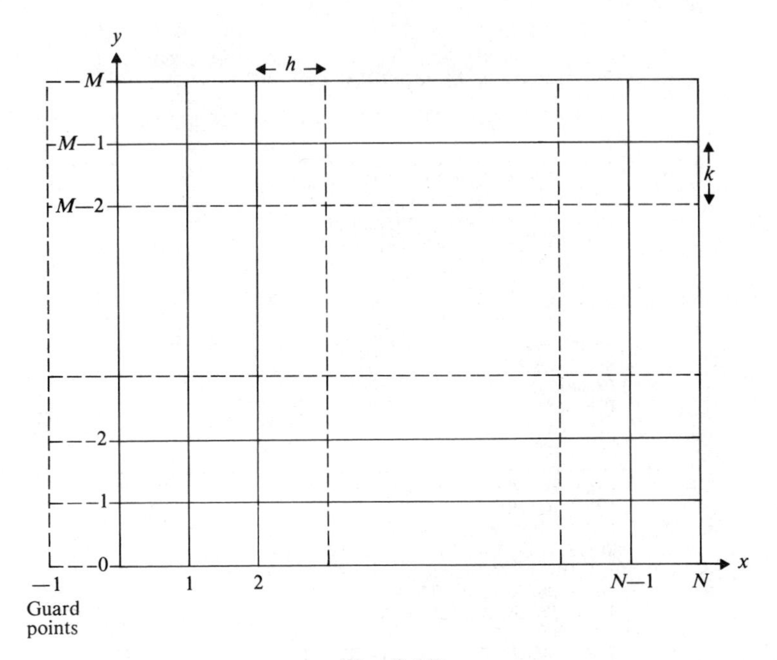

Fig. 7.17

travelling in the **same** direction. To increase the accuracy of the method we must reduce the mesh size h.

If, instead of prescribing function values along a boundary, the normal derivatives are given – e.g. $\frac{\partial \phi}{\partial x}$ might be prescribed along Oy – then we introduce a mesh line of guard points as shown in Fig. 7.17. We use (7.70) in the form

$$\phi(-h, y_j) = \phi(h, y_j) - 2h\frac{\partial \phi}{\partial x}(0, y_j)$$

to determine function values along that mesh line ($x = -h$) which is then treated as a fictitious boundary and the solution completed as above. Equation (7.74) or (7.75) has to be satisfied at all internal mesh points and also along Oy, leading to $N(M-1)$ equations in $N(M-1)$ unknowns, which are again solved by the Gauss–Seidel method as described.

The following program solves Poisson's equation over a rectangular region.

```
      PROGRAM POISSON
C     SOLVE POISSON'S EQUATION IN A REGION 0<=X<=XMAX,
C     0<=Y<=YMAX; ON A RECTANGULAR MESH M*N
C
C     DEFINE FUNCTION SOURCE(X,Y) AS A FUNCTION SUBROUTINE
C
C     I RUNS ALONG X WITH M+1 MAX VALUE
C     J RUNS ALONG Y WITH N+1 MAX VALUE
      DIMENSION V(25,25),S(25,25)
      DATA KYES,KNO,KX,KY/1HY,1HN,1HX,1HY/
      DATA NTTY,KHYF,E1/5,1H-,.0E-4/
      WRITE(NTTY,9000)
      READ(NTTY,9010)XMAX,YMAX
      WRITE(NTTY,9020)
      READ(NTTY,9030)M,N
      NPLUS1=N+1
      MPLUS1=M+1
      DX=XMAX/M
      DY=YMAX/N
      DXSQ=DX*DX
      DYSQ=DY*DY
C---------------------------------------------------
C     READ IN BOUNDARY VALUES
      DO 30 J=1,NPLUS1,N
C     ASK IF THEY ARE CONSTANT
10    WRITE(NTTY,9040)KX,J
      READ(NTTY,9050)KANS
      IF (KANS.EQ.KNO)GOTO 20
      IF (KANS.NE.KYES)GOTO 10
C
C     THE BOUNDARY IS CONSTANT-READ IT
      WRITE(NTTY,9060)KX,J
      READ(NTTY,9070)C
20    DO 30 I=1,MPLUS1
C     IF IT IS NOT CONSTANT READ IT IN
      IF (KANS.EQ.KYES)GOTO 30
      WRITE(NTTY,9080)I,J
      READ(NTTY,9090)C
30    V(I,J)=C
C
C-----------------------
      DO 60 I=1,MPLUS1,M
C     ASK IF THESES ARE CONSTANT
40    WRITE(NTTY,9040)KY,I
      READ(NTTY,9050)KANS
      IF (KANS.EQ.KNO)GOTO 50
      IF (KANS.NE.KYES)GOTO 40
C
C     THE BOUNDARY IS CONSTANT-READ IT
      WRITE(NTTY,9060)KY,I
      READ(NTTY,9070)C
C     THIS LOOP IS 2 TO N SINCE (1,1) AND (M+1,N+1)
C     WERE ASSIGNED ABOVE
50    DO 60 J=2,N
      IF (KANS.EQ.KYES)GOTO 60
C     THE BOUNDARY IS NOT CONSTANT- READ IT IN
      WRITE(NTTY,9080)I,J
      READ(NTTY,9090)C
60    V(I,J)=C
C
C---------------------------------------------------
C     AVERAGE BOUNDARY VALUE BECOMES INITIAL GUESS
```

```
      STEMP=0.
      DO 70 I=1,MPLUS1
70    STEMP=STEMP+V(I,1)
      DO 80 J=2,N
80    STEMP=STEMP+V(1,J)+V(MPLUS1,J)
      STEMP=STEMP/(2.*M+2.*N)
C
C     NOW ASSIGN IT TO V(I,J)- ALSO CALCULATE SOURCE FUNCTION
      DO 90 I=2,M
      X=(I-1)*DX
      DO 90 J=2,N
      Y=(J-1)*DY
      S(I,J)=SOURCE(X,Y)
90    V(I,J)=STEMP
C-----------------------------------------------------------
C     SOLUTION PROCEDURE
      DO 110 ITER=1,1000
      ERROR=0.
      DO 100 I=2,M
      DO 100 J=2,N
      U=DYSQ*(V(I-1,J)+V(I+1,J))+DXSQ*(V(I,J-1)+V(I,J+1))
      U=U-DXSQ*DYSQ*S(I,J)
      U=U/(2.*(DXSQ+DYSQ))
      IF (ABS(U-V(I,J)).LE.ERROR)GOTO 100
      ERROR=ABS(U-V(I,J))
100   V(I,J)=U
110   IF (ERROR.LT.E1) GOTO 200
C-----------------------------------------------------------
C     IF WE REACH THIS POINT NO CONVERGENCE AFTER 1000 ITERATION
      WRITE(NTTY,9130)
      STOP
C-----------------------------------------------------------
C     THE SOLUTION HAS CONVERGED. PRINT OUT THE RESULT
C
200   NSTEP=AMAX0(1,(12+MPLUS1)/13)
C     NSTEP IS CALCULATED SO THAT THERE ARE 13 COLUMNS ACROSS PAGE
C     N1=LENGTH OF LINE OF HYPHENS
      N1=10*MPLUS1)/NSTEP+3
      WRITE(NTTY,9100)(I,I=1,MPLUS1,NSTEP)
        WRITE(NTTY,9110)(KHYF,I=1,N1)
      DO 210 J=NPLUS1,1,-1
210   WRITE(NTTY,9120)J,(V(I,J),I=1,MPLUS1,NSTEP)
      STOP
C-----------------------------------------------------------
9000  FORMAT(1H/,41HX LIES BETWEEN 0 AND XMAX; Y LIES BETWEEN,
     1/,1H ,28H0 AND YMAX. TYPE XMAX,YMAX :,$)
9010  FORMAT(2F8.2)
9020  FORMAT(1H ,45HTHERE ARE M MESH INTERVALS IN THE X DIRECTION,
     1/,1H ,36HAND N IN THE Y DIRECTION. TYPE M,N :,$)
9030  FORMAT(2I2)
9040  FORMAT(1H ,26HDO YOU WISH THE VALUES ON ,
     1A1,1H=,I2,11H CONSTANT  ,$)
9050  FORMAT(A1)
9060  FORMAT(1H ,21HWHAT IS THE VALUE ON ,A1,1H=,I2,2H :,$)
9070  FORMAT(F8.2)
9080  FORMAT(1H ,10HWHAT IS V(,I2,1H,,I2,3H): ,$)
9090  FORMAT(G)
9100  FORMAT(1H/,30HSOLUTION OF POISSON'S EQUATION,///,
     11H ,3H   ,13(5H     ,I3,2H !))
9110  FORMAT(1H ,130A1)
9120  FORMAT(1H ,I2,1H!,13F10.4)
9130  FORMAT(1H ,20H***NO CONVERGENCE***)
```

```
      END
      FUNCTION SOURCE(X,Y)
      SOURCE=SIN(3.141592654*X)
      RETURN
      END
```

§7.19 Problems

1. Show that the result of the uniqueness theorem in §7.3 is unaffected by any sudden discontinuities in the dielectric constant.

2. A given series of electrodes are maintained at given potentials. Prove that the potential function is unique.

3. Show that the analysis of §7.3 applies equally well with regard to the magnetostatic potential arising from a given set of permanent magnets.

4. S is a closed equipotential surface completely surrounding a system of charges whose algebraic sum is Q. The potential of S is V. Show that if the surface S is regarded as one plate of a condenser, of which infinity is the other, the capacity is Q/V. If S_1 and S_2 are two such surfaces at potentials V_1 and V_2, show that the capacity of the condenser formed by S_1 and S_2 is $Q/(V_1 - V_2)$.

5. The space between two equipotential surfaces ϕ_1 and ϕ_2 which completely surround a system of electrostatic charges in vacuum is filled with a medium of constant dielectric constant K. Show that at all points inside the inner surface ϕ_1 the potential is increased by $\{(K-1)/K\}(\phi_1 - \phi_2)$.

6. Show that if the angle XOY in Fig. 7.3 is made equal to π/n, where n is an integer, the number of charges in the image system of a charge at A inside XOY, is $2n$. Show also that unless n is an integer the method of images breaks down.

7. A line charge e is placed at a distance a from an infinite conducting plane at zero potential. Show that at a point P on the plane the density of induced charge is $-ea/\pi r^2$, where r is the shortest distance from P to the line charge.

8. Two equal charges e are placed a distance a apart, and each of them is $a/2$ from an infinite conducting plane at zero potential.

Prove that the force between the charges is $3e^2/8\pi\varepsilon_0 a^2$. If the sign of one of the charges is reversed, why is the force not simply reversed also?

9. The charge e_1 at A in Fig. 7.4 is taken further away from the conducting sphere and simultaneously increased in size in such a way that as $f \to \infty$, $e_1/f^2 \to E_0$. Deduce that if an uncharged conducting sphere of radius a is placed in a uniform field E_0, the image of the field is a doublet at the centre of magnitude $E_0 a^3$. This is the result of §7.7.

10. A point charge e is placed inside a spherical cavity of radius a cut out of a conducting block of metal at zero potential. If the distance of the charge from the centre of the cavity is f, show that the force on it is $e^2af/4\pi\varepsilon_0(a^2-f^2)^2$.

11. An electric dipole of moment p is at a distance f from the centre of a conducting sphere of radius a kept at zero potential. The dipole points away from the sphere. Prove that the image is a dipole of moment pa^3/f^3 and a charge pa/f^2 at the inverse point.

12. An electric dipole of moment p is held at a distance a from an infinite conducting plane at zero potential. Show that the image is an equal dipole. If the original dipole makes an angle θ with the normal to the plane, show that the mutual potential energy is $-\dfrac{p^2}{32\pi\varepsilon_0 a^3}(1+\cos^2\theta)$. Deduce that if it is free to turn about its centre, it will take up a position perpendicular to the plane, and the period of small oscillations, about this direction is $4\pi\sqrt{(2a^3A)}/p$, where A is the moment of inertia.

13. The whole of space for which $x>0$ is filled with a dielectric of uniform dielectric constant K, and the rest is vacuum. P is the point $(-a, 0, 0)$ and P′ is the point $(a, 0, 0)$. A charge e is placed at P. Show that in the vacuum the potential is the same as that due to a charge e at P and e' at P′, and in the dielectric it is the same as that due to a charge e'' at P, where

$$e' = -\frac{K-1}{K+1}e, \qquad e'' = \frac{2}{K+1}e.$$

Use the fact that the pull on the dielectric is equal and opposite to the pull on the charge e to show that the dielectric is attracted towards the charge with a force $\dfrac{K-1}{K+1}\dfrac{e^2}{16\pi\varepsilon_0 a^2}$. Explain this in terms of tensions in the tubes of force.

14. Show that the result in question 13 may be used to determine the image system when a small permanent magnet is held some distance away from an infinite block of material of permeability μ.

15. Show how to find the image system for a line charge parallel (i) to an infinite dielectric medium, and (ii) to an uncharged conducting cylinder at zero potential. By removing the line charge to infinity in (ii) show how to find the potential for a conducting cylinder in a uniform field perpendicular to its axis.

16. A current I flows in a straight wire parallel to a circular cylinder of permeability μ. Show that inside the cylinder the magnetic induction is $2\mu/(\mu+\mu_0)$ times as big as it would be if the cylinder were removed, and is everywhere in the same direction. Deduce that if the cylinder is placed in any given two-dimensional field, the induction inside it is unaffected in direction, but is multiplied by a factor $2\mu/(\mu+\mu_0)$ in magnitude.

17. A uniform current I flows in a straight wire parallel to a semi-infinite block of material of permeability μ. Show that the potential is given by an image system similar to that in question 13.

18. A telegraph wire is of radius a and its height above the earth is h. Prove by means of (7.18) that the capacity per unit length is $2\pi\varepsilon_0/\ln(2h/a)$.

19. A charge e is placed a distance f from the centre of an insulated conducting sphere of radius a. Show that the least positive charge which must be given to the sphere so that the surface density is everywhere positive, is $ea\{(f+a)/(f-a)^2-1/f\}$.

20. Show that the least charge Q that can be given to a conducting sphere of radius a so that when the system is placed in a uniform field E_0 no part of the sphere is negatively charged is $Q=12\pi a^2\varepsilon_0 E_0$.

21. A conducting cylinder of radius a and infinite length is placed with its axis perpendicular to a uniform field E_0. The cylinder is

at zero potential. Calculate the potential at all points and deduce that the greatest surface density of induced charge is $2\varepsilon_0 E_0$.

22. Show that (7.29) gives for the field inside a small circular cavity in a dielectric the result stated at the end of §6.8.

23. A charge e is placed at a point P outside an uncharged conducting sphere of radius a. The distance of P from the centre O of the sphere is f. Show that the potential at any point whose distance from P is R may be put in the form

$$\phi = \frac{e}{4\pi\varepsilon_0}\left[\frac{1}{R} - \sum_{n=1}^{\infty} \frac{a^{2n+1}}{f^{n+1} r^{n+1}} P_n(\cos\theta)\right]$$

Show that this may be transformed to

$$\phi = \frac{e}{4\pi\varepsilon_0}\left[\frac{1}{R} - \frac{a}{fR_1} + \frac{a}{fr}\right]$$

where R_1 is the distance from the inverse point of P with respect to the sphere.

24. A point charge e is placed a small distance c from the centre of a spherical cavity of radius a in an infinite dielectric. Show that the charge experiences a force approximately equal to $\dfrac{(K-1)}{2K+1}\dfrac{e^2 c}{2\pi\varepsilon_0 a^3}$ away from O.

25. A spherical hole of radius a is cut out of an infinite block of uniform conductivity. At large distances the current flow is uniform and in the z direction. Show that the potential function is $\phi = A\left(r + \dfrac{a^3}{2r^2}\right)\cos\theta$. Deduce that the lines of flow are

$$r^3 - a^3 = Cr \operatorname{cosec}^2\theta.$$

26. Solve the potential equations for a uniformly magnetized sphere.

27. A sphere of radius a and permeability μ is placed in a uniform magnetic field H_0. Prove that the field inside the sphere is uniform with value $3\mu_0 H_0/(\mu + 2\mu_0)$.

28. Verify that the solution in (7.30) for the potential due to a small magnet in a spherical cavity satisfies all the conditions (i)–(v) of §7.9.

29. A magnetic particle of moment m is at the centre of a spherical hole of radius a cut in an infinite block of matter of permeability μ. A uniform field H_0 exists at infinity, and the magnetic particle points at an angle α with the direction of H_0. Show that the field inside the cavity consists of the field due to the magnet m together with a constant part whose direction makes an angle β with H_0, where

$$\cot\beta = \cot\alpha + \frac{3\mu a^3 H_0}{2(\mu-\mu_0)m\sin\alpha}$$

30. A small magnet of moment m is at the centre of a spherical shell bounded by spheres of radii a and b and filled with material of permeability μ. Show that outside the shell the field is the same as that of a magnet of moment m', where

$$9\mu\mu_0^2 b^3 m = m'\{(\mu+2\mu_0)(2\mu+\mu_0)b^3 - 2(\mu-\mu_0)^2 a^3\}$$

31. A and B are the points $z=\pm a$. Show that the function $w=\dfrac{1}{2\pi\varepsilon_0}\ln\dfrac{z-a}{z+a}$ gives equipotential surfaces $r_1/r_2=\text{constant}$, where r_1 and r_2 are distances from A and B. Deduce that the capacity C per unit length between two parallel cylinders of radius R a distance $2D$ apart, is given by

$$\frac{1}{C} = \frac{1}{\pi\varepsilon_0}\ln\left(\{D+\sqrt{(D^2-R^2)}\}/R\right) = \frac{1}{\pi\varepsilon_0}\cosh^{-1}(D/R)$$

32. A line charge e is at a distance h from an infinite conducting plane at zero potential. Show that one half of the charge induced on the plane lies within $\sqrt{2}h$ from the line charge.

33. Show that the capacity per unit thickness of a two-dimensional condenser in which the plates are at potentials V_1 and V_2 is $\varepsilon_0[\psi]/(V_1-V_2)$ where $[\psi]$ denotes the increment in ψ in going once completely round one of the plates.

34. Show that $w=-E_0\left(z-\dfrac{a^2}{z}\right)$ gives the potential distribution for a conducting cylinder of radius a at zero potential in a uniform field E_0. (Cf. problem 21). Interpret the expression for w in terms of images.

35. Show that $w = iAz^n$ (n not necessarily integral) gives the potential distribution for two intersecting planes OX, OR making an angle π/n, if both are at zero potential. Prove that with this potential function, $\phi = -Ar^n \sin n\theta$. Find the equations of the field lines, and deduce from (7.42) that the charge density at any point on the planes is $nAr^{n-1}/4\pi$. Draw diagrams for $n = 2$ and $n = \frac{2}{3}$, and notice the bunching of tubes of force near the tip in the one case, and the absence in the other. This bunching explains why a lightning conductor is given a pointed tip.

36. ABCD is the region bounded on two sides, AB,CD by the parabolas $y^2 = 4b(b - x)$, $y^2 = 4c(c - x)$, and on the other two sides AD,BC by the x-axis and the parabola $y^2 = 4a(x + a)$. Show that the transformation $\zeta = z^{\frac{1}{2}}$ maps ABCD on to a rectangle in the ζ-plane. Hence prove that if AB and CD are taken as electrodes, and if the strip is of uniform thickness t, its resistance is $(b^{\frac{1}{2}} - c^{\frac{1}{2}})/\sigma t a^{\frac{1}{2}}$.

37. Show that the total current leaving an electrode per unit thickness of material is $\sigma[\psi]$, where $[\psi]$ denotes the change in ψ on going round the boundary of the electrode in a clockwise direction.

38. A thin strip of metal of thickness t lies in the xy-plane, and is bounded by the lines $y = \pm\frac{\pi}{2}$. Equal currents I enter and leave by small circular electrodes of radius δ at the points $(\pm a, 0)$. Show that the potential function w is

$$w = -\frac{I}{2\pi\sigma t} \ln \left(\frac{e^{2z} - e^{2a}}{e^{2z} - e^{-2a}} \right) + \text{constant}$$

Deduce that the resistance is $-\dfrac{I}{\pi\sigma t} \ln (\delta \operatorname{cosech} 2a)$

39. Show that if n is an integer, (7.63) may be interpreted as showing that if two planes at zero potential intersect at an angle π/n, the image system of a line charge e between them consists of $n - 1$ line charges $+e$, and n lines $-e$.

40. If the strip in Fig. 7.14a is of width $2b$, show that the transformation (7.64) needs to be replaced by $\zeta = e^{\pi z/2b}$, and that

$$w = -\frac{I}{2\pi\sigma} \ln \left(\frac{e^{\pi z/b} - e^{\pi a/b}}{e^{\pi z/b} - e^{-\pi a/b}} \right)$$

41. A condenser is formed by taking OX, OR (Fig. 7.13a) as one plate, and a small circular cylinder with centre P and radius δ as the other. Show that its capacity per unit length perpendicular to the xy plane is $-2\pi\varepsilon_0/\ln\left\{\dfrac{n\delta}{2r}\operatorname{cosec} n\alpha\right\}$, where $z_0 = re^{i\alpha}$.

42. What happens if you use the transformation $\zeta = z^{2n}$ in §7.16, Fig. 7.13a, instead of $\zeta = z^n$?

43. Show that in the transformation $\zeta = f(z)$ a line doublet of moment m at z_0, making an angle α with the x-axis, transforms into a line doublet of moment $m|f'(z_0)|$ at $\zeta_0 = f(z_0)$ making an angle $\alpha + \arg f'(z_0)$ with the ξ-axis. Deduce that if the original doublet lies between the two planes of Fig. 7.13a, which are kept at zero potential,

$$w = 2mn\,|z_0^{n-1}|\left\{\frac{e^{i\theta}}{z^n - z_0^n} - \frac{e^{-i\theta}}{z^n - \bar{z}_0^n}\right\}$$

where $\theta = \alpha + (n-1)\arg z_0$.

8 Electromagnetic induction

§ 8.1 Introduction

In Chapter 5 we defined the magnetic field **B** in terms of the force $\mathbf{v}\times\mathbf{B}$ (newtons) felt by a unit charge moving with velocity $\mathbf{v}$(m. $\sec^{-1}$); the units of **B** are weber m^{-2}. We went on to discuss the magnetostatic field associated with the flow of a steady current. We must now consider magnetic fields which change with time. Just as in Chapter 5 in which the essential ingredients for the development of electrodynamics came from the experiments of Oersted, Ampère and others, we must look to experiments for the new information that will allow us to extend the theory to time-dependent **B** fields. The rapid advances made by Ampère and others during the 1820s had left the science of electrodynamics, as Ampère called it, ready for the next step forward. It was taken almost simultaneously in Britain and the United States in 1831 by Faraday and Henry, working quite independently.

What Faraday and Henry discovered was that when a closed circuit moved across a magnetic field, a current flowed even though there were no batteries present. The same effect resulted from varying the magnetic field and keeping the loop of wire still. In either case the current lasted only so long as the circuit was moving, or the field changing. It made no difference whether the magnetic field was caused by a permanent magnet or an electrical circuit. To this phenomenon Faraday gave the name **electromagnetic induction**. In a series of brilliant experiments he showed that the induced current flowed whenever the number of tubes of induction (i.e. the flux Φ) through the circuit was altered, and that the magnitude of the effect depended only on the rate of change of Φ. Regrettably a theoretical account of electrodynamics allows no space for a description of Faraday's experiments (see [1]) but simply a summary of the results in what has become

known as the flux rule, which may be stated as follows:
(a) When the magnetic flux threading a conducting circuit is changed **in any way** an electromotive force (emf) is induced in the circuit and its magnitude is equal to the rate of change of flux.

Thus if we denote the emf by $\mathscr{E}$ and the flux by Φ we may express this mathematically by

$$|\mathscr{E}| = \frac{d\Phi}{dt} \tag{8.1}$$

So far we have not specified the direction of the emf. To do so we must supplement the information in the flux rule by a second law known as **Lenz's law**:
(b) The induced emf acts in such a way as to oppose the change in flux. That is, it acts to generate currents in the circuit whose associated magnetic effect would counteract the external change.

We may combine (8.1) and Lenz's law and state that the emf associated with a change in magnetic flux through a circuit is

$$\mathscr{E} = -\frac{d\Phi}{dt} \tag{8.2}$$

Lenz's law is merely a particular case of a very general physical principle, known as **Le Chatelier's Principle** which states that a physical system always reacts to oppose any change that is imposed from outside. In this case the change is an alteration in Φ; the reaction is an **induced emf** or **back emf** which should oppose the change of Φ.

Figure 8.1 presents a simple picture of how Lenz's law works. Figure 8.1a represents a situation in which the flux through the circuit C is decreasing. By Lenz's law we know that the induced emf acts in this case to produce in the circuit a current whose associated magnetic field is so directed to increase the flux through C. Thus the induced current in Fig. 8.1a flows in an anticlockwise sense and the emf $\mathscr{E} > 0$. In Fig. 8.1b, on the other hand, the flux through C is increasing so that the induced current now flows in a clockwise direction and $\mathscr{E} < 0$.

§ 8.2 Motion of a conductor in a uniform magnetic field

In this section we shall see how induction may be understood as a consequence of the $\mathbf{v} \times \mathbf{B}$ force on charges moving in a magnetic

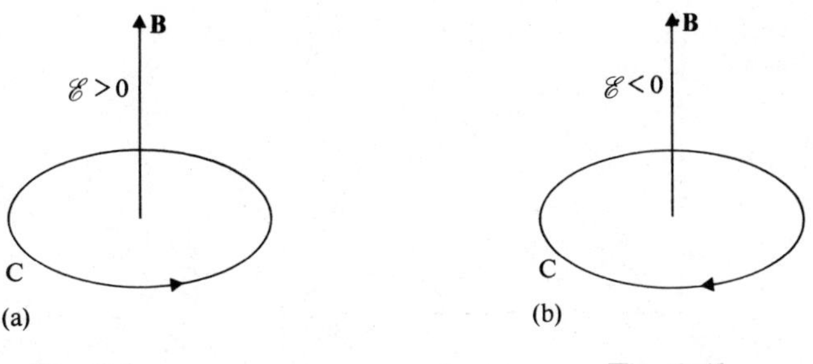

Fig. 8.1a *Fig. 8.1b*

field. Suppose we take a strip of conductor and move it with constant velocity **v** at right angles to a magnetic field **B** (Fig. 8.2). We shall further suppose that **B** is spatially uniform and constant in time. As the conducting strip moves across the magnetic field a particle of charge e experiences a force $\mathbf{F}=e\mathbf{v}\times\mathbf{B}$ in the positive or negative x direction depending on whether the particle is positively or negatively charged. For a metallic conductor the free charges are electrons with $e<0$, i.e.

$$\mathbf{F}=-|e|\mathbf{v}\times\mathbf{B}=-|e|\mathbf{E}' \tag{8.3}$$

As a consequence one end of the strip will become negatively, and the other positively, charged and this separation of charge naturally causes an electrostatic field **E** to be established in the negative x direction. The magntide of **E** is such that the resultant force on the electron vanishes, i.e. $\mathbf{E}+\mathbf{E}'=0$.

Next let us suppose that the conducting strip PQ moves freely with velocity **v** along the conducting bars, shown in Fig. 8.3, a distance x_0 apart and connected at one end by a conducting arm RS. Now, of course, a current I flows in a clockwise sense. The charges in the strip feel a force along the strip equal to $\mathbf{E}'=\mathbf{v}\times\mathbf{B}$ per unit charge. The electromotive force is therefore

$$\mathscr{E}=\oint \mathbf{E}'\cdot d\mathbf{s}=-vBx_0$$

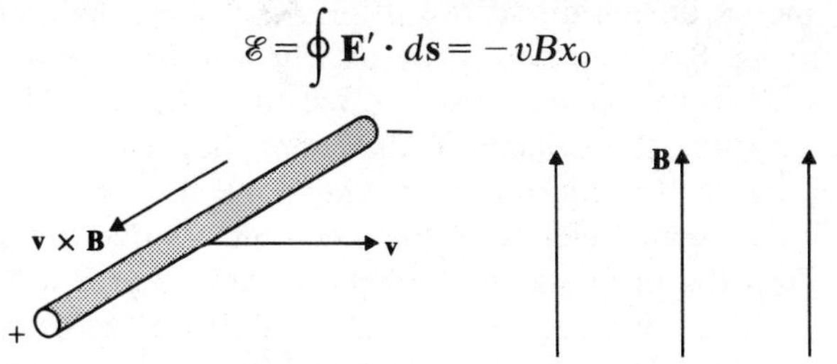

Fig. 8.2

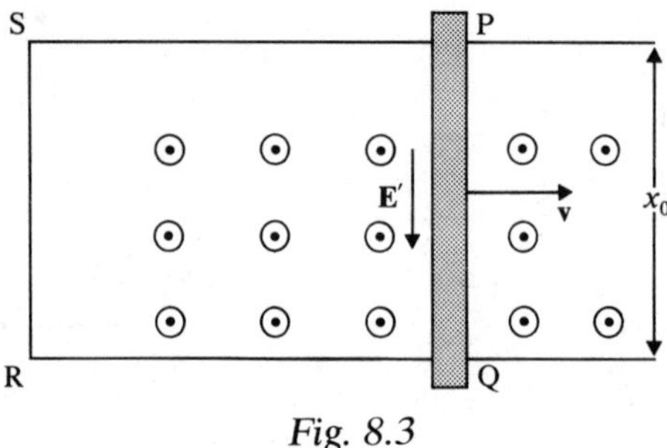

Fig. 8.3

since the only contribution round the circuit PQRS comes from the arm PQ. This is just the result we get from the flux rule:

$$\mathscr{E} = -\frac{d\Phi}{dt} = -\frac{d}{dt}[Bx_0y]$$

$$\text{i.e. } \mathscr{E} = -vBx_0$$

So in this particular case in which the circuit PQRS is changing with time we have derived the emf from the $\mathbf{v}\times\mathbf{B}$ force on the charges in the moving strip. We must resist concluding from this that we have "proved" Faraday's law! Faraday's result holds **regardless** of what brings about the change in flux $\Phi = \int_S \mathbf{B}\cdot d\mathbf{S}$. We happen to have used an argument which applies only to flux changes brought about by a change in S.

§ 8.3 The induction law for a moving circuit

In the preceding section we considered a moving circuit with a particular configuration. It is instructive to present an argument for an arbitrary moving circuit. Once again the magnetic field in which it moves does not vary with time. Consider, then, an element PQ (Fig. 8.4) represented by $d\mathbf{s}$ which moves with velocity $\mathbf{v}$. In time dt it will move a distance $\mathbf{v}dt$ to a position P′Q′. Suppose also that the velocity of the conduction electrons relative to the wire is $\mathbf{u}$. If no current is flowing $\mathbf{u}=0$, but when current does flow the actual velocity of the electrons relative to the field is $\mathbf{v}+\mathbf{u}$. Thus the magnetic field exerts on each charge a force $e(\mathbf{v}+\mathbf{u})\times\mathbf{B}$. The component of this along PQ is $e\{(\mathbf{v}+\mathbf{u})\times\mathbf{B}\}\cdot\hat{\mathbf{s}}$ where $\hat{\mathbf{s}}$ is a unit vector in the direction PQ. Since $\mathbf{u}$ and $\hat{\mathbf{s}}$ are parallel, $(\mathbf{u}\times\mathbf{B}\cdot\hat{\mathbf{s}}=0$ and the component force along PQ is

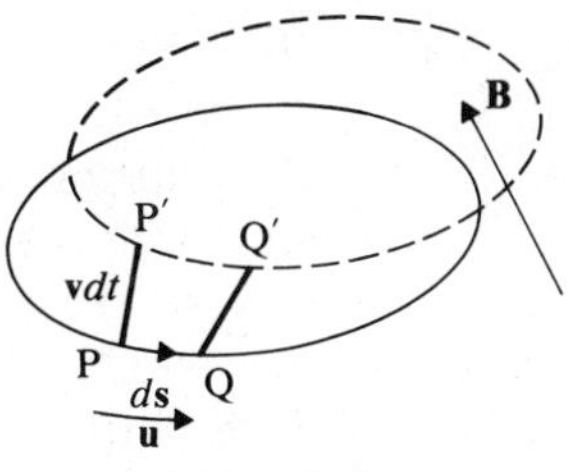

Fig. 8.4

$e(\mathbf{v}\times\mathbf{B})\cdot\hat{\mathbf{s}}$. This means that in addition to any forces due to batteries which may be present, there is an electric field $\mathbf{E}=\mathbf{v}\times\mathbf{B}$ induced in the wire; its component along the wire is $\mathbf{v}\times\mathbf{B}\cdot\hat{\mathbf{s}}$. The corresponding emf is simply the line integral of this field round the circuit i.e.

$$\mathscr{E}=\oint(\mathbf{v}\times\mathbf{B})\cdot\hat{\mathbf{s}}\,ds$$

$$=\oint(\mathbf{v}\times\mathbf{B})\cdot d\mathbf{s} \tag{8.4}$$

We now compute the change in the flux through a surface S spanning the circuit as it moves during the interval dt. At time $t+dt$ the original surface S has been increased by a band of which PP′Q′Q is an element. Now $\mathbf{v}\,dt\times d\mathbf{s}$ is the vector area PP′Q′Q so that $(\mathbf{v}\,dt\times d\mathbf{s})\cdot\mathbf{B}$ represents the flux of B across this element. Thus over the entire band

$$d\Phi=\oint\mathbf{B}\cdot(\mathbf{v}\,dt\times d\mathbf{s}) \tag{8.5}$$

$$\text{i.e.}\ \frac{d\Phi}{dt}=\oint\mathbf{B}\cdot(\mathbf{v}\times d\mathbf{s})$$

$$=-\oint(\mathbf{v}\times\mathbf{B})\cdot d\mathbf{s} \tag{8.6}$$

by applying the rule governing triple scalar products. Comparing (8.4) and (8.6) we see

$$\mathscr{E}=-\frac{d\Phi}{dt}$$

just as in the special case considered in §8.2. However we have succeeded in making a considerable generalization to a circuit of any shape moving arbitrarily (e.g. $\mathbf{v}$ need not be uniform so that the circuit might deform as it moves.)

§ 8.4 A simple dynamo

The law of induction, (8.2), is the principle on which dynamos, or machines for creating an emf, are based. Consider, for example, an almost closed circular loop of wire PQ of area A (Fig. 8.5) which can rotate in air about an axis XY. We shall suppose that there is a constant magnetic field **B** perpendicular to the plane of the paper. Then when the normal to the loop makes an angle θ with the field, the flux $\Phi = AB\cos\theta$, so that the emf induced between P and Q when the coil rotates with angular velocity ω, is

$$\mathscr{E} = -\frac{d\Phi}{dt} = AB\omega \sin\theta \tag{8.7}$$

If P and Q were joined to complete the circuit, this emf would cause a current to flow round the wire. But if by some device such as brushes, or sliding contacts on the axle XY, we connect PQ to the terminals of an outside circuit, this emf may be used to create a current in that circuit. On account of the $\sin\theta$ term in (8.7) the emf is **alternating** with frequency $\omega/2\pi$. A **direct** emf may be obtained by a **commutator** which interchanges the roles of P and Q in the outside circuit at every half-revolution. The emf thus obtained still fluctuates between 0 and $AB\omega$ and commercial dynamos therefore have a large number of similar coils, each slightly displaced relative to its neighbours, so that their phases are staggered. In this way fluctuations are smoothed out, and the resulting combination may be made almost steady.

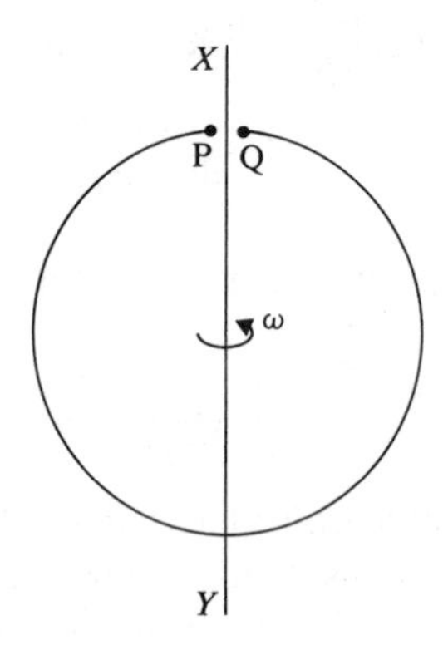

Fig. 8.5

§ 8.5 Faraday's disc

Faraday's disc (Fig. 8.6) is a simple generator and is an interesting example in which the flux through the circuit appears not to change. There is nevertheless an emf generated in the circuit. A thin conducting disc rotates with angular velocity ω about an axle. There is a uniform magnetic field $\mathbf{B}$ perpendicular to the plane of the disc. A and C are electrical connections to the axle and rim of the disc respectively. On joining A and C by a conducting wire it is found that a current flows, indicating that A and C are at different potentials.

What is meant by the flux through the circuit in this case? The "circuit" consists of the wire joining A and C together with the part lying in the rotating disc itself. Thus part of the circuit passes through moving material even though it does not move with the material as in the example in §8.2. The flux through this "circuit" is constant but the $\mathbf{v}\times\mathbf{B}$ force generates an emf. If we consider an electron at a point P(OP$=r$), the force on this electron is $\pm er\omega\mathrm{B}$ in the radial direction (the sign depending on the direction of $\mathbf{B}$). Thus the work done in taking a charge e from A to C is

$$\pm\int_a^b er\omega\mathrm{B}\,dr=\pm\tfrac{1}{2}e\omega\mathrm{B}\,(b^2-a^2)$$

where a and b are the radii of axle and disc respectively. The potential difference between A and C is therefore

$$V=\pm\tfrac{1}{2}\omega\mathrm{B}\,(b^2-a^2)$$

and this gives rise to the observed current when A and C are connected by a conductor.

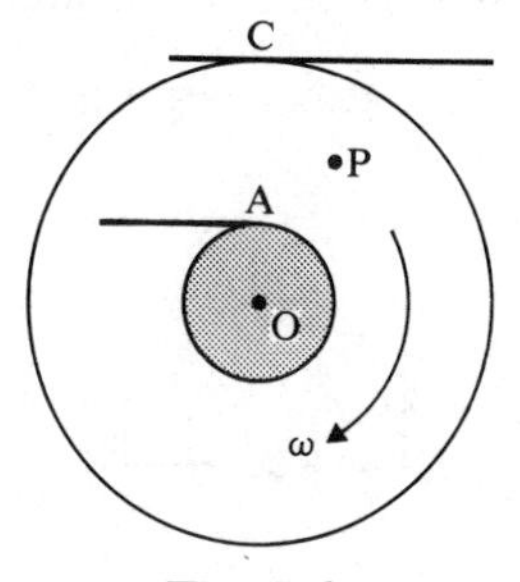

Fig. 8.6

§ 8.6 Self-inductance and mutual inductance

We now turn to situations in which the circuits are stationary but the magnetic fields change with time so that $\Phi = \Phi(t)$. Consider for example an electrical circuit in which the current is time-dependent. Then we may write

$$\frac{d\Phi}{dt} = \frac{d\Phi}{dI}\frac{dI}{dt} = L\frac{dI}{dt} \tag{8.8}$$

where $L = d\Phi/dI$ is known as the **self-inductance**. Thus we may express the back emf of (8.2) as

$$\mathscr{E} = -L\frac{dI}{dt} \tag{8.9}$$

The unit of self-inductance is the **henry**. A circuit (or part of a circuit) in which an emf of 1 volt is generated by a current I varying at a rate of 1 ampere per second has a self-inductance of 1 henry. L is a function of the geometry of the conductor. Obviously a wire coiled to form a solenoid has a very much larger self-inductance than when unwound. The symbol used in circuits for inductance is in fact just a solenoid ⅏.

If we consider more than one circuit then we may generalize (8.8) to

$$\frac{d\Phi_{kl}}{dt} = \frac{d\Phi_{kl}}{dI_l}\frac{dI_l}{dt} = M_{kl}\frac{dI_l}{dt} \tag{8.10}$$

where $M_{kl} = d\Phi_{kl}/dI_l$ is the **mutual inductance** between circuits k and l. The unit is again the henry. Note that $M_{kk} = d\Phi_{kk}/dI_k = L_k$.

Let us apply the law of induction to the simple circuit in Fig. 8.7, in which the resistance is R, the self-inductance L and the emf of the battery, $\mathscr{E}$. When the current flowing is I, the effective forward emf is that of the battery **less** the back emf, i.e. $\mathscr{E} - L\dfrac{dI}{dt}$.

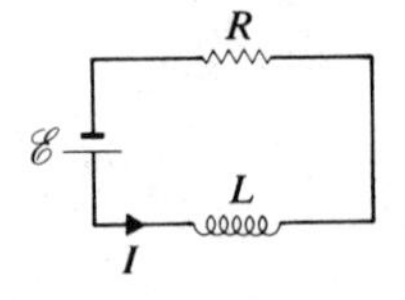

Fig. 8.7

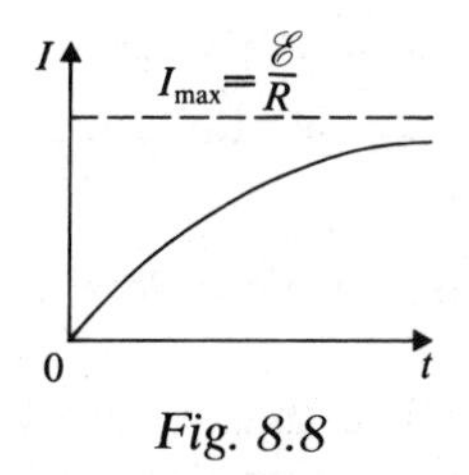

Fig. 8.8

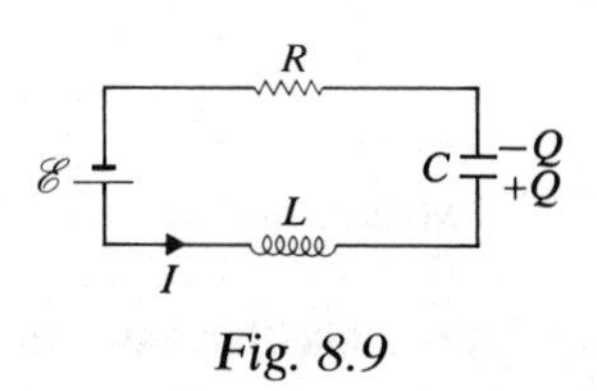

Fig. 8.9

Ohm's law then gives the differential equation for I:

$$\mathscr{E} - L\frac{dI}{dt} = RI \tag{8.11}$$

If we suppose that at $t=0$ the circuit is closed by a switch so that $I(0)=0$, the solution of (8.11) is

$$I = \frac{\mathscr{E}}{R}(1 - e^{-Rt/L}) \tag{8.12}$$

Thus the current builds up to its peak value $\mathscr{E}/R$ as shown in Fig. 8.8. If we write $\tau = L/R$, $I_{max} = \mathscr{E}/R$ then

$$I = I_{max}(1 - e^{-t/\tau}) \tag{8.13}$$

where τ is the **time constant** of the circuit.

Another important circuit is that in Fig. 8.9 which, in addition to the battery, resistance and inductance now includes a condenser of capacity C and and is known as an LCR circuit. We shall suppose that at time t the current flowing in the circuit is $I(t)$ and that $\pm Q$ are the charges on the condenser plates. The potential drop across the condenser is Q/C and so the net forward emf is now $\mathscr{E} - L\frac{dI}{dt} - \frac{Q}{C}$. The differential equation for the current now reads

$$\mathscr{E} - L\frac{dI}{dt} - \frac{Q}{C} = RI \tag{8.14}$$

in which the coefficients L, R, C are constants. However, Q is clearly time-dependent and

$$\frac{dQ}{dt} = I \tag{8.15}$$

Combining (8.14), (8.15) produces

$$L\frac{d^2I}{dt^2}+R\frac{dI}{dt}+\frac{I}{C}=\frac{d\mathscr{E}}{dt} \tag{8.16}$$

Two special cases of (8.16) are worth discussing in more detail:

(a) **Free oscillations** In the first case we set $\mathscr{E}=0$ so that we are concerned only with the free oscillations of charge within the circuit. Equation (8.16) becomes

$$L\frac{d^2I}{dt^2}+R\frac{dI}{dt}+\frac{I}{C}=0 \tag{8.17}$$

and provided $R^2<4L/C$ the solution is

$$I=A_e^{-Rt/2L}\cos(nt+\varepsilon) \tag{8.18}$$

A and ε are arbitrary constants depending only on initial conditions and

$$n^2=\frac{1}{LC}-\frac{R^2}{4L^2} \tag{8.19}$$

Thus there are oscillations of period $T=2\pi/n$ but the amplitude decays exponentially with time. If the resistance is small, $n^2\simeq(LC)^{-1}$ and so the period of free oscillations becomes

$$T=2\pi\sqrt{LC} \tag{8.20}$$

(b) **Forced oscillations** In this case we suppose that $\mathscr{E}=\mathscr{E}_0\cos\omega t$, i.e. the applied emf is an alternating one of frequency $\omega/2\pi$. In this case (8.16) becomes

$$L\frac{d^2I}{dt^2}+R\frac{dI}{dt}+\frac{I}{C}=-\omega\mathscr{E}_0\sin\omega t \tag{8.21}$$

Introducing the operator $D\equiv\dfrac{d}{dt}$ gives

$$I=-\frac{\omega\mathscr{E}_0\sin\omega t}{LD^2+RD+1/C}=\frac{-\omega\mathscr{E}_0\sin\omega t}{-\omega^2L+RD+1/C}$$

$$\text{i.e. } I=-\frac{\left[-RD+\left(\frac{1}{C}-\omega^2L\right)\right]\omega\mathscr{E}_0\sin\omega t}{\left[\omega^2R^2+\left(\frac{1}{C}-\omega^2L\right)^2\right]}$$

$$=(\mathscr{E}_0/Z)\cos(\omega t-\delta) \tag{8.22}$$

where

$$Z^2 = R^2 + \left(\omega L - \frac{1}{\omega C}\right)^2, \qquad \tan\delta = \left(\omega L - \frac{1}{\omega C}\right)\Big/ R \quad (8.23)$$

This alternating current has the same frequency as the applied emf $\mathscr{E}_0 \sin \omega t$ and its amplitude $\mathscr{E}_0/Z$ is independent of time. It is, however, out of phase with the emf. It represents a forced oscillation and is a particular integral of the differential equation (8.21). The other part of the solution – the complementary function (8.18) – represents oscillations that die away with time on account of damping. Their frequency is quite independent of the applied emf $\mathscr{E}$ and they represent free oscillations or **transients**. When the circuit is first joined, both parts of the total current are effective, but in a short time the transients decay to a negligible amplitude, leaving only the forced oscillations. If all the other quantities remain constant, the amplitude of the forced oscillations has its maximum value when

$$\omega^2 = 1/LC \quad (8.24)$$

which is the **resonance frequency** of the circuit.

§ 8.7 Two circuits – the transformer

Suppose that we have two coils with self-inductances L_1, L_2 and mutual inductance M, which form parts of two circuits as in Fig. 8.10. On account of the mutual induction we describe these circuits as **inductively coupled**.

Let I_1, I_2 be the currents in the two circuits. Then the flux of B through the first coil is $L_1 I_1 + M I_2$. When I_1 and I_2 change this gives rise to a back emf $L_1 \dfrac{dI_1}{dt} + M\dfrac{dI_2}{dt}$. Consequently the net forward emf in this circuit is

$$\mathscr{E}_1 - L_1 \frac{dI_1}{dt} - M\frac{dI_2}{dt}$$

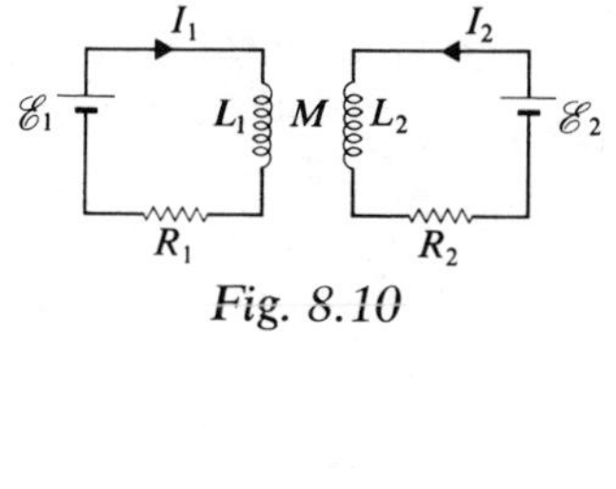

Fig. 8.10

and Ohm's law now gives the differential equation

$$\mathscr{E}_1 - L_1\frac{dI_1}{dt} - M\frac{dI_2}{dt} = R_1 I_1 \tag{8.25}$$

Similarly from the other circuit

$$\mathscr{E}_2 - L_2\frac{dI_2}{dt} - M\frac{dI_1}{dt} = R_2 I_2 \tag{8.26}$$

Corresponding equations are written in a similar way for any number of linked circuits. They are easily solved if $\mathscr{E}_1, \mathscr{E}_2 \ldots$ are known and the initial conditions stated.

There is one particular case of (8.25) and (8.26) which is of especial interest. Let us suppose that $\mathscr{E}_1$ is an alternating emf and both $\mathscr{E}_2$ and R_1 are zero. This means that the current in the second circuit is solely due to the emf in the first. We shall see that so far as the second circuit is concerned, the effect of the mutual inductance is to provide an emf of a different numerical value from the original emf in the first circuit.

In practice the two coils L_1 and L_2 may be supposed to consist of n_1 and n_2 turns of wire respectively wound close together on the same iron core, and are such that when unit current flows in any one of these turns a flux of **B** equal to Φ crosses this particular turn of wire. If the coils are wound closely, this flux will cross all the other turns in both coils. Thus unit current in the coil L_1 produces a total flux $n_1\Phi$, and this flux threads each of the n_1 turns of the first coil, and each of the n_2 turns of the second. Hence, from the definitions of L and M

$$L_1 = n_1^2\Phi, \qquad M = n_1 n_2 \Phi, \qquad L_2 = n_2^2\Phi \tag{8.27}$$

It follows that

$$L_1 L_2 = M^2 \tag{8.28}$$

This relation, sometimes called the **transformer condition,** will not be wholly satisfied in practice on account of leakage of lines of B; but if we assume that it is satisfied and put $\mathscr{E}_2 = R_1 = 0$, we can eliminate I_1 from (8.25) and (8.26) and obtain

$$\mathscr{E}_1 L_2 = -MR_2 I_2$$

Using (8.27) this becomes

$$R_2 I_2 = -\frac{n_2}{n_1}\mathscr{E}_1 \tag{8.29}$$

This shows that the potential R_2I_2 across the ends of the second coil is in a constant ratio n_2/n_1 to the emf in the first coil. In this way, by varying n_1/n_2, we are able to step up or down a given alternating emf. For this reason a device of this kind is called a transformer.

§ 8.8 Impedance and reactance

In this discussion we shall be concerned solely with the forced oscillations of §8.6, so that all transient effects will be neglected, and all oscillating quantities may be assumed to have the same frequency $\omega/2\pi$, though there may be certain phase differences between them. In such problems as these it is possible to simplify the working very considerably.

Let us first consider certain special cases of the general circuit shown in Fig. 8.9. In particular let us suppose that

$$\mathscr{E} = \mathscr{E}_0 \cos \omega t, \tag{8.30}$$

and that we connect, in turn, just one of the three quantities R, L and C to the emf $\mathscr{E}$. This may be achieved by supposing that the other two quantities have zero magnitude. The reader will have no difficulty in verifying that in these three cases the full solutions for the forced oscillations, (8.22), (8.23) reduce to

$$R \text{ alone:} \quad I = \frac{\mathscr{E}_0}{R} \cos \omega t \tag{8.31}$$

$$L \text{ alone:} \quad I = \frac{\mathscr{E}_0}{\omega L} \cos \left(\omega t - \frac{\pi}{2}\right) \tag{8.32}$$

$$C \text{ alone:} \quad I = \frac{\mathscr{E}_0}{(\omega C)^{-1}} \cos \left(\omega t + \frac{\pi}{2}\right) \tag{8.33}$$

We may describe the results (8.31)–(8.33) by saying that in this particular circuit

(i) an inductance L behaves like an effective resistance ωL, but the phase of the current lags $\pi/2$ behind the voltage,

(ii) a capacity C behaves like an effective resistance $1/\omega C$, but the phase of the current is $\pi/2$ in advance of the voltage,

(iii) a resistance R has no effect on the phase of the current.

We shall find a similar situation if we take R, L and C in pairs. The appropriate solutions now reduce to

$$\text{R and L together:} \quad I = \frac{\mathscr{E}_0}{Z}\cos(\omega t - \delta) \quad \text{where}$$

$$Z^2 = R^2 + \omega^2 L^2, \qquad \tan\delta = \omega L/R \tag{8.34}$$

$$\text{R and C together:} \quad I = \frac{\mathscr{E}_0}{Z}\cos(\omega t + \delta) \quad \text{where}$$

$$Z^2 = R^2 + 1/(\omega C)^2, \qquad \tan\delta = 1/\omega CR, \tag{8.35}$$

$$\text{L and C together:} \quad I = \frac{\mathscr{E}_0}{Z}\cos(\omega t - \pi/2) \quad \text{where}$$

$$Z = \omega L - 1/\omega C \tag{8.36}$$

We may describe the results (8.34)–(8.36) by saying that in each case there is an effective resistance Z, such that the maximum current is given by Ohm's law

$$I_{max} = \mathscr{E}_0/Z \tag{8.37}$$

and that there are phase differences which, except in the last case, are neither 0 nor $\pm\pi/2$. Evidently the actual phase differences in (8.34) and (8.35) represent a compromise between the effect of the pure resistance R which tends to leave the original phase unchanged, and that of L which tends to retard it, or of C which tends to advance it. Both Z and δ may be found graphically by drawing a right-angled triangle in which R is measured in one direction, whereas L and C are measured in a perpendicular direction. Z is then the magnitude of the vector sum of two components, and the magnitudes of these components are R, ωL and $1/\omega C$. Figure 8.11 shows the three diagrams obtained in this way. Note, however, that in Fig. 8.11c, Z is the sum of the two vectors ωL and $1/\omega C$. These are both perpendicular to the direction of R, but are opposed to one another. If $\omega L > 1/\omega C$ the resultant

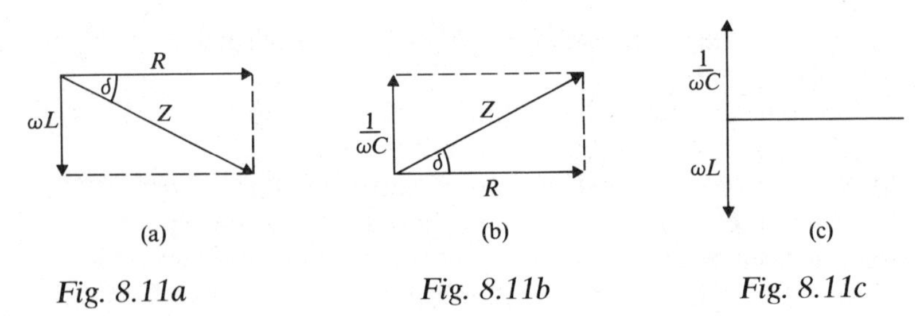

Fig. 8.11a Fig. 8.11b Fig. 8.11c

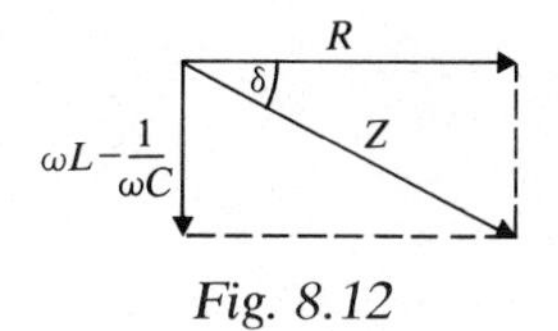

Fig. 8.12

points downward, but if $\omega L < 1/\omega C$ it points up. Let us call the direction of R the voltage line; then if Z is above the voltage line the phase of the current is in advance of the emf, and if Z lies below the voltage line the phase of the current is behind that of the emf. In each case the magnitude of the phase difference is given by the angle between Z and the voltage line.

We call Z the **impedance** of the circuit. Thus Z is the magnitude of the vector sum of the **ohmic resistance** R and the **inductive reactance** ωL, or the **capacitative reactance** $1/\omega C$, proper account being taken of the directions in which these two reactances must be measured. The quantity $1/Z$ is sometimes called the **admittance**.

Our argument has so far only been applied to the cases where either one or two of the three quantities are present. But it is equally valid when all three are present, as in Fig. 8.9. For, as Fig. 8.12 shows, the impedance Z of the circuit and the phase δ of the current are given by

$$Z^2 = R^2 + \left(\omega L - \frac{1}{\omega C}\right)^2, \qquad \tan\delta = \left(\omega L - \frac{1}{\omega C}\right)\Big/R$$

with

$$I = \frac{\mathscr{E}_0}{Z}\cos(\omega t - \delta), \qquad I_{\max} = \frac{\mathscr{E}_0}{Z} \tag{8.38}$$

This result is precisely that in (8.22), which was found by solving the appropriate differential equation. But the method just used is vastly easier to handle than the original differential equation. We shall show in §8.10 how it may be formalized and applied quite generally.

§ 8.9 LCR circuit program

It is instructive to write a simple program to display the characteristics of an LCR circuit on a graphics screen. The following program is an example which outputs the supply voltage together with the voltages across the various circuit elements.

```
      PROGRAM LCR
C     NOTE: ALL VARIABLES STARTING WITH I J OR L ARE REAL
      IMPLICIT REAL(I,J,L)
      DIMENSION V(4),KMESS(20)
      DATA KMESS/
    1 2HV ,2HAC,2HRO,2HSS,2H L,
    2 2HV ,2HAC,2HRO,2HSS,2H C,
    3 2HV ,2HAC,2HRO,2HSS,2H R,
    4 2HV ,2HSU,2HPP,2HLY,2H  /
C     NOTE: ON A PDP11/40 ONLY POSSIBLE TO STORE
C     2 CHARACTERS PER INTEGER WORD
      DATA NTTY/5/
C----------------------------------------------------
      FNY(K,Y)=Y+(FLOAT(K)*2.-1.)*VMAX
C     STATEMENT FUNCTION FOR SCALING
C----------------------------------------------------
C     NOW INPUT THE RELEVANT DATA
      WRITE(NTTY,9000)
      READ(NTTY,9010)R,L,C,V0,W,Q
C----------------------------------------------------
C     NOW TRY SCALING
C     Z=IMPEDANCE OF THE CIRCUIT
      Z=SQRT(R*R+(W*L-1./(W*C))**2)
      I=V0/Z
      V3=R*I
      V2=(Q+I/W)/C
      V1=L*W*I
C     VMAX WILL BE ESTIMATE OF LARGEST VOLTAGE
C     1.1*VMAX GIVES SOME ROOM BETWEEN GRAPHS
      VMAX=1.1*AMAX1(V0,V1,V2,V3,V4)
      DT=2.*3.14159/(100.*W)
C     NRUN=NUMBER OF TIME STEPS TO BE TAKEN
      NRUN=500
      TMAX=NRUN*DT
C     TMAX IS MAX. TIME ACHIEVED
C----------------------------------------------------
C     SCALE IS (IN X) -.2*TMAX TO TMAX.  -.2*TMAX TO GIVE ROOM FOR TITLES
C     SCAL IN Y IS 0. TO 8*VMAX. FOUR GRAPHS EACH BETWEEN -,+VMAX
      CALL INITAL
      TMIN=-.2*TMAX
      YMIN=0.
      YMAX=8.*VMAX
      CALL SCAL(TMIN,YMIN,TMAX,YMAX)
C----------------------------------------------------
C     INITIAL CONDITIONS
C     INITIAL CURRENT IS ALWAYS 0.  INITIAL CHARGE IS Q
C     J=DI/DT
      I=0.
      TIME=0.
      Q=Q
      V(4)=V0*SIN(W*TIME)
      V(3)=I*R
      J=(1./L)*(V(4)-I*R-Q/C)
      V(2)=Q/C
      V(1)=J*L
C----------------------------------------------------
C     DRAW AXES
      CALL POINT(0.,VMAX)
      CALL VECT(TMAX,0.)
      CALL POINT(0.,3.*VMAX)
      CALL VECT(TMAX,0.)
      CALL POINT(0.,5*VMAX)
```

```
      CALL VECT(TMAX,0.)
      CALL POINT(0.,7.*VMAX)
      CALL VECT(TMAX,0.)
      PLOT=2
      DO 30 K=4,1,-1
      VAX=FNY(K,0.)
      K1=K*5-4
C     VAX IS Y COORD OF K'TH AXES. K1 IS ADDRESS IN ARRAY KMESS
C     OF RELEVANT TITLE
      CALL POINT(TMIN,VAX)
30    CALL TTEXT(10,KMESS(K1))
C---------------------------------------------------------
C     INTEGRATE
C     PLOT POINT ON SCREEN INITIALLY  AND THEN EVERY 5 STEPS
C
      DO 60 N=1,NRUN
C
      PLOT=PLOT-1
      IF (PLOT .NE. 1)GOTO 50
C-------------------
      PLOT=5
      DO 40 K=1,4
      P1=FNY(K,V(K))
      CALL POINT(TIME,P1)
40    CONTINUE
C-------------------
50    CONTINUE
      VA=VO*SIN(W*TIME)
      J=(VA-I*R-Q/C)/L
      QNEW=Q+I*DT
      INEW=I*J*DT
      TIME=TIME+DT
      VA=VO*SIN(W*TIME)
      JNEW=(VA+INEW*R-QNEW/C)/L
      Q=Q+(I+INEW)*.5*DT
      I=I+(J+JNEW)*.5*DT
C     STORE VALUES FOR PLOTTING LATER
      V(4)=VA
      V(3)=I*R
      V(2)=Q/C
      V(1)=L*JNEW
60    CONTINUE
      CALL GREND
      STOP
C---------------------------------------------------------
9000  FORMAT(1H ,21HWHAT ARE R,L,C,V,W,Q ,$)
9010  FORMAT(6F8.2)
      END
```

The voltages across the various elements are shown in Fig. 8.13, reproduced from a graphics screen. Notice the departure from sinusoidal characteristics at early times caused by the transients discussed in §8.6. After a time transient effects decay leaving only the forced oscillations. The program is easily modified to examine, say, a square-wave applied voltage

$$V_A = +V_0, \sin \omega t > 0; \qquad V_A = -V_0, \sin \omega t < 0.$$

Further modifications might include nonlinear circuit elements such as a non-Ohmic resistance from a vacuum tube or semiconductor diode [12].

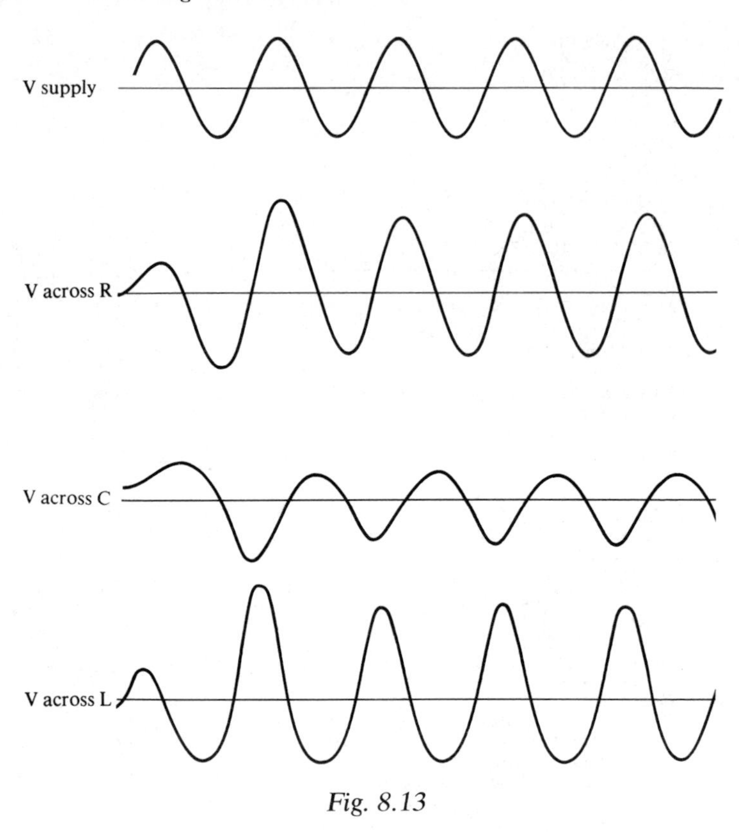

Fig. 8.13

§ 8.10 Complex algebra and circuit theory

The method developed in §8.8 rests on simple properties of the Argand diagram and complex numbers. Instead of putting $\mathscr{E} = \mathscr{E}_0 \cos \omega t$, we shall find it convenient to use

$$\mathscr{E} = \mathscr{E}_0 e^{i\omega t} \tag{8.39}$$

We may interpret this as meaning that we are always to take the real part of all expressions involving $e^{i\omega t}$. But it will be quite good enough to work right through all our problems with functions of the type (8.39), without distinguishing between real and imaginary parts; then, on completing the calculations, we take the real part. A similar type of expression will hold for the current; let us put

$$I = I_0 e^{i\omega t} \tag{8.40}$$

but I_0 may not necessarily be purely real, as we shall see in a moment. Indeed, our primary object is to discover the modulus and argument of I_0. The type of substitution represented in (8.39) and (8.40) is possible because we have decided to consider only forced oscillations in which all fluctuating quantities have the same frequency $\omega/2\pi$.

Let us consider first the circuit containing $\mathscr{E}$, L and R, all in series. Ohm's law for this circuit gives

$$\mathscr{E} - L\frac{dI}{dt} = RI \tag{8.41}$$

Substituting the values of $\mathscr{E}$ and I given by (8.39) and (8.40) we see that (8.41) is satisfied if $\mathscr{E}_0 - Li\omega I_0 = RI_0$, that is, if

$$I_0 = \frac{\mathscr{E}_0}{R + i\omega L} \tag{8.42}$$

Thus I_0 is a complex quantity. To separate real and imaginary parts let us put

$$R + i\omega L = Ze^{i\delta} \tag{8.43}$$

It will be recognized at once that Z and δ thus defined are in fact precisely the same as in (8.34), so that Z is the impedance and δ is the phase lag of current behind emf. Combining (8.40), (8.42) and (8.43)

$$I = \frac{\mathscr{E}_0}{Z} e^{i(\omega t - \delta)} \tag{8.44}$$

Taking the real part of this equation we reproduce in detail the formulae in (8.34). The complex number which we introduced in (8.43) and which we have seen gives us both the impedance and the phase is often written $\mathbf{Z}$, and is called the **vector impedance**. When drawn on an Argand diagram its modulus is the impedance Z and its argument, the phase δ. In fact

$$\mathbf{Z} = Ze^{i\delta} \tag{8.45}$$

$\mathbf{Z}$ may be regarded as composed of the ohmic resistance (or real part) R, and the inductive reactance (or complex part) $i\omega L$. The relation of all this to the Argand diagram is soon recognized in Fig. 8.11.

The method just described is quite general in its application. Thus, suppose that we have the circuit of Fig. 8.9 containing L, R and C all in series, the fundamental equations are (8.14) and (8.15):

$$\mathscr{E} - L\frac{dI}{dt} - \frac{Q}{C} = RI, \qquad I = \frac{dQ}{dt} \tag{8.46}$$

To solve these, we put

$$\mathscr{E} = \mathscr{E}_0 e^{i\omega t}, \qquad I = I_0 e^{i\omega t}, \qquad Q = Q_0 e^{i\omega t} \tag{8.47}$$

It follows that

$$I_0 = i\omega Q_0 \tag{8.48}$$

$$\left\{R + i\omega L + \frac{1}{i\omega C}\right\} I_0 = \mathscr{E}_0 \tag{8.49}$$

The vector impedance is

$$\mathbf{Z} = R + i\omega L + \frac{1}{i\omega C} \tag{8.50}$$

and

$$I = \frac{\mathscr{E}}{Z} \tag{8.51}$$

It is easily recognized that these equations lead to precisely the same results as in (8.38), which is known to be the correct solution.

We could, however, apply the same analysis to any other circuit, or part of a circuit. For as we saw in Chapter 4 the current distribution in electrical networks is governed by Kirchhoff's laws; these state that no current piles up at any junction of wires, and that the potential drop round any closed contour is zero, making allowance for batteries that may be present. With alternating currents of frequency $\omega/2\pi$, all these conditions will resemble those of (8.46), with perhaps more terms according to the complexity of the contour, and continuity equations for the current. Now all currents and charges will vary with the time according to the law $e^{i\omega t}$, so that we may always replace quantities such as $L\frac{dI}{dt}$ by $i\omega LI$, and $\frac{Q}{C}$ by $\frac{I}{i\omega C}$. Thus each inductance may be treated as a complex resistance $i\omega L$ and each condenser as a complex resistance $\frac{1}{i\omega C}$; after this we may proceed to solve the equations for the currents by using Ohm's law, just as in (8.51). The final current, or currents, will be given by complex expressions, whose moduli are the maximum values and whose arguments determine the phases.

We have just seen that a resistance R and an inductance L in series may be regarded as a single vector impedance $\mathbf{Z}$, where $\mathbf{Z} = R + i\omega L$. In a similar way we may treat other combinations, so that a complete network may be broken up into a set of impedances $\mathbf{Z}_1, \mathbf{Z}_2 \ldots$. This is often easier than treating each part separately, for it is easy to show that impedances in series are compounded together according to the law

$$\mathbf{Z} = \mathbf{Z}_1 + \mathbf{Z}_2 + \ldots \tag{8.52}$$

and if they are in parallel

$$\mathbf{Z}^{-1} = \mathbf{Z}_1^{-1} + \mathbf{Z}_2^{-1} + \ldots \tag{8.53}$$

It is important to remember that in both expressions we must use the vector impedance $\mathbf{Z}$ and not its magnitude Z.

§ 8.11 Generalized law of induction

In this section we derive a remarkable generalization of the law of induction due to Maxwell. Consider a stationary circuit in which there are no sources of emf; then

$$\oint \mathbf{E} \cdot d\mathbf{s} = -\frac{d\Phi}{dt} \tag{8.54}$$

This equation applies, so far, only for integration round the closed wire circuit, $\mathbf{E}$ and $d\mathbf{s}$ both being directed along the wire. But the tangential component of $\mathbf{E}$ is continuous at the surface of the wire where the medium changes, and so (8.54) must also hold for a circuit situated adjacent to the wire. Indeed we may expect it to hold for any closed curve since the fact that it holds for at least one curve not coincident with the wire shows that the electrical forces (represented by $\mathbf{E}$) which are brought into existence by a change of magnetic flux cannot depend on whether there is an actual flow of current. The flow of current is an effect in this case and not a cause. The induced or back emf is independent of whether a current does – indeed even whether it can – flow. So we shall assert the truth of (8.54) for any closed curve whatever. Though we have shown that Maxwell's generalization is plausible, it is a hypothesis nevertheless, though one that has been abundantly justified by the results which it predicts.

If we write $\Phi = \int \mathbf{B} \cdot d\mathbf{S}$, where the integration is over any surface spanning the curve along which $\oint \mathbf{E} \cdot d\mathbf{s}$ is evaluated, then (8.54) is equivalent to

$$\oint \mathbf{E} \cdot d\mathbf{s} = -\frac{d}{dt} \int \mathbf{B} \cdot d\mathbf{S}$$

For circuits which are stationary we may attribute any change in flux to the time dependence of **B** and in such cases

$$\oint \mathbf{E} \cdot d\mathbf{s} = -\int \frac{\partial \mathbf{B}}{\partial t} \cdot d\mathbf{S} \tag{8.55}$$

So for any surface S spanning an arbitrary contour

$$\int \nabla \times \mathbf{E} \cdot d\mathbf{S} = -\int \frac{\partial \mathbf{B}}{\partial t} \cdot d\mathbf{S}$$

which is only possible if

$$\nabla \times \mathbf{E} = -\frac{\partial \mathbf{B}}{\partial t} \tag{8.56}$$

This extremely important result following from Maxwell's generalization of the law of induction is another of the set of equations – known as **Maxwell's equations** – governing the behaviour of the electromagnetic field.

Note that if we write $\mathbf{B} = \nabla \times \mathbf{A}$ where **A** is the vector potential, (8.56) becomes

$$\nabla \times \left\{ \mathbf{E} + \frac{\partial \mathbf{A}}{\partial t} \right\} = 0$$

$$\text{i.e. } \mathbf{E} = -\frac{\partial \mathbf{A}}{\partial t} - \nabla \phi \tag{8.57}$$

where ϕ is an arbitrary scalar function. We may quickly give it identity however by noting that if there is no change in the magnetic field $\frac{\partial \mathbf{A}}{\partial t} = 0$ so that $\mathbf{E} = -\nabla \phi$. In other words ϕ is simply the electrostatic potential of Chapters 2 and 3 and (8.57) expresses the two sources of electric field – magnetic fields which change with time and distributions of charge. Throughout the remainder of the book we shall retain ϕ to denote the electric potential.

In deriving (8.56) we assumed that the circuit was at rest. What happens if this is not so? Is Maxwell's equation (8.56) still valid? As we might expect, the answer to this requires careful handling of the expression $\dfrac{d}{dt}\displaystyle\int_S \mathbf{B}\cdot d\mathbf{S}$ since not only $\mathbf{B}$ but S itself is now time-dependent. To analyse this we make use of the convective or total derivative, familiar from fluid dynamics, i.e. $\dfrac{D}{Dt}\equiv\dfrac{\partial}{\partial t}+\mathbf{v}\cdot\boldsymbol{\nabla}$. This allows for the fact that the flux may be changed by changing the magnetic field or by changing the position (or the shape) of the circuit.

$$\Phi=\int_S \mathbf{B}\cdot d\mathbf{S}=\oint \mathbf{A}\cdot d\mathbf{s}$$

so that

$$\frac{D\Phi}{Dt}=\oint\left[\frac{D\mathbf{A}}{Dt}\cdot d\mathbf{s}+\mathbf{A}\cdot\frac{D}{Dt}(d\mathbf{s})\right] \qquad (8.58)$$

Writing $\dfrac{D}{Dt}(d\mathbf{s})=(d\mathbf{s}\cdot\boldsymbol{\nabla})\mathbf{v}$ (problem 29) this becomes

$$\frac{D\Phi}{Dt}=\oint\left[\frac{D\mathbf{A}}{Dt}\cdot d\mathbf{s}+\mathbf{A}\cdot(d\mathbf{s}\cdot\boldsymbol{\nabla})\mathbf{v}\right]$$

The vector identity (problem 30)

$$\mathbf{A}\cdot(d\mathbf{s}\cdot\boldsymbol{\nabla})\mathbf{v}\equiv d\mathbf{s}\cdot[(\mathbf{A}\cdot\boldsymbol{\nabla})\mathbf{v}+\mathbf{A}\times(\boldsymbol{\nabla}\times\mathbf{v})]$$

enables us to express (8.58) as

$$\frac{D\Phi}{Dt}=\oint\left[\frac{\partial\mathbf{A}}{\partial t}+\boldsymbol{\nabla}(\mathbf{v}\cdot\mathbf{A})-\mathbf{v}\times\boldsymbol{\nabla}\times\mathbf{A}\right]\cdot d\mathbf{s}$$

$$=-\oint[\mathbf{E}+\mathbf{v}\times\mathbf{B}+\boldsymbol{\nabla}(\phi-\mathbf{v}\cdot\mathbf{A})]\cdot d\mathbf{s}$$

by using (8.57). It follows that

$$\frac{D\Phi}{Dt}=-\oint[\mathbf{E}+\mathbf{v}\times\mathbf{B}]\cdot d\mathbf{s} \qquad (8.59)$$

Let us now think of the rate of change of flux as measured by an observer who moves **with** the circuit; such an observer is said to be in the **rest frame.** Let us use $\mathbf{E}'$ to denote the electric field in this coordinate frame, so that

$$\frac{d\Phi}{dt} = -\oint \mathbf{E}' \cdot d\mathbf{s} \tag{8.60}$$

and, from (8.59), it follows that

$$\mathbf{E}' = \mathbf{E} + \mathbf{v} \times \mathbf{B} \tag{8.61}$$

Thus in a moving conductor $(\mathbf{E} + \mathbf{v} \times \mathbf{B})$ acts as an **effective** electric field. The electric field as seen by the moving observer is made up of two terms, the first due to the time dependence of the magnetic fields and the other to the motion **across** the magnetic field.

§ 8.12 Energy in the magnetic field

In Chapter 5 a number of parallels between the magnetostatic and electrostatic fields were drawn but these did not extend to comparisons of the energy density of the two fields. The reason for avoiding this was simply that one cannot introduce the concept of energy in magnetostatics since to set up a magnetic field from a steady current requires first switching on the current so that there is an interval over which the field is established. Since the field during this interval is time-dependent it follows that induced emfs are present and these cause the current source to do work.

Let us begin with an analogy. We have seen in considering the *LCR* circuit in §8.6 that the charge Q satisfies the differential equation

$$L\frac{d^2Q}{dt^2} + R\frac{dQ}{dt} + \frac{Q}{C} = \mathscr{E}(t) \tag{8.62}$$

The form of this equation is identical to that describing the oscillations of a mechanical system consisting of a load of mass m, a spring of stiffness k and a dashpot with coefficient of damping D. If we consider such a system driven by a force $F(t)$ then, corresponding to (8.62), we have

$$m\frac{d^2x}{dt^2} + D\frac{dx}{dt} + kx = F(t) \tag{8.63}$$

By analogy therefore we may pair the variables in the two systems as follows:

$$L \text{ inductance} \to m \qquad \mathscr{E} \text{ emf} \to F$$
$$C \text{ capacity} \to k^{-1} \qquad Q \text{ charge} \to x$$
$$R \text{ resistance} \to D \qquad I \text{ current} \to \frac{dx}{dt}$$

Using this identification it is tempting to form the electromagnetic equivalent of kinetic energy $\frac{1}{2}m\dot{x}^2$, i.e. $\frac{1}{2}LI^2$ and treat this as an expression for the electromagnetic energy. We shall show this is indeed true but we should resist carrying the analogy further to identify, for example, LI as the "electromagnetic momentum" associated with the circuit.

By analogy with the electrostatic case we will begin with a "discrete" argument in terms of electrical circuits and then generalize the result in terms of field variables **B** and **H**. If magnetic monopoles existed then of course the argument would parallel precisely that in the electrostatic case. Let us consider the work done by the emf $\mathscr{E}$ in moving an increment of charge dQ through a circuit, i.e.

$$dW = \mathscr{E}dQ = \mathscr{E}I\,dt = \left(\frac{d\Phi}{dt} + RI\right)I\,dt$$

in which the first term represents the work done to overcome the back emf in the circuit, while the second arises from ohmic heating. If we disregard this irreversible effect then

$$dW = I\,d\Phi \tag{8.64}$$

and we may regard this as a measure of the change in the magnetic energy of the system. We may generalize this to n circuits, i.e.

$$dW = \sum_{k=1}^{n} I_k\,d\Phi_k$$

This is a perfectly general expression which is valid irrespective of the way in which the $d\Phi_k$ arise. In the case in which the $d\Phi_k$ are produced by changes in the currents flowing in the circuits then

$$d\Phi_k = \sum_{l=1}^{n} \frac{d\Phi_{kl}}{dI_l}\,dI_l = \sum_{l=1}^{n} M_{kl}dI_l$$

and this may be equated to the change in the magnetic energy of the system (assuming that no mechanical work is done, e.g. in moving the circuits themselves). Next suppose that the currents in all the circuits attain their equilibrium values simultaneously at some time $t=t_0$ so that at any instant in the interval $0 \leq t \leq t_0$ each I_k will be some fraction f of its ultimate value, i.e. $I_k(t)=fI_k(t_0)$. Moreover $d\Phi_k(t)=\Phi_k(t_0)\,df$ so that

$$\begin{aligned} W &= \sum_{k=1}^{n} I_k(t_0)\Phi_k(t_0)\int_0^1 f\,df \\ &= \frac{1}{2}\sum_{k=1}^{n} I_k\Phi_k \end{aligned} \tag{8.65}$$

Using (8.10) this gives

$$\begin{aligned} W &= \frac{1}{2}\sum_{k=1}^{n}\sum_{l=1}^{n} M_{kl}I_kI_l \\ &= \tfrac{1}{2}[L_1I_1^2+L_2I_2^2+\ldots+L_nI_n^2] \\ &\quad +[M_{12}I_1I_2+\ldots+M_{1n}I_1I_n \\ &\quad +\ldots\ldots+M_{n-1,\,n}I_{n-1}I_n] \end{aligned} \tag{8.66}$$

in which we have used $M_{kl}=M_{lk}$, $M_{kk}=L_k$.

For the case of two circuits

$$W=\tfrac{1}{2}(L_1I_1^2+L_2I_2^2)+M_{12}I_1I_2 \geq 0$$
$$\text{i.e. } L_1L_2 \geq M_{12}^2$$

While (8.66) is a perfectly general result for a system in which the elements (circuits) cannot move and in which the magnetic media are **linear** it is inconvenient in the same sense as (3.5) was inconvenient in electrostatics. As in the electrostatic case, there is a ready generalization which allows us to get rid of the scaffolding of circuits used to establish (8.66). We might have written

$$\Phi_k = \int_{S_k} \mathbf{B}\cdot d\mathbf{S} = \oint_{C_k} \mathbf{A}\cdot d\mathbf{s}_k$$

where $\mathbf{A}$ is the vector potential. Thus (8.65) becomes

$$W=\frac{1}{2}\sum_k \oint_{C_k} I_k\mathbf{A}\cdot d\mathbf{s}_k \tag{8.67}$$

If we now take C_k to be simply a closed mathematical circuit in a conducting medium rather than an actual physical circuit made of conducting material then we may make the transition from the discrete to the continuous case quite simply by replacing $I_k\, d\mathbf{s}_k$ by $\mathbf{j}\, d\mathbf{r}$ where $\mathbf{j}$ is the current density and by letting $\sum_k \oint_{C_k} \rightarrow \int_V$ through choosing a set of circuits C_k at infinitesimal separations. Then (8.67) becomes

$$W = \frac{1}{2} \int_V \mathbf{j} \cdot \mathbf{A}\, d\mathbf{r} \tag{8.68}$$

and using Ampère's law $\boldsymbol{\nabla} \times \mathbf{H} = \mathbf{j}$ this transforms to

$$W = \frac{1}{2} \int_V \mathbf{H} \cdot (\boldsymbol{\nabla} \times \mathbf{A})\, d\mathbf{r} - \frac{1}{2} \int (\mathbf{A} \times \mathbf{H}) \cdot d\mathbf{S}$$

Arguments by now familiar show that the surface integral vanishes so that

$$W = \frac{1}{2} \int \mathbf{B} \cdot \mathbf{H}\, d\mathbf{r} \tag{8.69}$$

This is the magnetic analogue of (6.21). As in the electric case we may define the energy density in a magnetic field by $\frac{1}{2} \mathbf{B} \cdot \mathbf{H}$ or, equivalently since $\mu \mathbf{H} = \mathbf{B}$ by $\frac{1}{2} \mu H^2$ or $B^2/2\mu$.

It is important that we recognize that the assumption of linear magnetic materials has been made. For nonlinear materials such as ferromagnets we have to modify the analysis given above.

§ 8.13 Hydromagnetics

In §8.11 we discussed the motion of a conductor in a magnetic field. A case of particular interest arises whenever the conductor is a fluid – whether a conducting liquid or a plasma. When such a fluid moves in a magnetic field, electric fields are induced in the fluid and currents flow. The magnetic field is then capable of acting on these currents through the $\mathbf{j} \times \mathbf{B}$ force and this interaction will in general modify the fluid motion itself. This interaction between field and flow is the concern of magnetohydrodynamics or hydromagnetics.

Consider first the kinematic aspect of hydromagnetics in which the effect of the magnetic forces on the motion of the conducting fluid is neglected. The field equations are

$$\nabla \times \mathbf{H} = \mathbf{j} \tag{8.70}$$

$$\nabla \times \mathbf{E} = -\frac{\partial \mathbf{B}}{\partial t} \tag{8.71}$$

$$\mathbf{j} = \sigma(\mathbf{E} + \mathbf{U} \times \mathbf{B}) \tag{8.72}$$

$$\mathbf{B} = \mu \mathbf{H} \tag{8.73}$$

in which $\mathbf{U}$ is the fluid velocity. Supposing σ and μ are spatially uniform it follows from (8.70)–(8.73) that

$$\frac{\partial \mathbf{B}}{\partial t} = \eta \nabla^2 \mathbf{B} + \nabla \times (\mathbf{U} \times \mathbf{B}) \tag{8.74}$$

in which $\eta = (\sigma\mu)^{-1}$. This equation is identical in structure to the hydrodynamic equation describing the evolution of the **vorticity** $\boldsymbol{\omega}$ of an incompressible fluid with kinematic viscosity ν. By analogy, therefore, η is known as the **magnetic viscosity**. The first term on the right-hand side of (8.74) represents the effect of **diffusion**, the second that of **convection** of the magnetic field.

Let us consider the diffusion and convection terms separately. If the fluid is at rest then (8.74) reduces to

$$\frac{\partial \mathbf{B}}{\partial t} = \eta \nabla^2 \mathbf{B} \tag{8.75}$$

i.e. a diffusion equation describing the way in which the magnetic field changes by diffusion through the electrically conducting fluid. The rate of change is characterized by the parameter η, i.e. by the electrical conductivity and the magnetic permeability of the fluid; thus η is also known as the **magnetic diffusivity**. From (8.75) we see that the field decays in a characteristic time τ where

$$\tau \sim \frac{L^2}{\eta} = \sigma\mu L^2 \tag{8.76}$$

L being a measure of the scale on which B varies.

Of course (8.75) applies with equal validity to a solid conductor. If we imagine a block of conductor in an external magnetic field which varies with time as $e^{i\omega t}$, then the magnetic field penetrates the conductor and induces an electric field within it. It follows since $\mathbf{j} = \sigma \mathbf{E}$ that there will also be an induced current distribution. These are called **eddy currents.** In many cases as in the core of a transformer or in an electric motor these eddy currents represent wasted energy, so the metal is usually laminated with a thin film of insulation between the layers. This serves to inhibit the undesirable eddy currents, though it does not completely eliminate them. The diffusion equation tells us that the oscillating magnetic field can only penetrate a distance determined by (8.76), i.e.

$$L \sim \left(\frac{2\pi}{\sigma\mu\omega}\right)^{1/2} \tag{8.77}$$

This limit to the penetration of the field is known as the **skin effect**. Note that for perfect conductors the field cannot penetrate at all.

Let us now return to (8.74) and consider a fluid which is a perfect conductor. Then

$$\frac{\partial \mathbf{B}}{\partial t} = \nabla \times (\mathbf{U} \times \mathbf{B}) \tag{8.78}$$

The analogy between (8.78) and the hydrodynamic analogue

$$\frac{\partial \boldsymbol{\omega}}{\partial t} = \nabla \times (\mathbf{U} \times \boldsymbol{\omega})$$

where $\boldsymbol{\omega}$ is the vorticity leads us to expect hydromagnetic counterparts to the Kelvin–Helmholtz theorems in fluid dynamics. In fact there are such theorems which state that:

(a) the magnetic flux through any closed contour moving with a perfectly conducting fluid is constant
(b) fluid elements which lie initially on a magnetic field line continue to lie on a field line in a perfectly conducting fluid.

The first result follows at once from (8.59), i.e.

$$\frac{D\Phi}{Dt} = -\oint [\mathbf{E} + \mathbf{U} \times \mathbf{B}] \cdot d\mathbf{s} \tag{8.79}$$

Since from Ohm's law

$$\mathbf{j} = \sigma(\mathbf{E} + \mathbf{U} \times \mathbf{B})$$

perfect conductivity requires that

$$\mathbf{E} + \mathbf{U} \times \mathbf{B} \equiv 0$$

Thus Φ is constant for any closed contour moving with the fluid.

To prove the second theorem we start from (8.78) and expand

the right-hand side to get

$$\frac{\partial \mathbf{B}}{\partial t} = (\mathbf{B} \cdot \nabla)\mathbf{U} - (\mathbf{U} \cdot \nabla)\mathbf{B} - \mathbf{B}(\nabla \cdot \mathbf{U})$$

giving

$$\frac{D\mathbf{B}}{Dt} = (\mathbf{B} \cdot \nabla)\mathbf{U} - \mathbf{B}(\nabla \cdot \mathbf{U}) \tag{8.80}$$

The continuity or mass conservation equation for the fluid

$$\frac{\partial \rho}{\partial t} + \nabla \cdot (\rho \mathbf{U}) = 0$$

in which ρ is the fluid density, allows us to replace the last term in (8.80) to give

$$\frac{D\mathbf{B}}{Dt} = (\mathbf{B} \cdot \nabla)\mathbf{U} + \frac{\mathbf{B}}{\rho}\frac{D\rho}{Dt}$$

or

$$\frac{D}{Dt}\left(\frac{\mathbf{B}}{\rho}\right) = \left(\frac{\mathbf{B}}{\rho} \cdot \nabla\right)\mathbf{U} \tag{8.81}$$

Moreover since (problem 29)

$$\frac{D}{Dt}(d\mathbf{s}) = (d\mathbf{s} \cdot \nabla)\mathbf{U} \tag{8.82}$$

we see that the behaviour of the quantities $\mathbf{B}/\rho$ and $d\mathbf{s}$ are described by the same equation. If at $t=0$, $\mathbf{B}$ and $d\mathbf{s}$ are parallel so that

$$\varepsilon \mathbf{B}_0 = \rho_0 \, d\mathbf{s}_0$$

where ε is infinitesimal, then from (8.81), (8.82)

$$\frac{D}{Dt}\left(d\mathbf{s} - \varepsilon \frac{\mathbf{B}}{\rho}\right) = 0 \quad \text{at} \quad t = 0$$

It is sometimes inferred from this that $[d\mathbf{s} - \varepsilon \mathbf{B}/\rho]$ remains zero for all t. While the conclusion is correct it may not be inferred directly; a rigorous discussion beyond the scope of this treatment is required to show that indeed

$$d\mathbf{s} = \varepsilon \frac{\mathbf{B}}{\rho}$$

for all time. Thus a fluid element which lies initially on a magnetic field line continues to lie on a field line in a perfectly conducting fluid.

These two theorems are sometimes summed up in the statement that the magnetic field in a perfectly conducting fluid behaves as if it were **frozen** into the fluid. In general the behaviour of the field in a conducting fluid is compounded of the two effects, convection and diffusion. On physical grounds we would expect that while the field lines are still carried along with the fluid, they are no longer frozen into it but may now "leak" through it. From (8.74) we see that if L is a characteristic length then convective effects are dominant over diffusion provided

$$LU \gg \eta$$

By analogy with the Reynolds number in fluid dynamics we may introduce a **magnetic Reynolds number** R_M defined by

$$R_M = \frac{LU}{\eta} \tag{8.83}$$

Thus convection is dominant over diffusion provided that $R_M \gg 1$.

§ 8.14 Example

Consider a particle of charge e, mass m moving in an axially symmetric magnetic field which is time-dependent. Suppose that the time variation is slow in the sense that $\dot{B}/B \ll \Omega$ where $\Omega = |e|B/m$ is the Larmor frequency. Show that for such fields the quantity $W_\perp/B$ is an approximate constant of the motion where $W_\perp = \frac{1}{2}mv_\perp^2$.

In §5.11 for constant uniform magnetic fields we showed that $W_\perp$ and $W_\parallel(=\frac{1}{2}mv_\parallel^2)$ are constants of the motion. Assume that $\mathbf{B}$ does not vary spatially. Its time dependence now means that there is an induced azimuthal electric field so that the equation of motion reads

$$m\ddot{\mathbf{r}} = e[\mathbf{E} + \dot{\mathbf{r}} \times \mathbf{B}]$$

i.e.

$$m\mathbf{v}_\perp \cdot \ddot{\mathbf{r}} = e\mathbf{v}_\perp \cdot [\mathbf{E} + \dot{\mathbf{r}} \times \mathbf{B}] \tag{8.84}$$

and since $\mathbf{E} \perp \mathbf{B}$ ($\mathbf{B}$ is axial), $\mathbf{v}_\parallel$ is constant and (8.84) reduces to

$$\frac{d}{dt}\left(\frac{1}{2}mv_\perp^2\right) = e\mathbf{E} \cdot \mathbf{v}_\perp$$

Therefore in completing a Larmor orbit the particle energy changes by

$$\delta(\tfrac{1}{2}mv_\perp^2) = e\oint \mathbf{E} \cdot d\mathbf{r}_\perp$$

$$= e\int \nabla \times \mathbf{E} \cdot d\mathbf{S}$$

where $d\mathbf{r}_\perp = \mathbf{v}_\perp \, dt$ and $d\mathbf{S}$ is an element of the surface enclosed by the orbit. Hence

$$\delta(\tfrac{1}{2}mv_\perp^2) = -e\int \frac{\partial \mathbf{B}}{\partial t} \cdot d\mathbf{S}$$

and since the magnetic field changes slowly

$$\delta(\tfrac{1}{2}mv_\perp^2) \simeq \pi r_L^2 \, e\dot{B}$$

where r_L is the Larmor radius. The negative sign disappears since for positive (negative) charges $e>0$ (<0) and $\mathbf{B} \cdot d\mathbf{S}<0$ (>0).

$$\therefore \quad \delta(\tfrac{1}{2}mv_\perp^2) = \frac{mv_\perp^2}{2}\frac{2\pi}{|\Omega|}\frac{\dot{B}}{B}$$

i.e.

$$\delta W_\perp = W_\perp \frac{\delta B}{B}$$

where δB is the change in magnitude of the magnetic field during one orbit. Thus

$$\delta\left(\frac{W_\perp}{B}\right) = 0 \tag{8.85}$$

so that $W_\perp/B$ is an approximate constant of the motion; it may be identified with the **magnetic moment** μ, of the particle (problem 33). Quantities which are approximately constant for systems which vary in some sense slowly – or adiabatically – are called **adiabatic invariants**. Another adiabatic invariant in this case is the magnetic flux through a Larmor orbit (problem 34).

It is possible (problem 35) to show that μ is also an adiabatic invariant in a spatially inhomogeneous axially symmetric magnetic

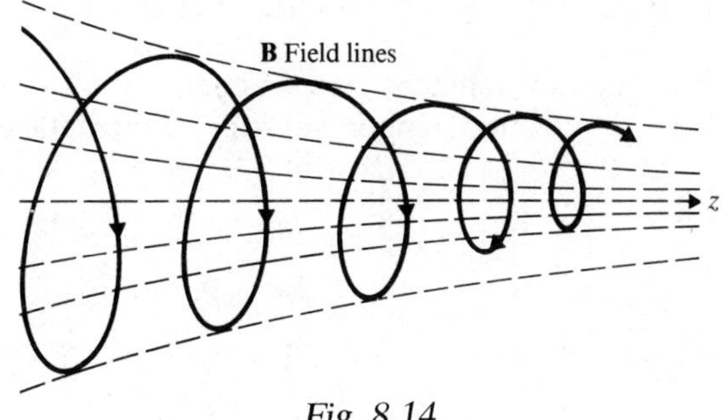

Fig. 8.14

field (Fig. 8.14). This result has important practical applications. For consider a particle moving in the field of Fig. 8.14 towards the region of increasing B. It follows from the invariance of $W_{\perp}/B$ that as B increases so must $W_{\perp}$. Since energy is conserved, this must be at the expense of $W_{\parallel}$. Thus at some point it may happen that $W_{\parallel}=0$ and the particle is unable to penetrate further into the field. The particle is reflected at this point and this field configuration is known as a **magnetic mirror**. If we arrange two mirror fields as in Fig. 8.15 we create a **magnetic bottle** or adiabatic mirror trap within which charged particles may be contained. Such mirror traps are used in confining plasmas in experiments on controlled thermonuclear reactions. The earth's magnetic field is an example of a natural mirror trap.

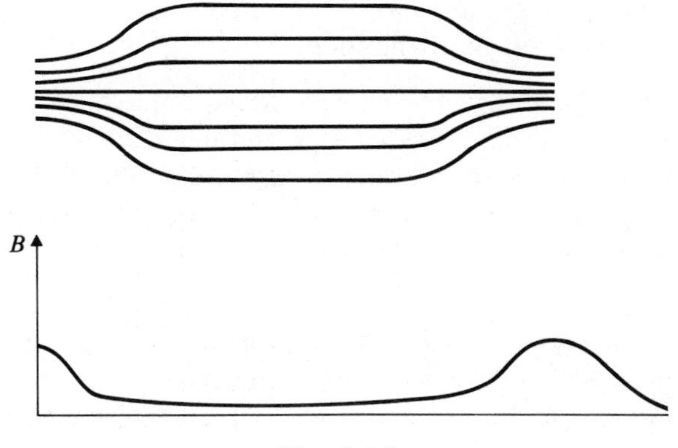

Fig. 8.15

§ 8.15 Summary of the field equations

At this stage in the development let us summarize all the information we have on electromagnetic fields in mathematical form:

$$\nabla \times \mathbf{D} = \rho \tag{8.86}$$

$$\nabla \times \mathbf{B} = 0 \tag{8.87}$$

$$\nabla \times \mathbf{E} = -\frac{\partial \mathbf{B}}{\partial t} \tag{8.88}$$

$$\nabla \times \mathbf{H} = \mathbf{j} \tag{8.89}$$

together with the constitutive relations

$$\mathbf{D} = \varepsilon \mathbf{E}; \qquad \mathbf{H} = \mathbf{B}/\mu; \qquad \mathbf{j} = \sigma \mathbf{E}$$

§ 8.16 Problems

1. A coil of area A, resistance R and induction L is rotated with constant angular velocity ω about a vertical axis in the earth's magnetic field, whose horizontal component is H. Show that the current I satisfies the equation

$$L\frac{dI}{dt} + RI = AH\omega \sin \omega t$$

By calculating the average rate of loss of energy show that the mean couple required to rotate the coil is $RA^2H^2\omega/2(R^2 + L^2\omega^2)$.

2. Calculate the emf induced in a circular loop if a spatially uniform magnetic field $\mathbf{B} = \mathbf{B}_0 \sin \omega t$ is directed at an angle θ to the normal of the plane of the loop.

3. A magnet of moment m can slide along the axis of a circular coil of wire of radius a, resistance R and negligible self-induction. Show that when the magnet is at a distance x from the coil the flux of B across the coil is $\Phi = \mu_0 m a^2/2(a^2 + x^2)^{3/2}$. Deduce that if the magnet is moved with constant velocity v the current in the coil is $I = 3\mu_0 m a^2 x v/R(a^2 + x^2)^{5/2}$.

4. A small search-coil consists of n turns of wire each of area A, and the two ends are connected to a galvanometer, the total resistance being R. The coil is placed perpendicular to a field which it is desired to measure and is suddenly withdrawn. If the

total charge that flows through the galvanometer is Q, show that the field is given by $B = QR/nA$. This is the principle of the **fluxmeter** by which we measure a field at any point.

5. A plane LCR circuit of area A rotates with angular velocity ω in a constant magnetic field $\mathbf{B}_0$ about an axis lying in the plane of the circuit and normal to $\mathbf{B}_0$. Calculate the average couple acting on the circuit.

6. Consider a square conducting loop of side L lying in the plane Oxz and moving in the positive z direction with uniform velocity u_0. In the laboratory frame there is a magnetic field $\mathbf{B} = B_0\hat{\mathbf{j}} \sin kz$ where $\hat{\mathbf{j}}$ is a unit vector along Oy. Find the emf in the rest frame of the loop.

If the impedance of the loop at the frequency $\omega' = ku_0$ is Z, show that the force required to maintain the loop at velocity u_0 is given by

$$F = -\frac{u_0 L B_0^2}{Z}[\sin(ku_0 t + kL) - \sin ku_0 t]^2$$

Comment on the unidirectional nature of the force for all values of kL.

7. A condenser is charged by means of a constant emf $\mathscr{E}$, the connecting wires having a resistance R. At $t = 0$ the emf is switched on. Show that at time t the charge on the condenser is $Q = C\mathscr{E}(1 - e^{-t/CR})$.

8. Verify that the time constant τ in (8.13) is such that in time τ the current has grown to about 0.63 of its maximum value.

9. The two plates of a condenser C carry charges $\pm Q_0$, and at $t = 0$ they are connected together through a coil of resistance R and inductance L. Show that the charge Q at any subsequent time is given by $L\dfrac{d^2Q}{dt^2} + R\dfrac{dQ}{dt} + \dfrac{Q}{C} = 0$. Solve this equation, given that $R^2 < 4L/C$.

10. Investigate the solutions of (8.17) for the free oscillations of a circuit in the cases $R^2 = 4L/C$ and $R^2 > 4L/C$. These are known as **critically damped** and **over-damped** circuits.

11. In the inductively coupled circuits of Fig. 8.10, $\mathscr{E}_2 = 0$ and $\mathscr{E}_1 = \mathscr{E}_0 \cos \omega t$. Show that there is a phase difference between I_1 and I_2, and that as $\omega \to \infty$ this phase difference tends to π.

12. In the inductively coupled circuits of Fig. 8.10, $\mathscr{E}_2=0$ and $\mathscr{E}_1=\mathscr{E}_0\cos\omega t$. Show that the current I_2 is given by the equation

$$(L_1L_2-M^2)\frac{d^2I_2}{dt^2}+(L_1R_2+L_2R_1)\frac{dI_2}{dt}+R_1R_2I_2=M\omega\mathscr{E}_0\sin\omega t.$$

Deduce that if the condition (8.28), is satisfied, then the amplitude of the oscillations of I_2 is $M\omega\mathscr{E}_0/S$, where

$$S^2=R_1^2R_2^2+\omega^2(L_1R_2+L_2R_1)^2$$

13. In the forced oscillations of a single circuit calculate the rate at which the emf $\mathscr{E}_0$ cos ωt is working, and show that the mean value is $\dfrac{\mathscr{E}_0^2}{2Z}\cos\delta$, Z and δ being given by (8.23). Verify that this is the same as the mean value of RI^2, so that conservation of energy is not violated. Deduce that if $R=0$ there is no consumption of energy even though there may be a large current. Explain this.

14. A coil of wire of m turns is closely wound on a circular ring of any form of cross-section, and a second coil of n turns is intertwined with the first. Show that the coefficients of mutual and self-inductance are

$$M_{12}=\frac{\mu_0mn}{2\pi}\int\frac{dS}{\rho},\qquad L_1=\frac{\mu_0m^2}{2\pi}\int\frac{dS}{\rho},\qquad L_2=\frac{\mu_0n^2}{2\pi}\int\frac{dS}{\rho}$$

where ρ is the distance of any point in the cross-section from the axis of the ring.

Deduce that $L_1L_2=M_{12}^2$. In practice $M_{12}^2<L_1L_2$ on account of flux leakage and the ratio $M_{12}/(L_1L_2)^{1/2}$ is called the **coupling factor**.

15. In the previous question suppose that the ring now has an iron core. Show that in this case

$$M_{12}=\frac{\mu mn}{2\pi}\int\frac{dS}{\rho}$$

If the cross-section is a circle of radius a, whose centre is R from the axis of the ring, deduce that $M_{12}=\mu mn[R-(R^2-a^2)^{1/2}]$.

16. Two equal circular loops of radius a lie opposite each other, a distance c apart. Show that the coefficient of mutual inductance

is

$$M_{12}=\frac{\mu_0 a^2}{2}\int_0^{2\pi}\frac{\cos\psi\,d\psi}{(c^2+2a^2-2a^2\cos\psi)^{1/2}}$$

If c is very large show that $M_{12}=\pi\mu_0 a^4/2c^3$. Deduce that if unit currents flow in the same direction round the coils, they attract each other with a force $3\pi\mu_0 a^4/2c^4$.

17. Show that the mutual inductance between a circle of radius a and an infinite straight line in the same plane is

$$M_{12}=\frac{\mu_0}{\pi}\int_{-a}^{+a}\frac{(a^2-x^2)^{1/2}}{x+f}\,dx$$

where f is the shortest distance from the centre of the circle to the straight line. Show that on integrating

$$M_{12}=\mu_0[f-(f^2-a^2)^{1/2}].$$

18. Use the program in §8.9 with the following parameters

$$V_0=0,\ Q=1\mu C,\quad L=1mH,\quad C=1\mu F,\quad 1\Omega<R<300\Omega$$

to verify that for $R<\sqrt{\frac{4L}{C}}$ there are damped oscillations and for $R>\sqrt{\frac{4L}{C}}$ there are no oscillations. (Put $\omega=10^4 s^{-1}$ rather than $\omega=0$ for computational reasons.) Estimate, from the graphs, the time constant of these circuits and check that they agree with the theoretical value.

19. Calculate the impedance and phase angle for the following series circuit; $L=1mH$, $C=1\mu F$, $\omega=10^4 s^{-1}$, $R=250\Omega$ and $V_0=2V$. Use the program to generate the solution (take $Q=0$) and from the output estimate the impedance and phase angle.

20. An alternating current $I_2\cos\omega t$ is superimposed on a direct current I_1. Show that the r.m.s. current is $(I_1^2+\frac{1}{2}I_2^2)^{1/2}$.

21. An emf represented by $\mathscr{E}_0+\mathscr{E}_1\cos\omega t$, where $\mathscr{E}_0$ and $\mathscr{E}_1$ are real constants, acts in a circuit whose vector impedance is $\mathbf{Z}=Ze^{i\delta}$. Show that the mean power expended in the circuit is

$$\{\mathscr{E}_0^2\sec\delta+\tfrac{1}{2}\mathscr{E}_1^2\cos\delta\}/Z.$$

22. Prove equations (8.52) and (8.53) for impedances in series and parallel.

Fig. 8.16a *Fig. 8.16b*

23. Show that the impedance of the system in Fig. 8.16a is $\alpha + i\omega\beta$, where $\alpha = R/(1+\omega^2C^2R^2)$, $\beta = L-(CR^2)/(1+\omega^2C^2R^2)$. Hence show that if ωCR is very much less than 1, it is possible to choose R and C so that there is zero resultant reactance. This device, which may be thought of as cancelling the inductance L, is useful in long-distance cables, where the presence of an inductance is undesirable.

24. Show that in the transformer of Fig. 8.16b the mutual inductance M is equivalent to an impedance $i\omega MI_2/I_1$ in circuit 1 and $i\omega MI_1/I_2$ in circuit 2.

25. Show that the forced oscillations in the transformer circuit Fig. 8.16c are given by

$$\mathscr{E} = i\omega L_1 I_1 + i\omega M I_2, \qquad 0 = RI_2 + i\omega L_2 I_2 + i\omega M I_1$$

Deduce that the peak value of the current in the secondary coil is $M\mathscr{E}_0/\{(L_1L_2-M^2)^2\omega^2+L_1^2R^2\}^{1/2}$.

26. Show that if the coefficient of coupling $M/(L_1L_2)^{1/2}$ in the last question is unity, then the power factor is $\omega L_2/(R^2+\omega^2L_2^2)^{1/2}$.

27. Show that the natural frequencies of the circuit in Fig. 8.16d are $\omega/2\pi$, where $C_1C_2(L_1L_2-M^2)\omega^4-(L_1C_1+L_2C_2)\omega^2+1=0$.

28. Figure 8.16e shows two circuits with a common capacity C. This is called **direct capacity coupling**. Show that the periods of free oscillation are $2\pi/\omega$ where

$$\{CC_1L_1\omega^2-(C+C_1)\}\{CC_2L_2\omega^2-(C+C_2)\}=C_1C_2$$

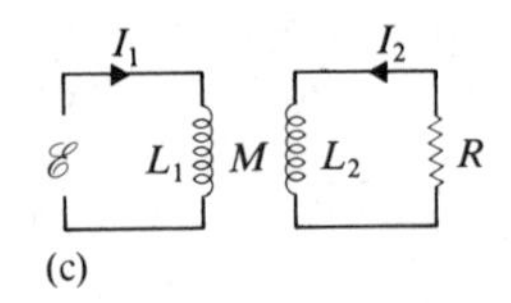

Fig. 8.16c

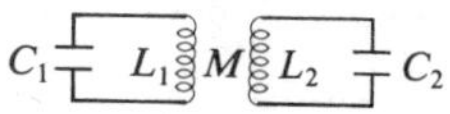

(d)

Fig. 8.16d

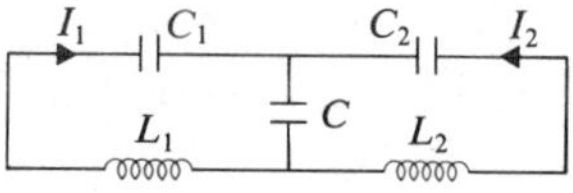

(e)

Fig. 8.16e

29. Show that $\frac{D}{Dt}(d\mathbf{s}) = (d\mathbf{s} \cdot \boldsymbol{\nabla})\mathbf{v}$.

30. Establish the identity:

$$\mathbf{A} \cdot (d\mathbf{s} \cdot \boldsymbol{\nabla})\mathbf{v} \equiv d\mathbf{s} \cdot [(\mathbf{A} \cdot \boldsymbol{\nabla})\mathbf{v} + \mathbf{A} \times (\boldsymbol{\nabla} \times \mathbf{v})]$$

31. Show that if we attempt to calculate the energy of the magnetic field due to a small magnet by integrating $\frac{1}{2}\mathbf{B} \cdot \mathbf{H}$ throughout all space we obtain an infinite value.

32. To avoid the difficulty in problem 31, suppose that each elementary magnet is a uniformly magnetized sphere of radius a. Show that $\frac{1}{2}\int \mathbf{B} \cdot \mathbf{H}\, d\mathbf{r}$ is now finite and determine its value.

33. Show that the adiabatic invariant $W_\perp/B$ may be identified with the magnetic moment of the charged particle.

34. Show that the magnetic flux through a Larmor orbit is an adiabatic invariant.

35. Show that the magnetic moment is an adiabatic invariant for the motion of a charge e in a spatially inhomogeneous magnetic field as well as for a time-dependent field. [**Hint:** Consider an axially symmetric field which increases slowly with z and show that $B_r \simeq -\frac{r_L}{2}\frac{\partial B}{\partial z}$ where r_L is the Larmor radius. From the Lorentz equation show to the same approximation, that

$$m\frac{dv_\parallel}{dt} = -\frac{er_L}{2}v_\perp\frac{\partial B}{\partial z}$$

and hence determine $\frac{dW_\parallel}{dt}$ where $W_\parallel = \frac{1}{2}mv_\parallel^2$.]

36. A small magnet $\mathbf{m}$ at the origin is rotating about its centre with an angular velocity $\boldsymbol{\omega}$, the magnitude of $\mathbf{m}$ remaining constant. Show that $\frac{d\mathbf{m}}{dt} = \boldsymbol{\omega} \times \mathbf{m}$. Hence using the expression (5.31)

for **A** and (8.57) for **E** show that the motion of the magnet gives rise to an electric field **E**, where

$$\mathbf{E} = \frac{\mu_0}{4\pi r^3}\{(\mathbf{r}\cdot\mathbf{m})\boldsymbol{\omega} - (\mathbf{r}\cdot\boldsymbol{\omega})\mathbf{m}\}$$

N.B. We know that moving charges (i.e. currents) give rise to a magnetic field. This is an example to show that moving magnets give rise to an electrostatic field.

37. A small magnet **m** has a constant velocity **v**. Show that at the moment when the vector distance from the magnet to a point P is **r** the electrostatic field at P due to the magnet's motion is

$$\mathbf{E} = \frac{\mu_0}{4\pi}\left[\frac{\mathbf{m}\times\mathbf{v}}{r^3} - \frac{3(\mathbf{v}\cdot\mathbf{r})(\mathbf{m}\times\mathbf{r})}{r^5}\right]$$

N.B. This is only valid if v is much less than the velocity of light.

38. In a betatron electrons move in a circular orbit in a vacuum chamber in a magnetic field **B** and are accelerated by the increasing flux linking the orbit. Show that for a stable circular orbit the magnetic field at the orbit must be exactly one half the average magnetic field over the duration of the acceleration.

39. An electron of charge e moves in free space with constant velocity **v**. Regarding the electron as a small conducting sphere of radius a and using the value of **B** from §5.14 problem 21, i.e. $\mathbf{B} = \frac{\mu_0}{4\pi}\frac{e\mathbf{v}\times\mathbf{r}}{r^3}$, show that the total magnetic energy associated with it is $\mu_0 e^2 v^2/12\pi a$. Deduce that if the "mass" m of the electron is supposed to be wholly electromagnetic then $m = \mu_0 e^2/6\pi a$. Using the values for m, μ_0, e in Appendix 1 show that the radius of the electron is approximately 3×10^{-15} m.

9

Maxwell's equations and electromagnetic waves

§ 9.1 Equations of electrodynamics to date

The equations of electrodynamics at this stage of development have been summarized in § 8.15. They represent essentially the equations of electrostatics and magnetostatics together with Maxwell's generalization of Faraday's law. They are not yet complete in that one of them requires an additional term. It was Maxwell who detected the omission and, in doing so, uncovered an enormously rich variety of new electromagnetic phenomena. The equation, which is incomplete at present, is not hard to find; it is (8.89). Since we accept the principle of conservation of charge as inviolable in classical electrodynamics, we know that $\nabla \cdot \mathbf{j} = -\dfrac{\partial \rho}{\partial t}$. However by taking the divergence of (8.89) we find that $\nabla \cdot \mathbf{j} = 0$. Now (8.89) arose from a consideration of static phenomena so it is hardly surprising that we need to generalize it to situations involving non-steady currents. If we write the continuity equation using (8.86) in the form

$$\nabla \cdot \mathbf{j} = -\frac{\partial \rho}{\partial t} = -\frac{\partial}{\partial t}(\nabla \cdot \mathbf{D}) \tag{9.1}$$

and consider a new current density defined by

$$\mathbf{J} = \mathbf{j} + \frac{\partial \mathbf{D}}{\partial t} \tag{9.2}$$

then the continuity equation may be written as $\nabla \cdot \mathbf{J} = 0$. Note that in arriving at this result we have used an **electrostatic** relation $\nabla \cdot \mathbf{D} = \rho$ in a non-stationary situation. This is an assumption and one that can only be justified by testing physical consequences of the resulting theory by experiment. What was previously just Ampère's law now becomes

$$\nabla \times \mathbf{H} = \mathbf{j} + \frac{\partial \mathbf{D}}{\partial t} \tag{9.3}$$

in which $\mathbf{j}$ is the **conduction current** and the new term $\dfrac{\partial \mathbf{D}}{\partial t}$ is known as the **displacement current**. Since $\mathbf{D} = \varepsilon_0 \mathbf{E} + \mathbf{P}$ we may write (9.3) as

$$\nabla \times \mathbf{H} = \mathbf{j} + \frac{\partial \mathbf{P}}{\partial t} + \varepsilon_0 \frac{\partial \mathbf{E}}{\partial t} \tag{9.4}$$

It is straightforward to interpret $\dfrac{\partial \mathbf{P}}{\partial t}$. When a dielectric is polarized by the action of an applied electric field the positive and negative charges separate, and while this separation is taking place, charges move so that a current must flow. This is the **polarization current** and is described by $\dfrac{\partial \mathbf{P}}{\partial t}$ (problem 1). Remember that we have already taken account of the effects of magnetic materials by writing (9.4) in terms of $\mathbf{H}$. If, for a moment, we revert to the fundamental magnetic field vector $\mathbf{B}$, then we should write (9.4) as

$$\nabla \times \mathbf{B} = \mu_0 \left[\mathbf{j} + \nabla \times \mathbf{M} + \frac{\partial \mathbf{P}}{\partial t} + \varepsilon_0 \frac{\partial \mathbf{E}}{\partial t} \right] \tag{9.5}$$

where $\mathbf{M}$ is the magnetization. The term, introduced by Maxwell in (9.3) and known as the displacement current, is named after the electric displacement $\mathbf{D}$. In (9.4) the term $\varepsilon_0 \dfrac{\partial \mathbf{E}}{\partial t}$ is sometimes called the **vacuum displacement current**.

The steps taken to complete the electrodynamic equations seem reasonable and consistent. It is possible to verify that they are in fact correct by applying an alternating emf to the plates of a condenser between which there is a uniform dielectric. In the (perfect) dielectric there is no conduction current $\mathbf{j}$ but according to (9.3) there should be a displacement current consisting of the polarization current together with the vacuum displacement current. In this, and in other ways to be discussed later, (9.3) is fully justified. Indeed the whole of the electromagnetic theory of light depends on the inclusion of the additional terms in (9.4).

In previous chapters we have neglected the displacement current. This is not only accurate for stationary conditions in which $\partial \mathbf{D}/\partial t = 0$ but is a very good approximation if the rate of change of $\mathbf{D}$ is not large. Such a state is known as **quasi-stationary** and all the equations and deductions in Chapter 8 depended, in fact, on this assumption. A study of numerical values shows that we may treat many problems in electrical engineering as quasi-stationary. However when we reach high frequencies such as those characteristic of optical phenomena, the displacement current predominates over the conduction current.

§ 9.2 Maxwell's equations

We may now summarize the fundamental equations of the electromagnetic field. They were first written as a set of equations by Maxwell in 1863 (though not in the form given here) and are known as **Maxwell's equations**:

$$\nabla \cdot \mathbf{D} = \rho \tag{9.6}$$

$$\nabla \cdot \mathbf{B} = 0 \tag{9.7}$$

$$\nabla \times \mathbf{E} = -\frac{\partial \mathbf{B}}{\partial t} \tag{9.8}$$

$$\nabla \times \mathbf{H} = \mathbf{j} + \frac{\partial \mathbf{D}}{\partial t} \tag{9.9}$$

in which **D** and **H** are defined by

$$\mathbf{D} = \varepsilon_0 \mathbf{E} + \mathbf{P} \tag{9.10}$$

$$\mathbf{H} = \frac{\mathbf{B}}{\mu_0} - \mathbf{M} \tag{9.11}$$

As formulated they are valid for media at rest; for moving media equation (9.9) requires some reinterpretation. Maxwell's equations need to be supplemented by the **constitutive relations**

$$\mathbf{D} = \varepsilon \mathbf{E} \tag{9.12}$$

$$\mathbf{H} = \mathbf{B}/\mu \tag{9.13}$$

$$\mathbf{j} = \sigma \mathbf{E} \tag{9.14}$$

and again Ohm's law (9.14) needs to be reinterpreted for moving conductors.

To these equations we may add those defining the potentials $\mathbf{A}$ and ϕ. We have already seen that these are

$$\mathbf{B} = \nabla \times \mathbf{A} \tag{9.15}$$

$$\mathbf{E} = -\frac{\partial \mathbf{A}}{\partial t} - \nabla \phi \tag{9.16}$$

Equation (9.15) does not completely define $\mathbf{A}$. In the stationary conditions of § 5.8 we adopted an additional condition $\nabla \cdot \mathbf{A} = 0$. When dealing with time-dependent currents it is more convenient to generalize this to

$$\nabla \cdot \mathbf{A} + \varepsilon\mu \frac{\partial \phi}{\partial t} = 0 \tag{9.17}$$

In stationary and quasi-stationary states this new condition is the same as the old and none of our earlier discussion is affected by the change. But the new form greatly simplifies the calculation of $\mathbf{A}$ and ϕ in non-stationary states (cf. Chapter 11).

§ 9.3 Decay of free charge

Several important deductions may be made from (9.6)–(9.17). Thus using (9.14) we may write (9.9) in the form

$$\nabla \times \mathbf{H} = \sigma \mathbf{E} + \frac{\partial \mathbf{D}}{\partial t}$$

Assuming that σ and ε are constant, we find, on taking the divergence of each side of this equation, that

$$\frac{\partial \rho}{\partial t} + \frac{\sigma}{\varepsilon}\rho = 0$$

so that

$$\rho = \rho_0 e^{-t/\tau} \tag{9.18}$$

where $\tau = \varepsilon/\sigma$ is known as the **relaxation time**. It follows from (9.18) that any original distribution of charge decays exponentially at a rate quite independent of any other electromagnetic disturbances that may be taking place simultaneously, and it allows us to set $\rho = 0$ in most of our problems. For metals τ is too small to measure with any accuracy (e.g. for copper $\tau \sim 10^{-18}$ seconds), but for dielectrics such as water the experimental value agrees excellently with (9.18). This result does not, of course, apply to charges at the surface of a conductor.

§ 9.4 The wave equation

Perhaps the most important deduction from Maxwell's equations is that they predict electromagnetic waves. Consider again the set of equations (9.6)–(9.14) and suppose that in the constitutive relations (9.12)–(9.14) ε, μ and σ are not functions of either space or time. Then taking the curl of (9.8) gives

$$\nabla\times(\nabla\times\mathbf{E})=-\nabla\times\frac{\partial\mathbf{B}}{\partial t}=-\mu\frac{\partial}{\partial t}(\nabla\times\mathbf{H})$$

which, using (9.9), gives

$$\nabla(\nabla\cdot\mathbf{E})-\nabla^2\mathbf{E}=-\mu\frac{\partial\mathbf{j}}{\partial t}-\varepsilon\mu\frac{\partial^2\mathbf{E}}{\partial t^2}$$

Using Ohm's law, (9.14), together with Gauss' equation (9.6) allows us to recast this as

$$\nabla^2\mathbf{E}-\varepsilon\mu\frac{\partial^2\mathbf{E}}{\partial t^2}-\sigma\mu\frac{\partial\mathbf{E}}{\partial t}=\frac{\nabla\rho}{\varepsilon} \tag{9.19}$$

If we consider a region in which there is no free charge then $\rho=0$ and we have a differential equation for the electric field

$$\nabla^2\mathbf{E}-\varepsilon\mu\frac{\partial^2\mathbf{E}}{\partial t^2}-\sigma\mu\frac{\partial\mathbf{E}}{\partial t}=0 \tag{9.20}$$

For non-conducting media $\sigma=0$ and (9.20) reduces to the standard **wave equation**, describing waves propagating with velocity $v=(\varepsilon\mu)^{-1/2}$. For conducting media, on the other hand, we must retain the third term in (9.20) but may then often neglect the second term so that we are left with a **diffusion equation**. If we compare the relative magnitudes of the two terms, assuming the time dependence of the electric field may be written as $\mathbf{E}(\mathbf{r},t)=\mathbf{E}(\mathbf{r})e^{-i\omega t}$, we find

$$\frac{|\sigma\mu\ \partial\mathbf{E}/\partial t|}{|\varepsilon\mu\ \partial^2\mathbf{E}/\partial t^2|}\sim\frac{\sigma}{\varepsilon\omega}=\frac{1}{\omega\tau} \tag{9.21}$$

where $\tau = \varepsilon/\sigma$ (cf. § 9.3). Thus if $\tau \gg T$, where $T = 2\pi/\omega$ is the period of the oscillation, then the term involving the conductivity in (9.20) is unimportant and we have effectively a wave equation. In the opposite limit $\tau \ll T$ the conductivity term is dominant and the electric field **E** is determined by the diffusion equation

$$\nabla^2 \mathbf{E} = \sigma\mu \frac{\partial \mathbf{E}}{\partial t} \tag{9.22}$$

Turning again to Maxwell's equations we may easily show that the magnetic field **H** satisfies

$$\nabla^2 \mathbf{H} - \varepsilon\mu \frac{\partial^2 \mathbf{H}}{\partial t^2} - \sigma\mu \frac{\partial \mathbf{H}}{\partial t} = 0 \tag{9.23}$$

Clearly the arguments presented in the case of the electric field apply to the magnetic field **H** as well. Neglecting the displacement current again produces a diffusion equation which has already been discussed in § 8.13 in discussing the hydromagnetics of a conducting fluid. In this chapter the discussion centres on the wave equation.

§ 9.5 Plane waves in a uniform non-conducting medium

In a uniform non-conducting medium the electric field **E** satisfies the wave equation

$$\nabla^2 \mathbf{E} = \varepsilon\mu \frac{\partial^2 \mathbf{E}}{\partial t^2} \tag{9.24}$$

We shall be particularly concerned with plane waves. A **plane wave** is one in which the wave amplitude is constant over all points of a plane drawn normal to the direction of propagation. Such a plane constitutes a wave-front, and this wave-front moves in a direction normal to itself with velocity of propagation $v = (\varepsilon\mu)^{-1/2}$ where $\nabla^2 \psi = \dfrac{1}{v^2}\dfrac{\partial^2 \psi}{\partial t^2}$, ψ representing any component of **E**. Thus, in a plane wave, amplitude is a function only of the perpendicular distance ξ from the origin to the wave-front (Fig. 9.1) and of the time i.e. $\psi = \psi(\xi, t)$. Plane waves are good approximations to actual waves in many situations, e.g. much of optics is

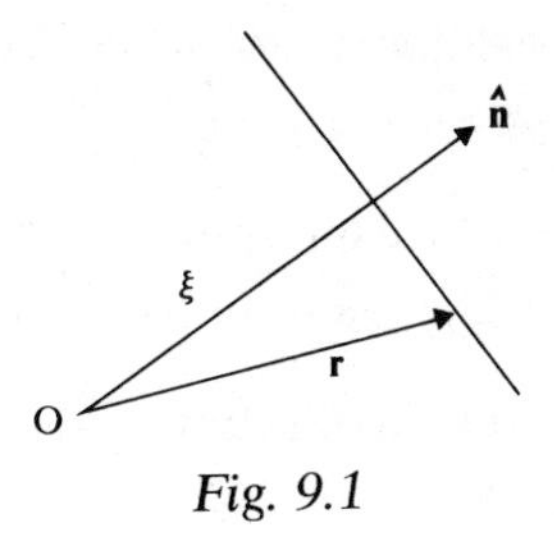

Fig. 9.1

based on the plane wave approximation, and equally, many radio waves far enough away from the antenna may be represented as plane waves.

In the wave equation

$$\frac{\partial^2 \psi}{\partial \xi^2} = \frac{1}{v^2} \frac{\partial^2 \psi}{\partial t^2} \tag{9.25}$$

let us assume a time dependence $e^{-i\omega t}$ so that (9.25) has the solution

$$\psi(\xi, t) = A_k e^{i(k\xi - \omega t)} + B_k e^{-i(k\xi + \omega t)} \tag{9.26}$$

in which A_k, B_k are constants (in general, complex) and the wave number $k = \dfrac{\omega}{v} = \dfrac{2\pi}{\lambda}$ where λ is the wave-length; $v = \omega/k$ is the **phase velocity** of the wave. Since the wave equation is linear we may invoke the principle of superposition and apply the Fourier integral theorem to construct the general plane-wave solution, i.e.

$$\begin{aligned} \psi(\xi, t) &= \int \{A_k e^{ik(\xi - vt)} + B_k e^{-ik(\xi + vt)}\} \, dk \\ &= f(\xi - vt) + g(\xi + vt) \end{aligned} \tag{9.27}$$

providing v is not itself a function of k (problem 4). Both f and g are arbitrary functions of their arguments; $f(\xi - vt)$ represents a wave-form propagating in the direction $\hat{\mathbf{n}}$ while $g(\xi + vt)$ corresponds to propagation in the opposite direction but with the same velocity v. The quantity ψ, being a component of the electric field, is of course real so that although it is convenient to work with forms like (9.26) we must always remember that by $A_k e^{ik(\xi - vt)}$ what in fact we mean is $Re[A_k e^{ik(\xi - vt)}]$. It is convenient to introduce the propagation vector **k** defined by

$$\mathbf{k} = k\hat{\mathbf{n}} \tag{9.28}$$

so that (9.26) may then be expressed

$$\psi(\mathbf{r}, t) = A_{\mathbf{k}} e^{i(\mathbf{k}\cdot\mathbf{r}-\omega t)} + B_{\mathbf{k}} e^{-i(\mathbf{k}\cdot\mathbf{r}+\omega t)}$$

Consider a plane wave propagating in the direction $\hat{\mathbf{n}}$ and represented by

$$\mathbf{E}(\mathbf{r}, t) = \mathbf{E}_0 e^{i(\mathbf{k}\cdot\mathbf{r}-\omega t)} \tag{9.29}$$

in which $\mathbf{E}_0$ is a constant, complex amplitude. Since **H** satisfies an identical wave equation, we may write

$$\mathbf{H}(\mathbf{r}, t) = \mathbf{H}_0 e^{i(\mathbf{k}\cdot\mathbf{r}-\omega t)} \tag{9.30}$$

Now not only do **E** and **H** represent solutions of the wave equation but in addition we require that they satisfy Maxwell's equations. It is important to recognize that this is not automatic since Maxwell's equations on their own do not completely determine the electromagnetic field. Thus for a uniform non-conducting medium (9.6) reduces to $\nabla\cdot\mathbf{E} = 0$ so that we require

$$\nabla\cdot\{\mathbf{E}_0 e^{i(\mathbf{k}\cdot\mathbf{r}-\omega t)}\} = 0$$

$$\text{i.e.}\quad \mathbf{k}\cdot\mathbf{E} = 0 \tag{9.31}$$

Similarly from (9.7) it follows that

$$\mathbf{k}\cdot\mathbf{H}_0 = 0 \tag{9.32}$$

so that both **E** and **H** are perpendicular to the propagation vector **k**. The electromagnetic wave is said to be **transverse** in the sense of waves on a string in which the displacement is perpendicular to the direction of propagation.

Next let us see what comes of requiring that the fields satisfy (9.8). Since

$$\nabla\times\mathbf{E} = \nabla[e^{i(\mathbf{k}\cdot\mathbf{r}-\omega t)}]\times\mathbf{E}_0 = i\mathbf{k}\times\mathbf{E}$$

Faraday's law gives

$$\mathbf{k}\times\mathbf{E} = \omega\mu\,\mathbf{H} \tag{9.33}$$

i.e. **H** is perpendicular to both **k** and **E** and the vectors $(\mathbf{E}, \mathbf{H}, \mathbf{k})$ constitute a right-handed orthogonal set as in Fig. 9.2.

We have identified the phase velocity v of the electromagnetic waves with the quantity $(\varepsilon\mu)^{-1/2}$. If we consider the propagation

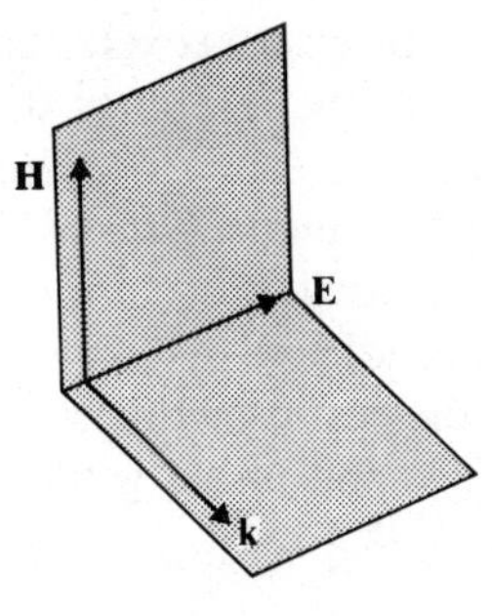

Fig. 9.2

of waves in free space then we conclude that their phase velocity is $(\varepsilon_0\mu_0)^{-1/2}$. Using the known values of ε_0, μ_0 we find that its magnitude is

$$c = 2.998 \times 10^8 \text{ m/sec.}$$

From the results of many very careful experiments it is known that the **velocity of light** in free space has this same value. We are thus led to conclude that light is simply a form of electromagnetic radiation, a view that has subsequently been completely verified. X-rays, ultra-violet and infra-red radiation, radio waves and microwaves are all electromagnetic, differing only in the order of magnitude of their wavelengths. All are transverse waves. A diagram of the electromagnetic spectrum is shown in Fig. 9.3.

In 1862 when Maxwell showed as a consequence of the field equations that electromagnetic waves may propagate there was no direct means of verifying his prediction, although experiments already carried out had resulted in an approximate value for the

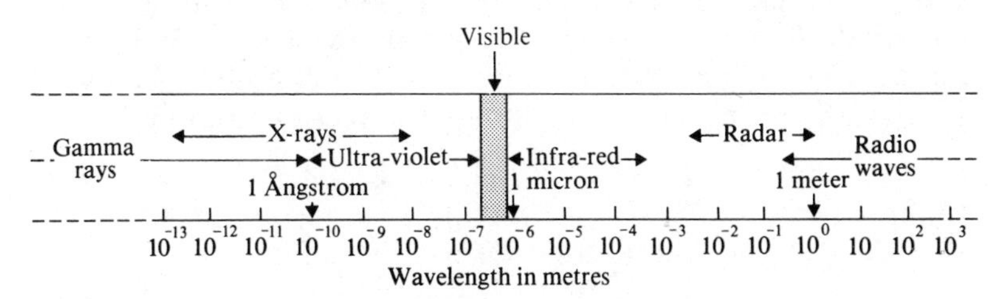

Fig. 9.3

speed of light, 3.1×10^8 m/sec. Maxwell remarked that it was scarcely possible "to avoid the inference that light consisted of transverse undulations of the same medium, which is the cause of electric and magnetic phenomena". In 1887 Hertz carried out experiments on electrical oscillations and verified directly the electromagnetic waves predicted by Maxwell's equation.

For propagation in dielectric media of permittivity ε and permeability μ, we identified the phase velocity of the waves with $(\varepsilon\mu)^{-1/2}$,

$$\text{i.e.} \quad v = \frac{c}{\sqrt{KK_m}} \tag{9.34}$$

where K is the dielectric constant and K_m the relative permeability. In most situations which concern us here we shall take $K_m = 1$, i.e. $\mu = \mu_0$, so that the phase velocity becomes

$$v = \frac{c}{\sqrt{K}} \tag{9.35}$$

Now in a medium whose refractive index is n it is known experimentally that the velocity of light is c/n. Hence if our original assumptions are valid $\varepsilon = n^2\varepsilon_0$. This result–known as Maxwell's relation–is approximately satisfied by many substances, but it fails because it does not take sufficiently detailed account of the atomic structure of the dielectric. It applies better for long waves (i.e. low frequency) than for short waves (high frequency).

§ 9.6 Polarization

The electric field of the wave discussed in § 9.5 is in the direction $\mathbf{E}_0$ which does not change with time. An electromagnetic wave with this property is said to be **linearly polarized** (or **plane polarized**). In general, however, to specify the electric field (or the magnetic field) in a plane wave we need two directions. Consider two linearly polarized waves with electric vectors given by

$$\mathbf{E}_1 = \mathbf{E}_{01} e^{i(\mathbf{k}\cdot\mathbf{r}-\omega t)}; \qquad \mathbf{E}_2 = \mathbf{E}_{02} e^{i(\mathbf{k}\cdot\mathbf{r}-\omega t)} \tag{9.36}$$

with

$$\mathbf{H}_1 = \frac{\mathbf{k}\times\mathbf{E}_1}{\omega\mu}; \qquad \mathbf{H}_2 = \frac{\mathbf{k}\times\mathbf{E}_2}{\omega\mu}. \tag{9.37}$$

The amplitudes $\mathbf{E}_{01}$, $\mathbf{E}_{02}$ are complex so that, in general, there will be a phase difference between the waves. The total electric field is found by superposition, i.e.

$$\mathbf{E}(\mathbf{r}, t) = (\mathbf{E}_{01} + \mathbf{E}_{02})e^{i(\mathbf{k}\cdot\mathbf{r}-\omega t)} \tag{9.38}$$

with $\mathbf{E}_{01} = |\mathbf{E}_{01}|e^{i\alpha}\hat{\mathbf{e}}_1$, $\mathbf{E}_{02} = |\mathbf{E}_{02}|e^{i\beta}\hat{\mathbf{e}}_2$. For simplicity consider a wave propagating along Ox and choose $\hat{\mathbf{e}}_1 = \hat{\mathbf{y}}$, $\hat{\mathbf{e}}_2 = \hat{\mathbf{z}}$ and so $|\mathbf{E}_{01}| \equiv E_{y0}$, $|\mathbf{E}_{02}| \equiv E_{z0}$. Then (9.38) becomes, on writing the phase difference $\delta = \beta - \alpha$

$$\mathbf{E}(x, t) = (E_{y0}\hat{\mathbf{y}} + E_{z0}e^{i\delta}\hat{\mathbf{z}})e^{i(kx-\omega t+\alpha)} \tag{9.39}$$

The electric vector at each point in space rotates in a plane normal to $\hat{\mathbf{x}}$ and its tip, as time evolves, describes an ellipse, Fig. 9.4. This is seen most easily for $\delta = \pm\pi/2$ when (9.39) becomes

$$\mathbf{E}(x, t) = (E_{y0}\hat{\mathbf{y}} \pm iE_{z0}\hat{\mathbf{z}})e^{i(kx-\omega t+\alpha)}$$

and taking the real part

$$\left.\begin{aligned} E_y(x, t) &= E_{y0}\cos(kx - \omega t + \alpha) \\ E_z(x, t) &= \mp E_{z0}\sin(kx - \omega t + \alpha) \end{aligned}\right\} \tag{9.40}$$

Then

$$\frac{E_y^2}{E_{y0}^2} + \frac{E_z^2}{E_{z0}^2} = 1 \tag{9.41}$$

The resultant **E** field in this case traces an ellipse in the clockwise sense $\delta = \pi/2$ since E_z leads E_y; when $\delta = -\pi/2$, E_y leads E_z and the ellipse is traced in the anticlockwise sense. In this case the

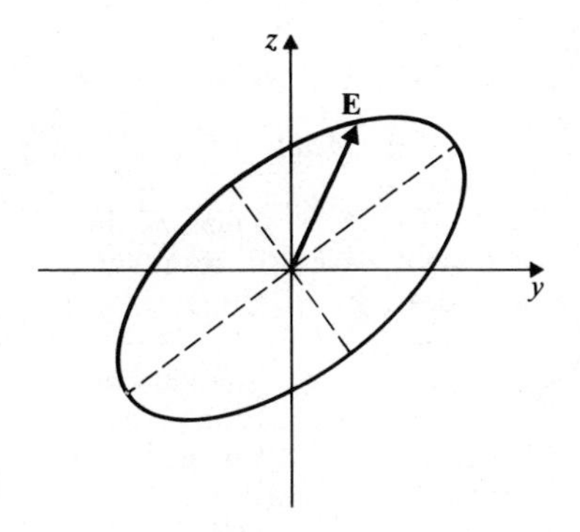

Fig. 9.4

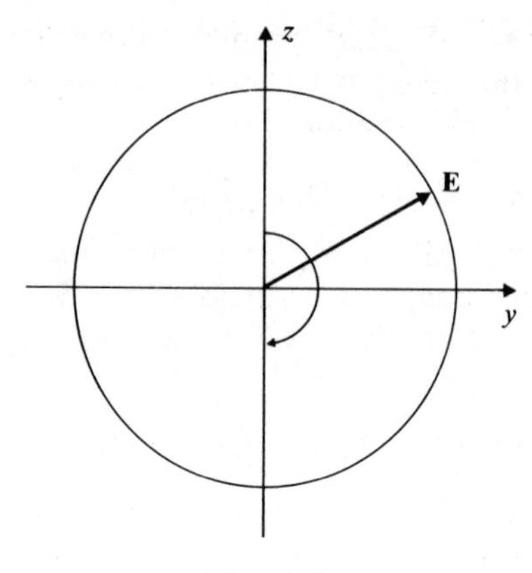

Fig. 9.5

electromagnetic wave is said to be **elliptically polarized**. The special case of linear (or plane) polarization is retrieved if $\delta = 0$ (problem 6). Another important special case arises when $\delta = \pi/2$ and $E_{y0} = E_{z0}$ so that (9.41) becomes $E_y^2 + E_z^2 = E_{y0}^2$; in this case the wave is said to be **circularly polarized**. To an observer looking at the approaching wave front the + sign in (9.40) corresponds to a wave in which **E** rotates in an anticlockwise sense; this wave is said to be **left circularly polarized**. Similarly the − sign (corresponding to clockwise rotation) give a **right circularly polarizrd** wave as in Fig. 9.5. Table 9.1 summarizes the various polarization states for different δ.

Table 9.1

Phase Difference δ	Component wave amplitudes	Locus of tip of resultant **E**	State of polarization
0	$E_{y0} \neq E_{z0}$	Straight line	Linear
$\pi/2$	$E_{y0} \neq E_{z0}$	Ellipse	Elliptic
$\pi/2$	$E_{y0} = E_{z0}$	Circle	Circular (RCP)
$-\pi/2$	$E_{y0} = E_{z0}$	Circle	Circular (LCP)
π	$E_{y0} \neq E_{z0}$	Straight line	Linear

§ 9.7 Electromagnetic energy and Poynting's theorem

Let us now return to the field equations (9.6)–(9.14). We recall from Chapter 8 that the expressions $\frac{1}{2}\int \mathbf{E}\cdot\mathbf{D}\,d\mathbf{r}$ and $\frac{1}{2}\int \mathbf{B}\cdot\mathbf{H}\,d\mathbf{r}$ are, respectively, the general expressions for the electrostatic and magnetostatic field energy. We now want to generalize these results to time-dependent situations.

We shall consider the rate of charge of energy in a given volume. We know that all electromagnetic effects may be attributed to charges–at rest or in motion–and that a moving charge experiences a force so that, as it moves, work is continually being done on it. In this section we shall calculate the rate at which the energy changes due to these forces exerted on each moving charge.

The force $\mathbf{F}$ on the moving charge e is expressed by

$$\mathbf{F}=e(\mathbf{E}+\mathbf{v}\times\mathbf{B}) \tag{9.42}$$

Thus the rate at which work is done on a **unit** charge is

$$[\mathbf{E}+\mathbf{v}\times\mathbf{B}]\cdot\mathbf{v}=\mathbf{E}\cdot\mathbf{v}$$

and consequently the total rate at which the electromagnetic forces are doing work in the given volume V is

$$\int_V (\mathbf{E}\cdot\mathbf{v})\rho\,d\mathbf{r}=\int_V \mathbf{E}\cdot\mathbf{j}\,d\mathbf{r}$$

where $\mathbf{j}$ is the free or conduction current density. It follows, using (9.9) that

$$\int_V (\mathbf{E}\cdot\mathbf{v})\rho\,d\mathbf{r}=\int_V \mathbf{E}\cdot\left(\boldsymbol{\nabla}\times\mathbf{H}-\frac{\partial\mathbf{D}}{\partial t}\right)d\mathbf{r} \tag{9.43}$$

The next step is to transform the first term in this expression using the vector identity

$$\begin{aligned}\boldsymbol{\nabla}\cdot(\mathbf{H}\times\mathbf{E})&=\mathbf{E}\cdot(\boldsymbol{\nabla}\times\mathbf{H})-\mathbf{H}\cdot(\boldsymbol{\nabla}\times\mathbf{E})\\&=\mathbf{E}\cdot(\boldsymbol{\nabla}\times\mathbf{H})+\mathbf{H}\cdot\frac{\partial\mathbf{B}}{\partial t}\end{aligned} \tag{9.44}$$

from (9.8). Substituting (9.44) into (9.43) gives

$$\int_V \mathbf{E}\cdot\mathbf{j}\,d\mathbf{r}=-\int_V\left[\boldsymbol{\nabla}\cdot(\mathbf{E}\times\mathbf{H})+\mathbf{H}\cdot\frac{\partial\mathbf{B}}{\partial t}+\mathbf{E}\cdot\frac{\partial\mathbf{D}}{\partial t}\right]d\mathbf{r} \tag{9.45}$$

The first integral may be transferred into a surface integral over the boundary of the volume V so that the rate at which the electromagnetic field is doing work on the charges is

$$-\int_S (\mathbf{E}\times\mathbf{H})\cdot d\mathbf{S}-\int_V \left[\mathbf{E}\cdot\frac{\partial\mathbf{D}}{\partial t}+\mathbf{H}\cdot\frac{\partial\mathbf{B}}{\partial t}\right] d\mathbf{r}$$

If we assume that ε and μ are constants with respect to t we then find that for **linear media**

$$-\frac{\partial}{\partial t}\int_V \frac{1}{2}[\mathbf{E}\cdot\mathbf{D}+\mathbf{B}\cdot\mathbf{H}]\, d\mathbf{r}=\int_V \mathbf{E}\cdot\mathbf{j}\, d\mathbf{r}+\int_S (\mathbf{E}\times\mathbf{H})\cdot d\mathbf{S} \tag{9.46}$$

The left-hand side of (9.46) is to be interpreted as the rate of decrease of electromagnetic energy; in other words **we take** $\frac{1}{2}[\mathbf{E}\cdot\mathbf{D}+\mathbf{B}\cdot\mathbf{H}]$ to be the electromagnetic energy density–a natural generalization of the earlier expressions derived for static fields, $\frac{1}{2}\mathbf{E}\cdot\mathbf{D}$ and $\frac{1}{2}\mathbf{B}\cdot\mathbf{H}$. This is an assumption for the form of the energy density for time-varying fields; it is perfectly plausible but an assumption nevertheless. What of the right-hand side? We earlier identified $\int_V \mathbf{E}\cdot\mathbf{j}\, d\mathbf{r}$ as the rate of working of the electromagnetic forces on the free charge within V. Explicitly from (4.16)

$$\int \mathbf{E}\cdot\mathbf{j}\, d\mathbf{r}=\int \frac{j^2}{\sigma}\, d\mathbf{r}-\int \mathbf{E}'\cdot\mathbf{j}\, d\mathbf{r} \tag{9.47}$$

where the $\mathbf{E}'$ represents an electromagnetic field. The first term on the right-hand side of (9.47) corresponds to the Joule heat loss and the second is (-1) times the rate at which the sources of emf in V do work. The remaining term in (9.46) corresponds to an energy term not previously considered. For static (or quasi-static) fields it can clearly be made vanishingly small by choosing S sufficiently large. However, anticipating a result to be established in Chapter 11, **radiation fields**, unlike static fields, have an r^{-1} dependence so that in general this term contributes to the energy balance equation. We shall in fact regard (9.46) as ensuring energy conservation by interpreting the surface integral on the right-hand side as a rate of flow, or **radiation**, of energy outward across the surface S which is the boundary of the volume V. The vector $\mathbf{E}\times\mathbf{H}$ is known as the **Poynting vector**, denoted by $\mathbf{S}$. While it seems reasonable to interpret $\mathbf{S}$ as the rate at which energy is radiated across unit area, this can lead to paradoxical results (problem 11) and in fact **only the integral of S** over a closed surface has general physical significance. In many cases, however, involving radiation from aerials or the propagation of light waves, in which the field is rapidly alternating about a zero mean, the vector $\mathbf{E}\times\mathbf{H}$ does represent the rate of flow across any isolated unit area.

§ 9.8 Energy flux in a plane wave

The time-averaged energy flux in a plane wave is determined by the Poynting vector $\mathbf{S}=\mathbf{E}\times\mathbf{H}$ defined in § 9.7. The appearance of $\mathbf{H}$ rather than $\mathbf{B}$ in the expression for $\mathbf{S}$ is one reason for preferring $\mathbf{H}$ to $\mathbf{B}$ in § 9.4. Since $\mathbf{S}$ is not a linear function we must be careful in computing it on account of having expressed the field vectors in terms of complex amplitudes. In many situations we shall only be interested in time-averaged–as opposed to instantaneous–values of the energy flux. Denoting time-averages by brackets $\langle\ \rangle$,

$$\begin{aligned}\langle\mathbf{S}\rangle &= \langle(Re\mathbf{E})\times(Re\mathbf{H})\rangle \\ &= \tfrac{1}{2}Re(\mathbf{E}\times\mathbf{H}^*)\end{aligned} \tag{9.48}$$

(problem 13). From (9.33) we have

$$\mathbf{H}^*=\frac{\hat{\mathbf{n}}\times\mathbf{E}^*}{v\mu} \tag{9.49}$$

$$\begin{aligned}\text{i.e.}\quad \mathbf{E}\times\mathbf{H}^* &= \{\mathbf{E}\times(\hat{\mathbf{n}}\times\mathbf{E}^*)\}/v\mu \\ &= |\mathbf{E}_0|^2/(v\mu)\,\hat{\mathbf{n}}\end{aligned}$$

$$\therefore\quad \langle\mathbf{S}\rangle=\frac{|\mathbf{E}_0|^2}{2v\mu}\,\hat{\mathbf{n}} \tag{9.50}$$

Recalling that $v=(\varepsilon\mu)^{-1/2}$, $(v\mu)^{-1}=\left(\dfrac{\varepsilon}{\mu}\right)^{1/2}$ so that

$$\langle\mathbf{S}\rangle=\frac{1}{2}\left(\frac{\varepsilon}{\mu}\right)^{1/2}|\mathbf{E}_0|^2\,\hat{\mathbf{n}} \tag{9.51}$$

Moreover the time-averaged energy density associated with the wave is (problem 14):

$$\langle W\rangle=\tfrac{1}{4}(\mathbf{E}\cdot\mathbf{D}^*+\mathbf{B}\cdot\mathbf{H}^*) \tag{9.52}$$

Writing $\mathbf{D}^* = \varepsilon \mathbf{E}^*$, $\mathbf{B} = \mu \mathbf{H}$ and using (9.49) this gives

$$\langle W \rangle = \frac{\varepsilon}{2} |E_0|^2 \tag{9.53}$$

From (9.51), (9.53)

$$\begin{aligned} \langle \mathbf{S} \rangle &= (\varepsilon\mu)^{-1/2} \langle W \rangle \hat{\mathbf{n}} \\ &= v \langle W \rangle \hat{\mathbf{n}} \end{aligned} \tag{9.54}$$

Thus for a plane wave the energy flux is in the direction of propagation of the wave, its magnitude being the product of the time-averaged energy density with the phase velocity of the wave. In other words energy flows with the same velocity as the wave itself. This result is **not** true in general. For a combination of two or more waves the velocity of energy transport given by $\mathbf{S}/W$ will in general be dependent on frequency and is of course the **group velocity** (cf. § 9.10).

§ 9.9 Plane waves in a conducting medium

In § 9.4 we saw that the components of $\mathbf{E}$ and $\mathbf{H}$ satisfied a modified wave equation (9.20) in which the additional term arising from the ohmic current density, $\mathbf{j}$, corresponds to a dissipation of energy. In this section we want to find plane wave solutions of Maxwell's equations for a conducting medium. Assume the field components vary harmonically with time, i.e. $\psi(\mathbf{r}, t) = \psi_0 e^{i(\mathbf{k} \cdot \mathbf{r} - \omega t)}$ and that the **longitudinal** amplitudes (i.e. those in the direction of $\mathbf{k}$) vanish so that we deal only with transverse waves as in § 9.5. Then from (9.20)

$$[k^2 - \varepsilon\mu\omega^2 - i\sigma\mu\omega]\psi_0 e^{i(\mathbf{k} \cdot \mathbf{r} - \omega t)} = 0$$

which requires

$$k^2 = \varepsilon\mu\omega^2 \left[1 + i\frac{\sigma}{\varepsilon\omega}\right] \tag{9.55}$$

In a conducting medium, therefore, the wave number k is complex, i.e. $k = \alpha + i\beta$, so that from (9.55)

$$\left.\begin{aligned} \alpha^2 - \beta^2 &= \varepsilon\mu\omega^2 \\ 2\alpha\beta &= \sigma\mu\omega \end{aligned}\right\} \tag{9.56}$$

Solving these equations simultaneously we find

$$\alpha = \omega\sqrt{\frac{\varepsilon\mu}{2}}\left[1+\left(1+\left(\frac{\sigma}{\varepsilon\omega}\right)^2\right)^{1/2}\right]^{1/2} \tag{9.57}$$

where the positive square root has been chosen (why?) and

$$\beta = \omega\sqrt{\frac{\varepsilon\mu}{2}}\left[-1+\left(1+\left(\frac{\sigma}{\varepsilon\omega}\right)^2\right)^{1/2}\right]^{1/2} \tag{9.58}$$

Observe that in the limit $\sigma \to 0$, $\alpha = \omega(\varepsilon\mu)^{1/2}$, $\beta = 0$.

The transverse components of the fields therefore behave as

$$\begin{aligned} e^{i(\mathbf{k}\cdot\mathbf{r}-\omega t)} &\equiv e^{i(k\xi-\omega t)} \\ &= e^{-\beta\xi}e^{i(\alpha\xi-\omega t)} \end{aligned} \tag{9.59}$$

with α,β given by (9.56). Since $\beta > 0$ we see from (9.59) that a plane wave cannot propagate in a conducting medium without attenuation; β is therefore called the **absorption coefficient**.

For good conductors, provided the frequency is not too high, $\sigma/\varepsilon\omega \gg 1$ so that (9.55) reduces to

$$k^2 \simeq i\omega\sigma\mu \tag{9.60}$$

and (9.58) becomes

$$\beta = \sqrt{\frac{\omega\sigma\mu}{2}} \tag{9.61}$$

The quantity $\beta^{-1} = \sqrt{2/\omega\sigma\mu}$ is a measure of the distance of penetration of an electromagnetic wave into a good conductor before its magnitude drops to $1/e$ times its surface value. This distance is known as the **skin depth** δ. We see from (9.61) that the higher the frequency (always provided $\sigma/\varepsilon\omega \gg 1$) or the better the conductor, the less the field penetrates.

§ 9.10 Plane electromagnetic waves in a uniform plasma

In §§ 9.5 and 9.9 we examined plane waves in uniform dielectrics and in conducting media respectively. We shall now take up the corresponding problem in a plasma. Whereas we left the permittivity ε and the electrical conductivity σ unspecified in the earlier discussions (since to determine them takes us again beyond the limits of **classical** electrodynamics), we shall see how, by simple arguments, it is possible to obtain an expression for the permittivity of a plasma. For simplicity we shall consider only **cold** plasmas in which it is assumed that the thermal motion of electrons and ions may be neglected. A fuller discussion of waves in plasmas may be found in [3].

Before dealing specifically with electromagnetic waves in a plasma, let us see what a generalization of the simple model of conductivity in § 4.2 to the time-dependent case gives. The Lorentz equation for an electron of charge *e*, mass *m*, may be written

$$m\frac{d\mathbf{v}}{dt}=e\mathbf{E}-m\nu\mathbf{v} \tag{9.62}$$

In the case of steady currents considered in Chapter 4 we had simply $e\mathbf{E}=m\nu\mathbf{v}$. The term $-m\nu\mathbf{v}$ is introduced into the equation of motion as a phenomenological representation of the momentum loss suffered by the electron as a result of collisions. We shall not be concerned with the motion of the positive ions as these, being massive in comparison with the electrons, are correspondingly sluggish in their response to an electric field. We have also neglected the magnetic force in the Lorentz equation. This is a good approximation for non-relativistic particle motion in the field of a plane wave since

$$\frac{|\mathbf{v}\times\mathbf{B}|}{|\mathbf{E}|}\sim\frac{|\mathbf{v}\times(\mathbf{k}\times\mathbf{E})|}{\omega|\mathbf{E}|}\sim\frac{v}{c}$$

Assuming a time dependence $e^{-i\omega t}$ in $\mathbf{E}$, $\mathbf{v}$ allows us to write (9.62) as

$$\mathbf{v}=\frac{e\mathbf{E}}{m(\nu-i\omega)} \tag{9.63}$$

and from this we may write for the corresponding current density $\mathbf{j}$

$$\mathbf{j}=n_0e\mathbf{v}=\frac{n_0e^2\mathbf{E}}{m(\nu-i\omega)} \tag{9.64}$$

We may regard this as the generalization of (4.9) to the non-stationary case.

Now plasmas in many instances may, to a very good approximation, be assumed to be collisionless. Neglecting ν in

(9.64) gives for the current density in a plasma

$$\mathbf{j} = i\frac{n_0 e^2}{m\omega}\mathbf{E} \tag{9.65}$$

We see that there is a phase difference between **j** and **E**. The plasma behaves as a dielectric, rather than a conducting, medium as may easily be seen from Maxwell's equation (9.9):

$$\nabla \times \mathbf{H} = \mathbf{j} - i\omega\varepsilon_0\mathbf{E} = -i\omega\varepsilon\mathbf{E}$$

Substituting from (9.65) for **j** gives

$$\varepsilon(\omega) = \varepsilon_0 - \frac{n_0 e^2}{m\omega^2}$$

$$\text{i.e.} \quad \varepsilon(\omega) = \varepsilon_0\left(1 - \frac{\omega_p^2}{\omega^2}\right) \tag{9.66}$$

where

$$\omega_p = \left(\frac{n_0 e^2}{\varepsilon_0 m}\right)^{1/2} \tag{9.67}$$

is the **plasma frequency**. It follows from (9.55) that

$$k^2 = \varepsilon\mu_0\omega^2 = \frac{\omega^2}{c^2}\left(1 - \frac{\omega_p^2}{\omega^2}\right)$$

$$\text{i.e.} \quad \omega^2 = \omega_p^2 + k^2c^2 \tag{9.68}$$

Thus for frequencies above the plasma frequency, k is real and waves propagate in the plasma without attenuation. The (ω, k) plot is shown in Fig. 9.6 and from this we see that the phase velocity of the wave, ω/k, is always greater than the velocity of light in vacuum. Since the phase velocity v_{ph} depends on frequency, a plasma is dispersive; (9.68) is the corresponding **dispersion relation**. The velocity of propagation of a wave packet, which disperses as it propagates, is given by the group velocity v_g,

$$v_g = \frac{d\omega}{dk} = \frac{kc^2}{\omega}$$

$$\text{i.e.} \quad v_g = \frac{c^2}{v_{ph}} < c \tag{9.69}$$

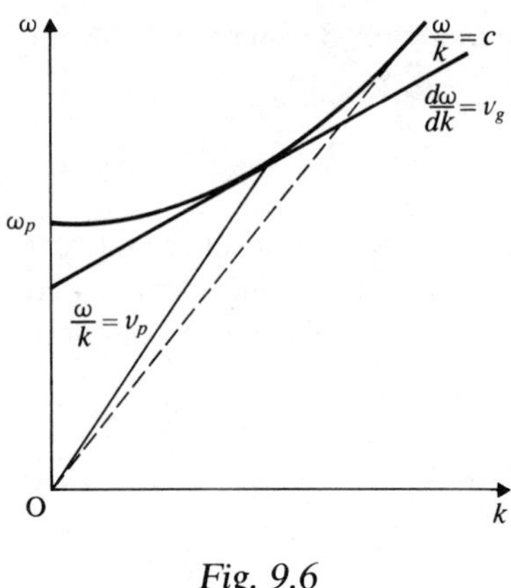

Fig. 9.6

Clearly when the plasma density becomes vanishingly small (9.68) gives the familiar free space result. What is the significance of the other limit, i.e. letting $k \to 0$ so that $\omega \to \omega_p$? When we considered conductors in electrostatics we saw that within the conductor $\mathbf{E} = 0$, not because the electric field is somehow excluded from the interior but simply because it is annulled by superposition of a second field. This field is produced by charges which have moved to the surface of the conductor under the action of the primary field. If this primary field is switched on at time $t = 0$, the electric field within the conductor is not **at first** zero because the electrons take a finite–if short–time to move to produce the response field, which exactly cancels the switched-on field. Some time τ is required to establish this equilibrium. Now if the external field is not static but alternating in time and reverses in a time short compared with τ, it follows that the electrons will not have time to respond and produce the field which cancels the primary field. Corresponding to τ is the frequency τ^{-1} which we may call a **cut-off frequency**. For electromagnetic waves such that $\omega\tau \gg 1$ the plasma will be transparent to the radiation since the plasma electrons will not have time enough to annul the electric field of the wave. For very high frequencies the wave will be unaware of the presence of the plasma and will behave as in free space, i.e. $\omega^2 \simeq k^2c^2$. It is easy to estimate the relaxation time τ and show that

$$\tau^{-1} = \left(\frac{n_0 e^2}{m\varepsilon_0}\right)^{1/2} = \omega_p$$

§ 9.11 Plane wave propagation in a uniform anisotropic plasma

In this section and the next we shall undertake two important developments of wave propagation in plasmas. In reality, whether we consider natural plasmas such as the ionosphere and the solar corona, or those created in the laboratory, in machines like the tokamaks mentioned in § 5.11, we find that they rarely conform to the simple model taken in the previous section. Plasmas, in general, are anisotropic and inhomogeneous, dissipative and dispersive. We can do little towards considering dissipative effects short of a statistical mechanical treatment and so we shall retain the cold plasma approximation. We may, however, within this approximation, generalize our discussion to include, in turn, wave propagation in anisotropic and inhomogeneous plasmas.

Anisotropic effects are important in practice since most natural plasmas such as in the ionosphere or the Crab nebula as well as those in fusion experiments are magnetized. This means that their properties are anisotropic and their response to an electromagnetic wave will differ according to the orientation of **k**, the propagation vector, to the magnetic field, $\mathbf{B}_0 = B_0\hat{\mathbf{z}}$. Corresponding to (9.62), but ignoring the dissipative term $-m\nu\mathbf{v}$, we now have the Lorentz equation

$$m\frac{d\mathbf{v}}{dt} = e[\mathbf{E} + \mathbf{v}\times\mathbf{B}_0] \tag{9.70}$$

Introducing the Larmor frequency $\Omega = |e|B_0/m$ and assuming a time dependence $e^{-i\omega t}$ in $\mathbf{v}$ gives

$$-i\omega\mathbf{v} = \frac{e\mathbf{E}}{m} - \Omega(\mathbf{v}\times\hat{\mathbf{z}})$$

Forming the current density as in (9.64) we find

$$\omega^2\mathbf{j} = i\omega\omega_p^2\varepsilon_0\mathbf{E} - i\omega\Omega(\mathbf{j}\times\hat{\mathbf{z}}) \tag{9.71}$$

Writing (9.71) in component form we may express $\mathbf{j}$ in terms of $\mathbf{E}$ as

$$\mathbf{j} = \varepsilon_0\omega_p^2 \begin{bmatrix} \dfrac{i\omega}{\omega^2-\Omega^2} & \dfrac{-\Omega}{\omega^2-\Omega^2} & 0 \\ \dfrac{\Omega}{\omega^2-\Omega^2} & \dfrac{i\omega}{\omega^2-\Omega^2} & 0 \\ 0 & 0 & \dfrac{i\omega}{\omega^2} \end{bmatrix} \begin{bmatrix} E_x \\ E_y \\ E_z \end{bmatrix} \tag{9.72}$$

If we now substitute this **tensor relation** for **j** into Maxwell's equation as in § 9.10 we get

$$j_\alpha - i\omega\varepsilon_0 E_\alpha = -i\omega\varepsilon_{\alpha\beta}E_\beta \tag{9.73}$$

with

$$j_\alpha = \sigma_{\alpha\beta}E_\beta \tag{9.72a}$$

giving the **permittivity tensor**

$$\varepsilon_{\alpha\beta} = \varepsilon_0\delta_{\alpha\beta} + i\frac{\sigma_{\alpha\beta}}{\omega} \tag{9.74}$$

which has components

$$\varepsilon_{\alpha\beta} = \begin{bmatrix} \varepsilon_{11} & \varepsilon_{12} & 0 \\ \varepsilon_{21} & \varepsilon_{22} & 0 \\ 0 & 0 & \varepsilon_{33} \end{bmatrix} \tag{9.75}$$

$$\left.\begin{aligned} \varepsilon_{11} &= \varepsilon_{22} = \varepsilon_0\left[1 - \frac{\omega_p^2}{\omega^2 - \Omega^2}\right] \\ \varepsilon_{12} &= -\varepsilon_{21} = -i\varepsilon_0\frac{\omega_p^2\Omega}{\omega(\omega^2 - \Omega^2)} \\ \varepsilon_{33} &= \varepsilon_0\left(1 - \frac{\omega_p^2}{\omega^2}\right) \end{aligned}\right\} \tag{9.76}$$

We see that the permittivity tensor is Hermitian. As we might have expected, the Larmor frequency Ω is an additional characteristic frequency in the case of a magnetized plasma. The singularity in ε_{11}, ε_{12} at $\omega = \Omega$ arises simply from a resonance between the frequency of the applied alternating electric field and that of an electron moving in a magnetic field and is not present in more realistic models of the plasma. Note too that ε_{33}, the permittivity of the plasma along the direction of the magnetic field, is simply that for an **isotropic** plasma, (9.66).

If we introduce terms R, L defined by

$$R = 1 - \frac{\omega_p^2}{\omega(\omega - \Omega)}; \qquad L = 1 - \frac{\omega_p^2}{\omega(\omega + \Omega)} \tag{9.77}$$

we may express

$$\left.\begin{aligned} \varepsilon_{11} &= \varepsilon_{22} = \tfrac{1}{2}(R + L) = S \\ \varepsilon_{12} &= -\varepsilon_{21} = \frac{i}{2}(R - L) = iD \end{aligned}\right\} \tag{9.78}$$

and if we set $\varepsilon_{33} = P$, the permittivity tensor becomes

$$\varepsilon_{\alpha\beta} = \varepsilon_0 \begin{bmatrix} S & iD & 0 \\ -iD & S & 0 \\ 0 & 0 & P \end{bmatrix} \tag{9.79}$$

We have already remarked on the significance of P. The meaning of the other elements is apparent once we adopt a coordinate frame in which the antisymmetric $\varepsilon_{\alpha\beta}$ is diagonalized. Introducing the unitary matrix

$$U = \frac{1}{\sqrt{2}} \begin{bmatrix} 1 & -i & 0 \\ 1 & i & 0 \\ 0 & 0 & \sqrt{2} \end{bmatrix} \tag{9.80}$$

and its inverse U^{-1} it is straightforward to show (problem 19) that $\varepsilon'_{\alpha\beta} = U_{\alpha\sigma}\varepsilon_{\sigma\rho}U^{-1}_{\rho\beta}$ is just

$$\varepsilon'_{\alpha\beta} = \begin{bmatrix} R & 0 & 0 \\ 0 & L & 0 \\ 0 & 0 & P \end{bmatrix} \tag{9.81}$$

Written in this form we see that the modes represented by R, L and P propagate independently of one another. It will become apparent that those corresponding to L and R are left and right circularly polarized waves.

To obtain the dispersion relation corresponding to (9.68) in the isotropic case, we write (9.8) as

$$\mathbf{k} \times \mathbf{E} = \omega\mu_0 \mathbf{H}$$

giving

$$\mathbf{k} \times (\mathbf{k} \times \mathbf{E}) = \omega\mu_0 (\mathbf{k} \times \mathbf{H})$$

Therefore from $(\mathbf{k} \times \mathbf{H})_\alpha = -\omega\varepsilon_{\alpha\beta}E_\beta$ we have

$$[\mathbf{k} \times (\mathbf{k} \times \mathbf{E})]_\alpha + \omega^2\mu_0\varepsilon_{\alpha\beta}E_\beta = 0$$

Writing $\mathbf{k}c/\omega = \mathbf{n}$ gives

$$[\mathbf{n} \times (\mathbf{n} \times \mathbf{E})]_\alpha + \varepsilon_0^{-1}\varepsilon_{\alpha\beta}E_\beta = 0 \tag{9.82}$$

If we choose $\mathbf{n}$ in the Oxz plane (Fig. 9.7) so that

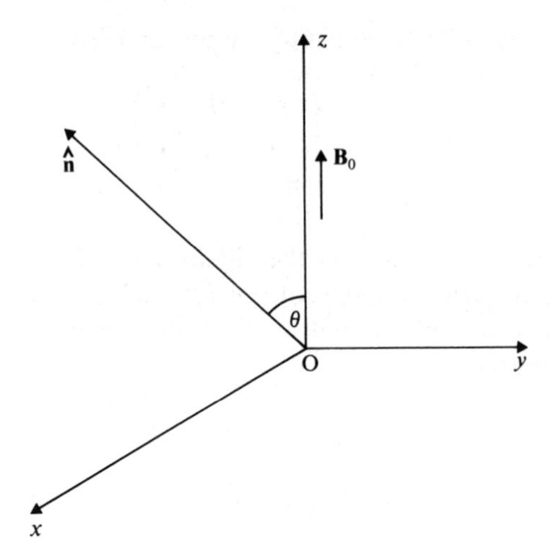

Fig. 9.7

$\mathbf{n}=(n\sin\theta, 0, n\cos\theta)$ then (9.28) takes the form

$$\begin{bmatrix} S-n^2\cos^2\theta & iD & n^2\sin\theta\cos\theta \\ -iD & S-n^2 & 0 \\ n^2\sin\theta\cos\theta & 0 & P-n^2\sin^2\theta \end{bmatrix}\begin{bmatrix} E_x \\ E_y \\ E_z \end{bmatrix}=0 \qquad (9.83)$$

and from this the general **dispersion relation** follows by setting the determinant of the matrix equal to zero. Thus for an anisotropic cold plasma (in which we have ignored the motion of the ions) we find the dispersion relation reduces to (problem 20)

$$An^4-Bn^2+C=0 \qquad (9.84)$$

with

$$\left.\begin{aligned} A &= S\sin^2\theta+P\cos^2\theta \\ B &= RL\sin^2\theta+PS(1+\cos^2\theta) \\ C &= PRL \end{aligned}\right\} \qquad (9.85)$$

The roots of the dispersion relation are determined by

$$n^2=\frac{B\pm[(RL-PS)^2\sin^4\theta+4P^2D^2\cos^2\theta]^{1/2}}{2A} \qquad (9.86)$$

This expression contains all the information about the propagation of waves in the plasma model we have chosen, but, as it stands, it conceals this very effectively! However, if we now add An^2 to each side of (9.84) and replace n^2 by (9.86) on the right-hand side we get (problem 21)

$$n^2 = 1 - \frac{2\omega_p^2(\omega^2 - \omega_p^2)}{2\omega^2(\omega^2 - \omega_p^2) - \omega^2\Omega^2 \sin^2\theta \pm \omega\Gamma} \tag{9.87}$$

with

$$\Gamma = [\omega^2\Omega^4 \sin^4\theta + 4\Omega^2(\omega^2 - \omega_p^2)^2 \cos^2\theta]^{1/2}$$

This is the **Appleton–Hartree formula** and was first derived to describe radio-wave propagation in the ionosphere. It is of course quite generally applicable within the limits of our approximation, i.e. to cold plasmas with the ion motion neglected (though it may readily be generalized to include the effects of the ions).

For propagation **along** the magnetic field we find

$$n_{R,L}^2 = 1 - \frac{\omega_p^2}{\omega(\omega \mp \Omega)} \tag{9.88}$$

corresponding to right and left circularly polarized waves (cf. § 9.6) and reducing to (9.68) in the limit $\Omega \to 0$. For propagation **orthogonal** to the magnetic field

$$n_X^2 = \frac{(\omega^2 - \omega_p^2)^2 - \omega^2\Omega^2}{\omega^2(\omega^2 - \omega_p^2 - \Omega^2)} \tag{9.89}$$

corresponding to the minus sign in (9.87) or

$$n_O^2 = 1 - \frac{\omega_p^2}{\omega^2} \tag{9.90}$$

corresponding to the plus sign. The subscripts X, O stand for **extra-ordinary** and **ordinary** and refer respectively to modes which depend on the magnetic fields for their propagation and to those which do not.

The various dispersion relations are presented as (ω, k) plots in Fig. 9.8. The shaded areas represent stop-bands for propagation. Beginning with the highest frequencies, i.e. those above the line $R = 0$ for which

$$\omega = \omega_2 = +\frac{\Omega}{2} + \left(\frac{\Omega^2}{4} + \omega_p^2\right)^{1/2}$$

we see from (9.88) and (9.89), (9.90) that both an R and L mode

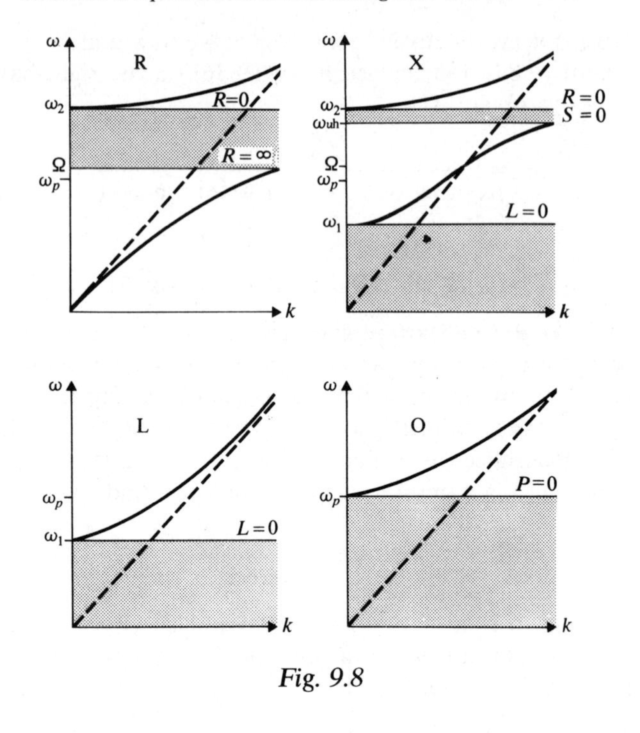

Fig. 9.8

and O and X modes can propagate. On crossing the line $R=0$ which represents a cut-off frequency both the R and X modes meet a stop-band but the others are unaffected. Continuing downwards in frequency we next meet the value given by

$$\omega^2=\omega_p^2+\Omega^2 \tag{9.91}$$

at which (9.89) is resonant. This value of ω is known as the **upper-hybrid resonance** (since it involves each of the two characteristic frequencies in a magnetized plasma, ω_p and Ω). Below this the X-mode again propagates but the R-mode is still forbidden. There is no further change in the dispersion characteristics until the **electron cyclotron resonance** is reached when $R=\infty$, i.e. at $\omega=\Omega$. Below this propagation is again possible in all four modes until the **plasma cut-off** at $P=0$, i.e. $\omega=\omega_p$. The O-mode now disappears as it must since no propagation is possible below ω_p in an isotropic plasma. At this point too the phase velocity of the X-mode increases above the speed of light in vacuum. Finally, continuing downwards one meets the **cyclotron cut-off** at $L=0$ which removes the L- and the X-modes. Only the R-mode is left.

§ 9.12 Plane wave propagation in an isotropic inhomogeneous plasma

In this section we shall consider some of the problems posed by inhomogeneous plasmas but will deal only with those which are isotropic. Let us derive a wave equation appropriate to a plasma in which the density, and hence the permittivity (9.66), is spatially varying. Let us suppose that $\varepsilon=\varepsilon(z)$. Then from Maxwell's equations, assuming a time dependence $e^{-i\omega t}$, we have

$$\nabla\times(\nabla\times\mathbf{E})=+i\omega\mu_0(\nabla\times\mathbf{H}) \tag{9.92}$$

$$\nabla\times\mathbf{H}=-i\omega\varepsilon(z)\mathbf{E} \tag{9.93}$$

giving

$$\nabla\times(\nabla\times\mathbf{E})-\omega^2\mu_0\varepsilon(z)\mathbf{E}=0 \tag{9.94}$$

To simplify the discussion we consider wave propagation **along** the density gradient. Then (9.94) reduces to

$$\frac{d^2E}{dz^2}+\frac{\omega^2}{c^2}\varepsilon'(z)E=0 \tag{9.95}$$

where $\varepsilon'(z)=\varepsilon(z)/\varepsilon_0$ and $E=E_x$ (or E_y). The equation for H is somewhat different in form but may be transformed into the form (9.95) (problem 22).

If we now look back to propagation in the homogeneous plasma in §9.10 we found there that $E=Ae^{i\phi(z)-i\omega t}$ with $\phi(z)=\frac{\omega_0}{c}\sqrt{\varepsilon'}z=kz$. Let us assume that in the case of (9.95) the electric field may be represented in this way though of course since now $\varepsilon=\varepsilon(z)$ we no longer know the form of $\phi(z)$. Dropping the time dependence we have

$$\frac{dE}{dz}=iA\phi' e^{i\phi}$$

$$\frac{d^2E}{dz^2}=\{iA\phi''-A(\phi')^2\}e^{i\phi}$$

with $\phi' = d\phi/dz$, $\phi'' = d^2\phi/dz^2$. Substituting in (9.95) gives

$$i\phi'' - (\phi')^2 + \frac{\omega^2}{c^2}\varepsilon'(z) = 0 \tag{9.96}$$

Clearly for homogeneous plasmas, $\phi'' \equiv 0$. We will suppose that for plasmas with densities which vary on a sufficiently long scale-length (the meaning of "long" to be determined later), the ϕ'' term may be regarded as small when compared with $\frac{\omega^2}{c^2}\varepsilon'(z)$, i.e.

$$\phi' = \pm\frac{\omega}{c}\sqrt{\varepsilon'}, \qquad \phi'' = \pm\frac{\omega}{c}\frac{d}{dz}(\sqrt{\varepsilon'}) \ll \frac{\omega^2}{c^2}\varepsilon'$$

Then in (9.96) we may write

$$\phi' = \left[\frac{\omega^2}{c^2}\varepsilon' \mp \frac{i\omega}{c}\frac{d}{dz}\sqrt{\varepsilon'}\right]^{1/2}$$

$$\simeq \mp\frac{\omega}{c}\sqrt{\varepsilon'} + \frac{1}{2\sqrt{\varepsilon'}}\frac{d}{dz}\sqrt{\varepsilon'}$$

$$\therefore \quad \phi(z) \simeq \mp\int^z \frac{\omega}{c}\sqrt{\varepsilon'}\,dz + i\ln(\varepsilon')^{1/4}$$

The electric field may now be written

$$E(z) = \frac{A}{(\varepsilon')^{1/4}}\exp\left(\pm i\frac{\omega}{c}\int^z \sqrt{\varepsilon'}\,dz\right) \tag{9.97}$$

corresponding to right (+) and left (−) travelling waves. This result is known as the WKBJ solution after the method evolved by Wentzel, Kramers, Brillouin and Jeffreys.

We may formalize the "slowly varying" condition mentioned above and show (problem 23) that the approximation is valid provided

$$\frac{c^2}{\omega^2}\left|\frac{3}{4}\left(\frac{1}{\varepsilon'}\frac{d}{dz}\sqrt{\varepsilon'}\right)^2 - \frac{1}{2(\varepsilon')^{3/2}}\frac{d^2}{dz^2}\sqrt{\varepsilon'}\right| \ll 1 \tag{9.98}$$

Thus the approximation is clearly invalid if $\varepsilon' \approx 0$.

As an example let us consider a linear density profile (Fig. 9.9a),

$$n(z) = n_0\frac{z}{L} \tag{9.99}$$

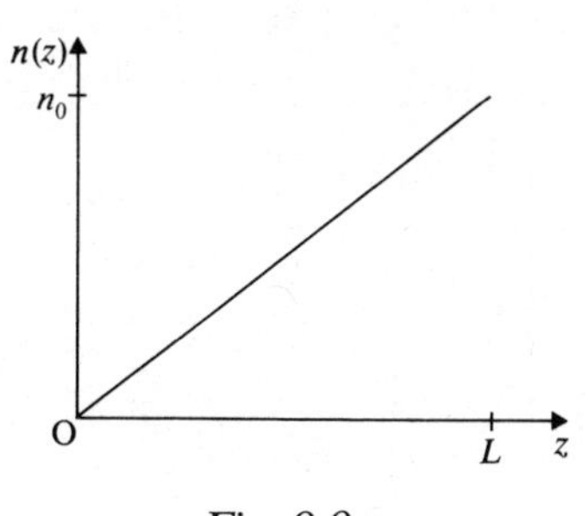

Fig. 9.9a

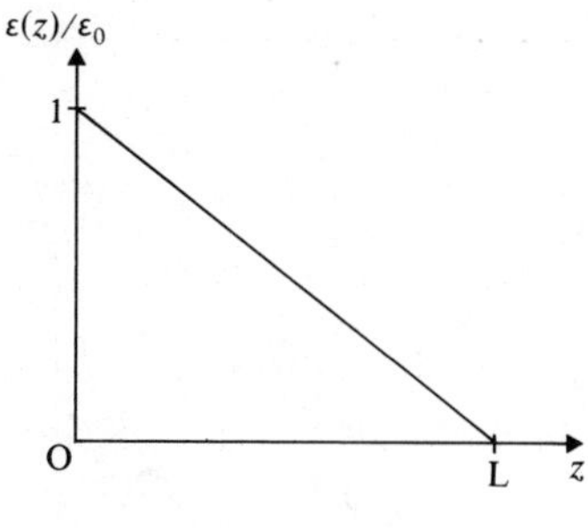

Fig. 9.9b

so chosen that at $z=L$, $\omega^2=\omega_p^2(L)$. Then

$$\varepsilon'(z)\equiv\frac{\varepsilon(z)}{\varepsilon_0}=1-\frac{z}{L} \tag{9.100}$$

and (9.95) is now

$$\frac{d^2E}{dz^2}+\frac{\omega^2}{c^2}\left(1-\frac{z}{L}\right)E=0 \tag{9.101}$$

with WKBJ solutions

$$E(z)=A\left(1-\frac{z}{L}\right)^{-1/4}\exp\left[\mp(2i/3)(\omega L/c)(1-z/L)^{3/2}\right] \tag{9.102}$$

If we write $\xi(z)=-\varepsilon'(z)=\dfrac{z}{L}-1$ then

$$E(\xi)=A(-\xi)^{-1/4}\exp\left(\mp\frac{2i}{3}\frac{\omega L}{c}(-\xi)^{3/2}\right) \tag{9.103}$$

which is valid sufficiently far away from $\xi=0$. Clearly for $z>L$, $\varepsilon'(z)<0$, i.e. $\xi>0$ and the exponential in (9.103) becomes $\exp\left(\mp\dfrac{2}{3}\dfrac{\omega L}{c}\xi^{3/2}\right)$. The solution for the electric field corresponding to the + sign is unbounded and we shall discard it as being physically unreasonable. Choosing the $-$sign represents a wave field which is evanescent in the region $\xi>0(z>L)$ as shown in Fig. 9.10.

Translating back to the physical variable we find that as the wave travels up the density gradient it reaches a point at which

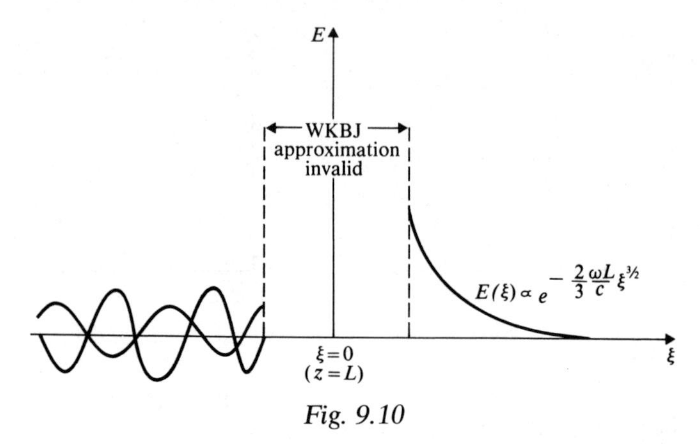

Fig. 9.10

$\omega = \omega_p(L)$–sometimes called the **critical frequency**–beyond which it cannot propagate, i.e. the electric field decays as in Fig. 9.10. Obviously we are not able to say anything about the wave field in the region $\xi \simeq 0$ (i.e. $\varepsilon' \simeq 0$) since here the WKBJ approximation fails (cf. (9.98)). For $\xi < 0$,

$$E(\xi) = A(|\xi|)^{-1/4} \exp\left(\mp \frac{2i}{3}\frac{\omega L}{c}(|\xi|)^{3/2}\right)$$

and this corresponds to a wave travelling towards the left (−) and one travelling in the opposite direction (+). One solution represents the incident wave while the other is a reflected wave.

In the region around $\xi = 0$ we must discard the WKBJ approximation and return to the differential equation. In the case of a linear density profile it so happens that we can find an exact solution to the differential equation. If we set

$$\zeta = -\left(\frac{L^2\omega^2}{c^2}\right)^{1/3}\left(1 - \frac{z}{L}\right)$$

Then (9.101) becomes (problem 24)

$$\frac{d^2E}{d\zeta^2} = \zeta E \tag{9.104}$$

which is known as **Stokes' differential equation** or **Airy's equation**. There are two linearly independent solutions which may be defined in a variety of ways. For our purposes it is convenient to

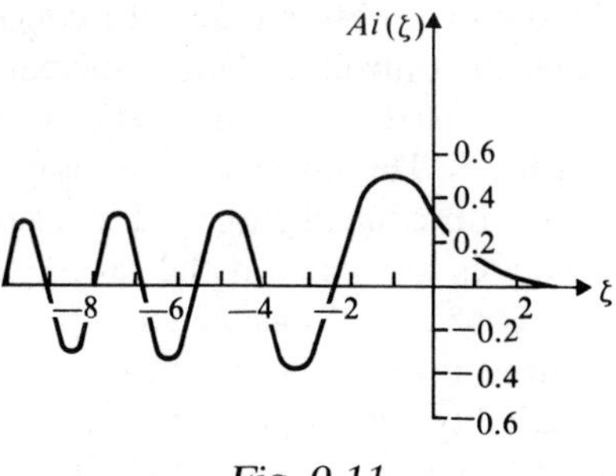

Fig. 9.11

introduce solutions in terms of the **Airy integral** [9]

$$Ai(\zeta) = \frac{1}{\pi}\int_0^\infty \cos\left(\zeta t + \tfrac{1}{3}t^3\right) dt \tag{9.105}$$

which, for large ζ, takes the forms

$$\begin{aligned} Ai(\zeta) &\sim \frac{1}{2\sqrt{\pi}}\zeta^{-1/4}e^{-2\zeta^{3/2}/3}, \qquad \zeta \to +\infty \\ &\sim \frac{1}{\sqrt{\pi}}|\zeta|^{-1/4}\sin\left(\tfrac{2}{3}|\zeta|^{3/2} + \frac{\pi}{4}\right), \qquad \zeta \to -\infty \end{aligned}$$

which correspond to the WKBJ solutions we have already found. The other Airy integral $Bi(\zeta)$ corresponds to the unphysical solution which we discarded.

Far from $\zeta = 0$ then, we retrieve the WKBJ solutions. Near to $\zeta = 0$ we must evaluate (9.105) numerically. $Ai(\zeta)$ is represented in Fig. 9.11; from this we see that near to $\zeta = 0$ there is a maximum in the electric field. Translating back to the physical variable we note that $\zeta = 0$ corresponds to $\omega = \omega_p(L)$, i.e. the critical frequency introduced above. The "swelling" of the electric field at $z = L$ corresponds to an accumulation of energy at this **critical layer**. The incident electromagnetic wave slows down as it approaches the critical layer and turns round at $z = L$.

§ 9.13 Interaction of intense laser radiation with plasmas

The process discussed in the previous section is of very great interest in studies of the interaction of intense radiation from high power lasers with matter. These studies are concerned with an alternative approach to controlled thermonuclear fusion via **inertial** containment as opposed to the **magnetic** containment mentioned in Chapter 5. The key idea is simply that the laser radiation on impact with a small target of solid deuterium ionizes the material and creates a plasma at the target surface. The energetic electrons in this plasma "boil off" and in so doing exert pressure on the un-ionized target material below the surface. Provided the laser pulse is suitably shaped so that it continues to ionize the target further in, so to speak, it is in principle possible to create such enormous pressures that the target material can be compressed to perhaps $10^4 \times$ liquid densities (Fig. 9.12). When this is achieved thermonuclear burn can be started in the central core and a "burn front" propagates outward, heating and igniting the dense fuel as it travels. In this scheme for fusion obviously a vital question concerns the interaction of the incident laser radiation with the plasma. The calculation presented in §9.12 is the simplest model possible of a very complex interaction but it illustrates some of the matters which concern physicists studying laser fusion.

We may follow our model a little further. We have seen that close to the critical layer the electric field is no longer homogeneous. In Chapter 8 (problem 25) we considered the behaviour of an electron in an inhomogeneous–though static–magnetic field; we might now ask how the motion of an electron in the field of an electromagnetic wave is affected by inhomogeneities in the field. This behaviour might give some insight as to how the plasma

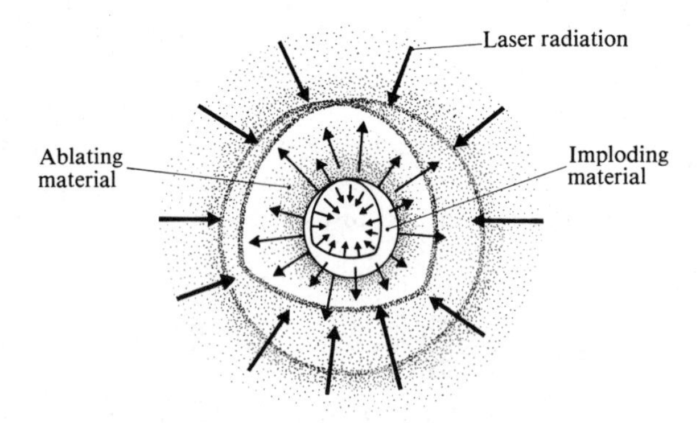

Fig. 9.12

itself behaves in this region. Consider then the motion of an electron in the field of a wave:

$$m\frac{d\mathbf{v}}{dt}=e[\mathbf{E}(\mathbf{r},t)+\mathbf{v}\times\mathbf{B}(\mathbf{r},t)] \tag{9.106}$$

and

$$\mathbf{E}(\mathbf{r},t)=\mathbf{E}_0(\mathbf{r})\cos\omega_0 t$$

$$\mathbf{B}(\mathbf{r},t)=-\frac{\nabla\times\mathbf{E}_0(\mathbf{r})}{\omega_0}\sin\omega_0 t$$

Thus the $\mathbf{v}\times\mathbf{B}$ term in (9.106) is $\sim(v/c)$ times smaller than the electric field and so to lowest order we may suppose that the electron follows $\mathbf{E}$, i.e.

$$m\frac{d\mathbf{v}^{(1)}}{dt}=e\mathbf{E}_0\cos\omega_0 t$$

$$\therefore\quad \mathbf{v}^{(1)}=\frac{e\mathbf{E}_0}{m\omega_0}\sin\omega_0 t=\mathbf{v}_0\sin\omega_0 t \tag{9.107}$$

where

$$\mathbf{v}_0\equiv\frac{e\mathbf{E}_0}{m\omega_0} \tag{9.108}$$

is known as the **quiver velocity**. For $v_0/c\ll 1$ we may use $\mathbf{v}^{(1)}$ given by (9.107) to solve (9.106) to next order. In doing so we must take account of the inhomogeneity of the field and expand $\mathbf{E}(\mathbf{r})$ about the **initial** position of the electron, i.e.

$$\mathbf{E}(\mathbf{r})\simeq\mathbf{E}(\mathbf{r}_0)+(\delta\mathbf{r}\cdot\nabla)\mathbf{E}+\dots$$

$$\delta\mathbf{r}=\int\mathbf{v}^{(1)}dt$$

$$=-\frac{e\mathbf{E}_0}{m\omega_0^2}\cos\omega_0 t$$

Then to second order,

$$m\frac{d\mathbf{v}^{(2)}}{dt}=e[(\delta\mathbf{r}\cdot\nabla)\mathbf{E}(\mathbf{r}_0)+\mathbf{v}^{(1)}\times\mathbf{B}(\mathbf{r}_0)]$$

$$=-\left(\frac{e^2}{m\omega_0^2}\right)[(\mathbf{E}_0\cdot\nabla)\mathbf{E}_0\cos^2\omega_0 t+\mathbf{E}_0\times\nabla\times\mathbf{E}_0\sin^2\omega_0 t]$$

Averaging both sides of this equation with respect to time we find an expression for the secular force acting on the electron, $\langle \mathbf{F} \rangle$, given by

$$\langle \mathbf{F} \rangle = -\left(\frac{e^2}{2m\omega_0^2}\right)[(\mathbf{E}_0 \cdot \nabla)\mathbf{E}_0 + \mathbf{E}_0 \times \nabla \times \mathbf{E}_0] \qquad (9.109)$$

The first term in (9.109) is due to the excursion of the electron in its oscillation into regions of differing field strength. For the case of an electromagnetic wave incident **normally** on an inhomogeneous plasma, this effect must vanish; it would of course contribute to the force on the electron if the wave were incident **obliquely** and polarized with its electric field in the plane of incidence. The second term is due to the magnetic field of the incident radiation which distorts the orbit of the electron and causes a **drift** of the electron in the direction of the wave. We may write (9.109) as

$$\langle \mathbf{F} \rangle = -\left(\frac{e^2}{4m\omega_0^2}\right)\nabla E_0^2$$

$$= -\left(\frac{e^2}{2m\omega_0^2}\right)\nabla\langle E^2 \rangle \qquad (9.110)$$

If we now sum over all the electrons in unit volume we get $\langle n_0 \mathbf{F} \rangle$ which we shall denote by $\langle \mathbf{F}_{\text{PM}} \rangle$, the **ponderomotive force**; i.e.

$$\langle \mathbf{F}_{\text{PM}} \rangle = -\left(\frac{n_0 e^2}{\varepsilon_0 m\omega_0^2}\right)\nabla\left\langle\frac{\varepsilon_0 E^2}{2}\right\rangle$$

$$= -\frac{\omega_p^2}{\omega_0^2}\nabla\left\langle\frac{\varepsilon_0 E^2}{2}\right\rangle \qquad (9.111)$$

Returning to the characteristic of the electric field in the region of the critical layer, namely, field swelling, we see (Fig. 9.13) that, on account of the strong gradients, the plasma may be forced out of this region. This leads to modification of the density profile by the laser radiation. This effect has been studied in detail on account of its importance in laser fusion studies. Figure 9.14 is a

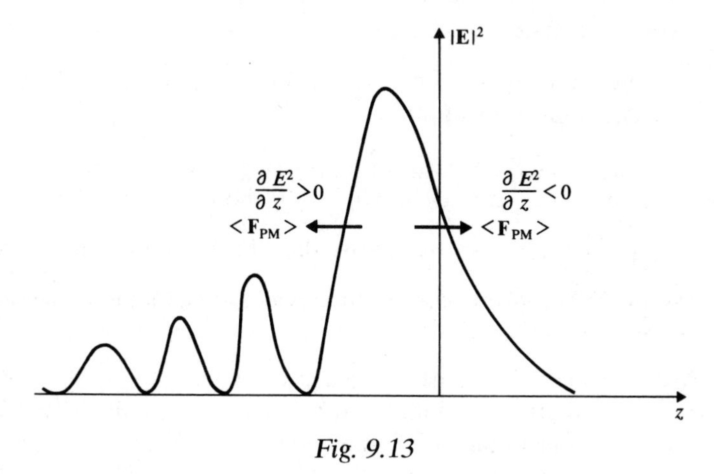

Fig. 9.13

plot of density as a function of distance in an inhomogeneous plasma. This is taken from a computer experiment in which the interaction of intense radiation with a plasma is simulated. Initially the density profile is linear. Fig. 9.14 shows clearly the modification to this profile in the region of the critical density.

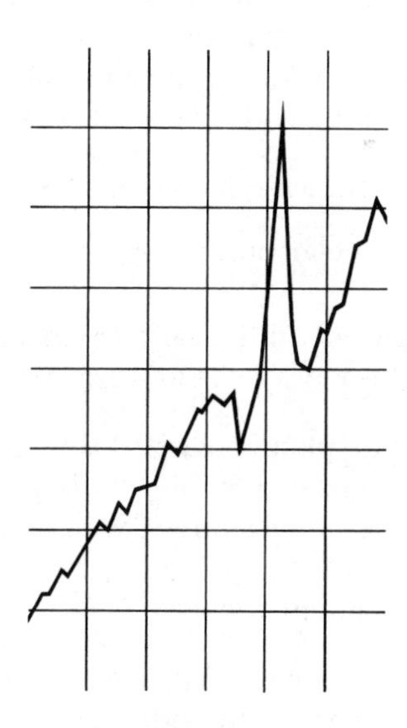

Fig. 9.14

§ 9.14 Problems

1. Show by an analysis similar to that in § 6.2 that the polarization current is given by $\partial\mathbf{P}/\partial t$.

2. Given $E_x = 0 = E_y$, $E_z = a \cos kx \cos \omega t$ and $\mathbf{H} = 0$ when $t = 0$ in vacuum, i.e. $\varepsilon = \varepsilon_0$, $\mu = \mu_0$, $\rho = 0$, $\sigma = 0$, show that $H_x = 0 = H_z$, $H_y = -a\left(\frac{\varepsilon_0}{\mu_0}\right)^{1/2} \sin kx \sin \omega t$. Verify that there is no mean flux of energy in this problem, which corresponds to standing or stationary waves.

3. A large sphere of radius b is made of material of conductivity σ and permittivity ε. At time $t = 0$ a charge Q_0 is uniformly distributed over the surface of a small concentric sphere of radius a. Show that at time t the charge Q on the inner sphere is given by $Q = Q_0 e^{-\sigma t/\varepsilon}$. Show further that the total Joule heat loss during the discharge is $\frac{Q_0^2}{2\varepsilon}\left(\frac{1}{a} - \frac{1}{b}\right)$.

4. Establish (9.27).

5. Show that the wave equation may be expressed directly in terms of any of the components of the electric field $\mathbf{E}$, the magnetic field $\mathbf{H}$ or the vector potential $\mathbf{A}$ in rectangular coordinates but **not** for any component in spherical polar coordinates. What happens in cylindrical polar coordinates?

6. Show that in § 9.6 the case $\delta = 0$ corresponds to plane polarization and $\delta = \frac{\pi}{2}$ to circular polarization.

7. Show that an arbitrarily elliptically polarized wave may be decomposed into a right and a left circularly polarized wave.

8. Two monochromatic plane polarized waves of the same frequency propagate along Oz. One is polarized in the $\hat{\mathbf{x}}$ direction with amplitude E_x, the other along $\hat{\mathbf{y}}$, with amplitude E_y. The phase of the second leads that of the first by ψ. Determine the polarization of the resultant wave.

9. Two monochromatic waves of the same frequency are circularly polarized, one right and the other left. They are in phase and propagate in the same direction. Determine the polarization

characteristics of the resultant for different values of the ratio E_1/E_2, E_1 and E_2 being the (real) amplitudes.

10. Show that for any finite system the Poynting vector tends to zero at infinity so fast that its integral over the sphere at infinity is zero, thus showing that the total electromagnetic energy remains constant.

11. The expression for the energy flux $\mathbf{E}\times\mathbf{H}$ was established for the case of a closed surface only. Verify that it is not true for an open surface in the case in which $\mathbf{E}$ and $\mathbf{H}$ are two constant perpendicular fields; but it is still true even in this case for any closed surface.

12. Starting with the total electromagnetic energy

$$U=\tfrac{1}{2}(\mathbf{E}\cdot\mathbf{D}+\mathbf{B}\cdot\mathbf{H})$$

in a given volume, show from Maxwell's equations that

$$\frac{\partial U}{\partial t}=-\int_S (\mathbf{E}\times\mathbf{H})\cdot d\mathbf{S}-\int_V \mathbf{E}\cdot\mathbf{j}\,d\mathbf{r}$$

13. Establish (9.48) for the time-averaged energy flux in a plane wave.

14. Establish (9.52) for the time-averaged energy density associated with a plane wave.

15. Show that the magnetic field energy $\frac{1}{2}\int_V \mathbf{B}\cdot\mathbf{H}\,d\mathbf{r}$ within a volume V may be written in the form

$$\frac{1}{2}\int\mathbf{A}\cdot\mathbf{j}\,d\mathbf{r}+\frac{1}{2}\int\mathbf{A}\cdot\frac{\partial\mathbf{D}}{\partial t}\,d\mathbf{r}-\frac{1}{2}\int(\mathbf{H}\times\mathbf{A})\cdot d\mathbf{S}.$$

Deduce that in a finite system of quasi-steady currents the total magnetic field energy is simply $\frac{1}{2}\int\mathbf{A}\cdot\mathbf{j}\,d\mathbf{r}$ which (for uniform μ) may be written as

$$\frac{\mu}{8\pi}\iint\frac{\mathbf{j}\cdot\mathbf{j}'}{|\mathbf{r}-\mathbf{r}'|}\,d\mathbf{r}\,d\mathbf{r}'.$$

16. Show that when $\sigma/\varepsilon\omega\gg 1$ the electric field in the conducting half-space $z>0$ may be expressed as

$$\mathbf{E}(z,t)=\mathbf{E}_0 e^{-\sqrt{\sigma\omega\mu/2}\,z}\cos\left(\sqrt{\frac{\sigma\omega\mu}{2}}\,z-\omega t\right)$$

Deduce that if **E** has only a component E_x then the only non-vanishing component of **B** is B_y and that B_y and E_x have a phase difference of $\pi/4$.

17. Obtain an expression for the group velocity of a uniform plane wave in a good conductor.

18. Determine the phase and group velocities in a medium of permittivity

$$\epsilon(\omega)=\varepsilon_0\left[1-\frac{\omega_p^2}{\omega(\omega-\Omega)}\right]$$

19. Establish (9.81)

20. Establish (9.84)

21. Obtain the Appleton–Hartree dispersion relation, (9.87)

22. Show that the differential equation satisfied by the wave field $\mathbf{H}=(0, H_y, 0)$ in an isotropic inhomogeneous plasma is

$$\frac{d^2H_y}{dz^2}-\frac{1}{\varepsilon(z)}\frac{d\varepsilon}{dz}\frac{dH_y}{dz}+\frac{\omega^2}{c^2}\frac{\varepsilon(z)}{\varepsilon_0}H_y=0$$

where $\varepsilon(z)$ is the plasma permittivity.

23. Establish the "slowly varying" condition (9.98) which validates the WKBJ approximation.

24. Show that the electric field of an electromagnetic wave propagating along a linear density profile satisfies (9.104).

25. Consider an electromagnetic wave incident obliquely on a plasma, the density of which varies slowly along Oz. Assuming the wave to be polarized so that the wave **H** field lies along Oy, show that this field satisfies the differential equation

$$\frac{d^2H_y}{dz^2}-\frac{1}{\varepsilon}\frac{d\varepsilon}{dz}\frac{dH_y}{dz}+q^2H_y=0$$

where $\varepsilon(z)$ is the dielectric function of the plasma and

$$q^2(z)=\frac{\omega^2}{c^2}\frac{\varepsilon(z)}{\varepsilon_0}-k^2\sin^2\theta$$

the x-dependence of the wave being represented by a factor $e^{-ikx\sin\theta}$.

Obtain WKBJ solutions to this wave equation and show that a condition for these to be good approximations to the wave **H** field is

$$\left|\frac{\varepsilon''}{2q^2\varepsilon}-\frac{q''}{2q^3}-\frac{3\varepsilon'^2}{4q^2\varepsilon^2}+\frac{3q'^2}{4q^4}\right| \ll 1$$

26. In the case of an electromagnetic wave propagating along a linear density profile show that the WKBJ condition reduces to $kL \gg 1$ and interpret this physically.

27. Laser light of wavelength 1.06 μm from a neodymium glass laser is incident on a plasma of varying density. Determine the critical density.

28. From (9.95) show that $\mathcal{I}m\left(E^*\dfrac{\partial E}{\partial z}\right)$ is constant. Interpret this result physically.

10

Electromagnetic waves in bounded media

§10.1 Boundary conditions

In Chapter 9, starting from Maxwell's equations, we derived wave equations for electric and magnetic fields and discussed plane wave solutions of these. In this chapter we extend the discussion to include problems of reflection and refraction at plane boundaries between media. By way of illustration we shall often consider optical wavelengths and show how optics is contained within the framework of Maxwell's electrodynamics; of course the theory is more generally applicable at other wavelengths, in particular to radio waves.

As observed in Chapter 9, Maxwell's equations on their own do not completely determine the electromagnetic field. In any particular situation one retrieves the physical solutions from an infinite set of solutions by satisfying appropriate boundary conditions. We saw in discussing the propagation of an electromagnetic wave in an inhomogeneous plasma in §9.12 that the only boundary condition in that case was the requirement that the fields vanish at infinity. We have already considered boundary conditions in previous chapters.

For convenience we summarize these in Table 10.1, for both dielectric and conducting boundaries

Table 10.1

Dielectrics	Conductors
$E_{t1}=E_{t2}$	$E_{t1}=E_{t2}$ $(=0$ for perfect conductors)
$D_{n1}=D_{n2}$	$D_{n1}-D_{n2}=\rho_s$
$H_{t1}=H_{t2}$	$H_1-H_{t2}=0$ (**or** $j_{s\perp}$ for perfect conductors)
$B_{n1}=B_{n2}$	$B_{n1}=B_{n2}$

For the time-dependent case the boundary conditions on the normal components of the fields **D** and **B** are not independent but are contained already in the electrodynamic field equations. This is a consequence of the fact that the divergence equations $\nabla \cdot \mathbf{D} = \rho$ and $\nabla \cdot \mathbf{B} = 0$ follow from the remaining Maxwell equations by applying the divergence operator to these. It follows therefore that the boundary conditions on D_n and B_n are automatically satisfied provided the conditions on E_t and H_t are met.

The idealization to perfect conductors may sometimes be made, in which case the skin depth becomes vanishingly small. Thus electric fields are excluded from inside a perfect conductor and the tangential component of the magnetic field **H** also vanishes within the conductor. It is not of course zero outside, so that $H_{t1} - H_{t2} = j_{s\perp}$ where $j_{s\perp}$ is the current per unit width, assumed to flow as a vanishingly thin current sheet. However, in general we should remember that the concept of perfect conductivity is not very useful at optical frequencies since for most metals it is a poor approximation.

Finally, as was done in §7.3 for static fields, it is possible to establish uniqueness in the time-dependent case by assuming two possible solutions to Maxwell's equations with identical tangential fields at the boundary. It is straightforward to show that the difference field satisfies Poynting's theorem and that for linear, isotropic, inhomogeneous media if we specify E_t and H_t on the boundary **and** the initial values of all fields then the fields within the solution space are **uniquely** specified at all subsequent times.

§10.2 Reflection and refraction at a dielectric boundary: Fresnel relations

In this section we shall consider a plane electromagnetic wave incident obliquely on the plane boundary separating two dielectrics. As mentioned in the previous section, we will use electromagnetic waves at optical frequencies to illustrate the general problem so that in this, and the following sections, we shall retrieve results familiar from elementary optics simply by applying the boundary conditions of §10.1 to the electric and magnetic fields satisfying Maxwell's equations. To begin with, we shall assume that both media (referred to as 1 and 2 from Fig. 10.1) are perfect dielectrics so that $\sigma = 0$ and no energy losses occur. The

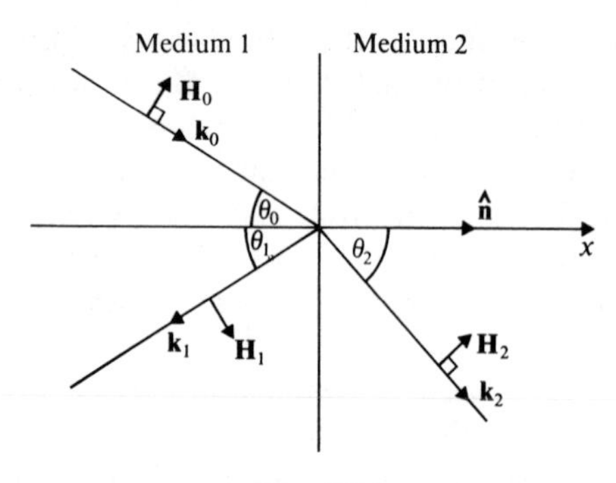

Fig. 10.1

geometry of the problem is represented in Fig. 10.1, the interface between the dielectrics being the plane $x=0$. Since we shall only need to satisfy the continuity of the tangential components of **E** and **H** at the boundary let us express these for the incident wave as

$$\mathbf{E}_0=\mathbf{E}_{00}e^{i(\mathbf{k}_0\cdot\mathbf{r}-\omega t)} \tag{10.1}$$

$$\mathbf{H}_0=\frac{\mathbf{k}_0\times\mathbf{E}_0}{\omega\mu_1} \tag{10.2}$$

where $\mathbf{k}_0$ is the arbitrary propagation vector of the wave. Similar expressions may be written for the wave reflected at the interface and for the wave in 2, i.e.

$$\mathbf{E}_1=\mathbf{E}_{10}e^{i(\mathbf{k}_1\cdot\mathbf{r}-\omega t)} \tag{10.3}$$

$$\mathbf{H}_1=\frac{\mathbf{k}_1\times\mathbf{E}_1}{\omega\mu_1} \tag{10.4}$$

$$\mathbf{E}_2=\mathbf{E}_{20}e^{i(\mathbf{k}_2\cdot\mathbf{r}-\omega t)} \tag{10.5}$$

$$\mathbf{H}_2=\frac{\mathbf{k}_2\times\mathbf{E}_2}{\omega\mu_2} \tag{10.6}$$

The boundary conditions must be satisfied at all points on the plane $x=0$ at all times, and so the phase factors of the field vectors must be equal; i.e. at $x=0$

$$\mathbf{k}_0\cdot\mathbf{r}=\mathbf{k}_1\cdot\mathbf{r}=\mathbf{k}_2\cdot\mathbf{r} \tag{10.7}$$

so that if we choose $\mathbf{k}_0$ in the Oxz plane it follows that

$$k_0\sin\theta_0=k_1\sin\theta_1=k_2\sin\theta_2$$

Moreover, since $k_0 = \omega\sqrt{\varepsilon_1\mu_1} = k_1$, $k_2 = \omega\sqrt{\varepsilon_2\mu_2}$, it follows that

$$\theta_0 = \theta_1 \tag{10.8}$$

and

$$k_0 \sin\theta_0 = k_2 \sin\theta_2$$

i.e.

$$\frac{\sin\theta_0}{\sin\theta_2} = \frac{k_2}{k_1} = \sqrt{\frac{\varepsilon_2\mu_2}{\varepsilon_1\mu_1}}$$

Since for non-magnetic materials we may take $\mu_2 = \mu_1$ this becomes

$$\frac{\sin\theta_0}{\sin\theta_2} = \sqrt{\frac{\varepsilon_2}{\varepsilon_1}} = \frac{n_2}{n_1} \tag{10.9}$$

The results (10.8), (10.9) are nothing other than the simple laws of geometric optics expressing the equality of the angles of incidence and reflection, and Snell's law which states that a light ray passing from one dielectric to another is refracted towards the normal to the interface if $\varepsilon_2 > \varepsilon_1$. To obtain information about the reflected and refracted radiation we must apply the boundary conditions summarized in §10.1 at $x = 0$. Omitting the factor $e^{-i\omega t}$ and taking $\hat{\mathbf{n}}$ to be the unit normal to the interface directed along Ox, these give

$$(\mathbf{E}_0 + \mathbf{E}_1) \times \hat{\mathbf{n}} = \mathbf{E}_2 \times \hat{\mathbf{n}} \tag{10.10}$$

$$(\mathbf{H}_0 + \mathbf{H}_1) \times \hat{\mathbf{n}} = \mathbf{H}_2 \times \hat{\mathbf{n}} \tag{10.11}$$

i.e.

$$(\mathbf{k}_0 \times \mathbf{E}_0 + \mathbf{k}_1 \times \mathbf{E}_1) \times \hat{\mathbf{n}} = (\mathbf{k}_2 \times \mathbf{E}_2) \times \hat{\mathbf{n}} \tag{10.11a}$$

since $\mu_1 = \mu_2$.

We must now consider two cases in which

(i) $\mathbf{E}$ is polarized perpendicular to the plane of incidence, i.e. the plane defined by the propagation vector of the wave, $\mathbf{k}_0$, and the normal to the interface, $\hat{\mathbf{n}}$.

(ii) $\mathbf{E}$ lies in the plane of incidence.

The general case is a linear combination of these two.

Case (i). In the first case we find on applying (10.10), (10.11a) that

$$E_{00} + E_{10} = E_{20} \tag{10.12}$$

$$(k_0 E_{00} \cos\theta_0 - k_1 E_{10} \cos\theta_0) = k_2 E_{20} \cos\theta_2 \tag{10.13}$$

From these we find that

$$\frac{E_{10}}{E_{00}}=\frac{\sqrt{\varepsilon_1}\cos\theta_0-\sqrt{\varepsilon_2}\cos\theta_2}{\sqrt{\varepsilon_1}\cos\theta_0+\sqrt{\varepsilon_2}\cos\theta_2} \tag{10.14}$$

$$\frac{E_{20}}{E_{00}}=\frac{2\sqrt{\varepsilon_1}\cos\theta_0}{\sqrt{\varepsilon_1}\cos\theta_0+\sqrt{\varepsilon_2}\cos\theta_2} \tag{10.15}$$

It is apparent from these expressions that while E_{20}/E_{00} is always positive, the sign of E_{10}/E_{00} depends on the nature of the dielectrics. If the dielectric in 2 has a larger refractive index than that in 1, it follows from (10.14) that $E_{10}/E_{00}<0$ so that reflection of an electromagnetic wave from such an interface results in a phase change of π in the electric field. In the opposite case of reflection from a medium of lower refractive index (i.e. $n_2<n_1$) there is no change of sign in $\mathbf{E}$.

The expressions in (10.14), (10.15) may be rewritten, using Snell's law, in the form

$$\frac{E_{10}}{E_{00}}=\frac{\sin(\theta_2-\theta_0)}{\sin(\theta_2+\theta_0)} \tag{10.16}$$

$$\frac{E_{20}}{E_{00}}=\frac{2\sin\theta_2\cos\theta_0}{\sin(\theta_2+\theta_0)} \tag{10.17}$$

These expressions are known as the **Fresnel relations.**

It is now a straightforward matter to use the Poynting vector to obtain information about the power transmitted and reflected at the interface. From §9.8 we have

$$\langle\mathbf{S}\rangle=\frac{1}{2}\operatorname{Re}(\mathbf{E}\times\mathbf{H}^*)$$

so that if we define the **reflection coefficient** as the energy flux reflected from the interface relative to that incident on this surface, we find

$$R_\perp=-\frac{\hat{\mathbf{n}}\cdot\langle\mathbf{S}_1\rangle}{\hat{\mathbf{n}}\cdot\langle\mathbf{S}_0\rangle}=\frac{|\mathbf{E}_1\times\mathbf{H}_1^*|}{|\mathbf{E}_0\times\mathbf{H}_0^*|}=\frac{|\mathbf{E}_{10}|^2}{|\mathbf{E}_{00}|^2} \tag{10.18}$$

where the subscript $\perp$ is added to distinguish the fact that $\mathbf{E}$ is polarized perpendicular to the plane of incidence. Similarly the

transmission coefficient $T_\perp$ is defined by

$$T_\perp = \frac{\hat{\mathbf{n}} \cdot \langle \mathbf{S}_2 \rangle}{\hat{\mathbf{n}} \cdot \langle \mathbf{S}_0 \rangle} = \frac{n_2 \cos\theta_2}{n_1 \cos\theta_0} \frac{|E_{20}|^2}{|E_{00}|^2} \tag{10.19}$$

In terms of (10.16), (10.17), together with Snell's law, we find from (10.18) and (10.19)

$$R_\perp = \frac{\sin^2(\theta_2 - \theta_0)}{\sin^2(\theta_2 + \theta_0)} \tag{10.20}$$

$$T_\perp = \frac{\sin 2\theta_0 \sin 2\theta_2}{\sin^2(\theta_2 + \theta_0)} \tag{10.21}$$

from which it is clear that $R_\perp + T_\perp = 1$, as it must, since we assumed at the outset that no energy losses occur.

For normal incidence these expressions take on simple forms:

$$R = \left(\frac{n_2 - n_1}{n_2 + n_1}\right)^2; \qquad T = \frac{n_2}{n_1}\left(\frac{2n_1}{n_2 + n_1}\right)^2 \tag{10.22}$$

which are analogous to the expressions obtained for waves on a string with a density discontinuity.

Case (ii). In this case **E** lies in the plane of incidence (Fig. 10.2) so that applying the boundary conditions on **E** and **H** gives

$$(E_{00} - E_{10}) \cos\theta_0 = E_{20} \cos\theta_2 \tag{10.23}$$

$$k_0 E_{00} + k_1 E_{10} = k_2 E_{20} \tag{10.24}$$

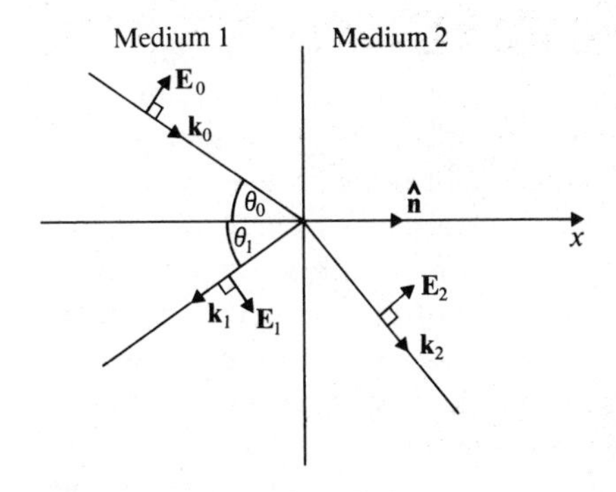

Fig. 10.2

From (10.23), (10.24) it follows that (problem 1)

$$\frac{E_{10}}{E_{00}}=\frac{\tan(\theta_0-\theta_2)}{\tan(\theta_0+\theta_2)} \tag{10.25}$$

and

$$\frac{E_{20}}{E_{00}}=\frac{2\sin\theta_2\cos\theta_0}{\sin(\theta_0+\theta_2)\cos(\theta_0-\theta_2)} \tag{10.26}$$

which are the Fresnel relations for this polarization. It is clear from (10.16) and (10.25) that there is an important distinction between the two states of polarization of **E**. From (10.16) we see that $(E_{10}/E_{00})_{\perp}\neq 0$ for a given $\theta_0(0<\theta_0<\pi/2)$ other than in the trivial case when $n_2=n_1$. However, it appears from (10.25) that $(E_{10}/E_{00})_{\parallel}$ and hence $R_{\parallel}$, will vanish when

$$\theta_0+\theta_2=\pi/2 \tag{10.27}$$

The angle of incidence determined by (10.27) is known as Brewster's angle, θ_B. Applying Snell's law for this angle of incidence gives

$$\frac{\sin\theta_B}{\sin(\pi/2-\theta_B)}=\frac{n_2}{n_1}$$

i.e.

$$\theta_B=\tan^{-1}\left(\frac{n_2}{n_1}\right) \tag{10.28}$$

Thus a wave polarized **in** the plane of incidence crosses an interface between two dielectrics without suffering reflection provided it is incident at the Brewster angle. If a wave is incident on the boundary at $\theta=\theta_B$ with both components of polarization present then only that component of the wave with **E** perpendicular to the plane of incidence will be reflected, i.e. the reflected wave will be linearly or **plane polarized** perpendicular to the plane of incidence. Thus θ_B is sometimes called the **polarizing angle**.

A simple physical interpretation of the behaviour of a wave incident at the Brewster angle can be given. Consider Fig. 10.3 in which an electromagnetic wave is incident from medium 1 (vacuum, say) on to medium 2 ($\varepsilon_2>\varepsilon_0$). The electrons in the material composing medium 2 are driven to oscillate at the frequency of the wave. We shall see in the following chapter that oscillating

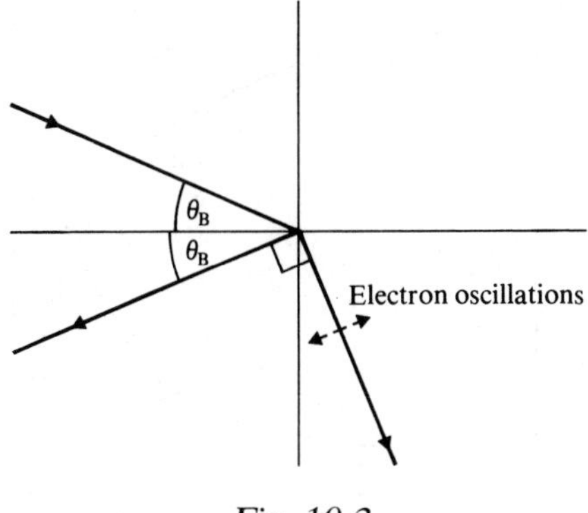

Fig. 10.3

electrons themselves **radiate** electromagnetic waves and the superposition of the incident radiation (electromagnetic wave) and that created by the oscillating electrons gives rise to the reflected and refracted waves. The direction of electron motion in the refracted wave is shown in Fig. 10.3 for a wave with **E** vector in the plane of incidence, and this direction is clearly parallel to the direction which would be followed by a reflected wave – if one existed. Now we shall see in Chapter 11 that electrons do not radiate in the direction of their motion and so the reflected wave which would be created by the reradiation from these electrons, does not materialize.

It is straightforward to calculate $R_{\parallel}$ and $T_{\parallel}$, as for the other polarization we found $R_{\perp}$, $T_{\perp}$ (cf. problem 10.2). Figure 10.4 plots the various coefficients in the case of an electromagnetic wave incident from a medium 1 **less dense** than medium 2.

One important application of the Brewster angle is to the design of a **Brewster window**, i.e. one with perfect transmission. At normal incidence just about 92 per cent (problem 7) of the incident intensity is transmitted through a glass window, i.e. about 4 per cent is lost at each surface. Typically in a gas laser with mirrors outside the windows one may have as many as 100 traverses of the window so that while a loss of 8 per cent per traverse might be tolerated, after a hundred or so there would be little light left! A way round this difficulty is found by adjusting the window so that the light is incident at the Brewster angle. The electric field component polarized parallel to the plane of incidence is transmitted perfectly and even after many traverses of

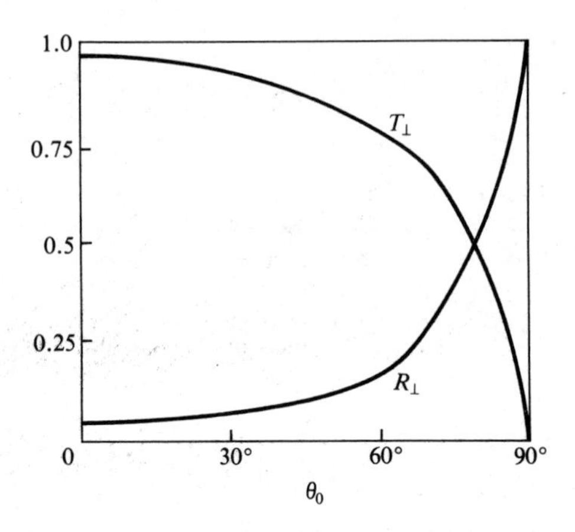

Fig. 10.4a

the window, has suffered negligible loss. On the other hand the field component polarized perpendicular to the plane of incidence is in part reflected and in part transmitted on each occasion so that after many traverses this component has been almost entirely eliminated from the beam. The light emitted by the laser is thus virtually 100 per cent linearly polarized.

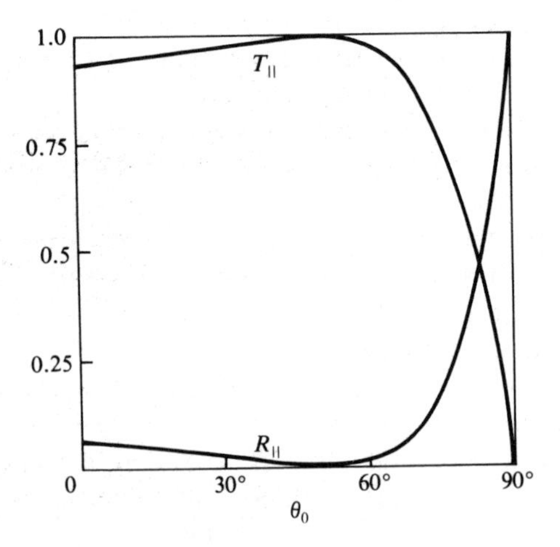

Fig. 10.4b

§10.3 Total reflection

In the previous section we considered radiation incident from medium 1 on medium 2 and generally we supposed that $n_2 > n_1$. We now turn to the opposite case when $n_1 > n_2$ and consider what takes place when θ_0 is increased steadily from zero. From Snell's law $\sin\theta_2 = (n_1/n_2)\sin\theta_0$, i.e. as θ_0 increases so does θ_2 until the point is reached at which $\theta_2 = \pi/2$. This happens whenever θ_0 attains a critical value θ_c for which $\sin\theta_c = n_2/n_1$. At this point the transmitted wave no longer propagates in medium 2 so that only a reflected wave remains. This phenomenon is known as **total reflection** (Fig. 10.5). If θ_0 is increased beyond the critical value θ_c, then clearly θ_2 becomes imaginary. This is not inconsistent with the analysis of §10.2 since we did not at any time suppose that the coefficients were real. Consider, for example, the case of an incident electromagnetic wave polarized so that **E** is normal to the plane of incidence. Then the incident and reflected wave amplitudes are described by (10.14). The electric field of the refracted wave is

$$\mathbf{E}_2 = \hat{\mathbf{y}} E_{20} e^{i(k_2 x \cos\theta_2 - k_2 z \sin\theta_2 - \omega t)} \tag{10.29}$$

and from the exponent we see that whereas $k_2 \sin\theta_2$ $(= k_1 \sin\theta_0)$ is real, the coefficient

$$k_2 \cos\theta_2 = (k_2^2 - k_1^2 \sin^2\theta_0)^{1/2} \tag{10.30}$$

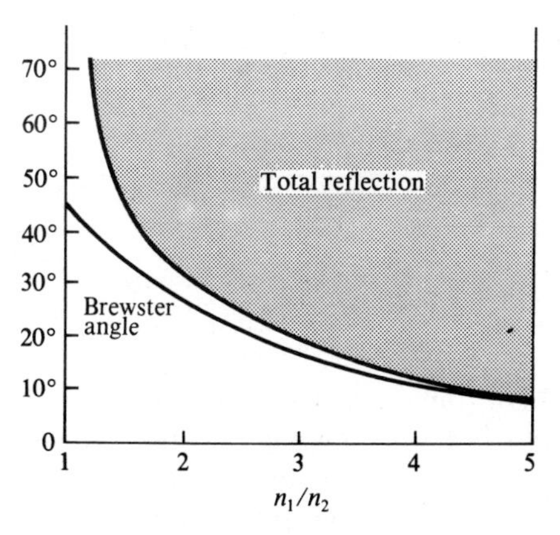

Fig. 10.5

i.e.

$$\cos\theta_2 = (1 - \sin^2\theta_0/\sin^2\theta_c)^{1/2} \tag{10.31}$$

is purely imaginary since the condition for total reflection to take place is $\theta_0 > \theta_c$. Thus writing

$$q = k_2\left(\frac{\sin^2\theta_0}{\sin^2\theta_c} - 1\right)^{1/2} \tag{10.32}$$

gives $k_2 \cos\theta_2 = \pm iq$ so that the refracted wave may now be written

$$\mathbf{E}_2 = \hat{\mathbf{y}} E_{20} e^{\pm qx} e^{-i(k_1 z \sin\theta_0 + \omega t)}$$

Since this expression applies to the region $0 \le x < \infty$, we must discard the positive sign, i.e.

$$\mathbf{E}_2 = \hat{\mathbf{y}} E_{20}{}^{-qx} e^{-i(k_1 z \sin\theta_0 + \omega t)} \tag{10.33}$$

So for $\theta > \theta_c$ the refracted wave propagates along the interface and is attentuated in the x direction, i.e. **into** the less dense dielectric (which is often air). To get some idea of the "penetration distance" $x_0 \sim q^{-1}$, consider light incident internally from glass to air at $\theta_0 = 45°$. For glass of refractive index 1.5, θ_c is given by $\sin^{-1}(2/3)$ so that the condition for total reflection is satisfied. Thus

$$x_0 \simeq \frac{1}{k_2}\left(\frac{\sin^2\theta_0}{\sin^2\theta_c} - 1\right)^{-1/2}$$

$$= \frac{\lambda_0}{2\pi}\left(\frac{1.5^2}{2} - 1\right)^{-1/2}$$

i.e.

$$x_0 \simeq 0.5\lambda_0$$

We see from this estimate that the electric fields are negligible at distances above a few wavelengths into the vacuum (medium 2).

Reflection and transmission coefficients are plotted in Fig. 10.6 for radiation incident internally from glass of refractive index 1.5 to air.

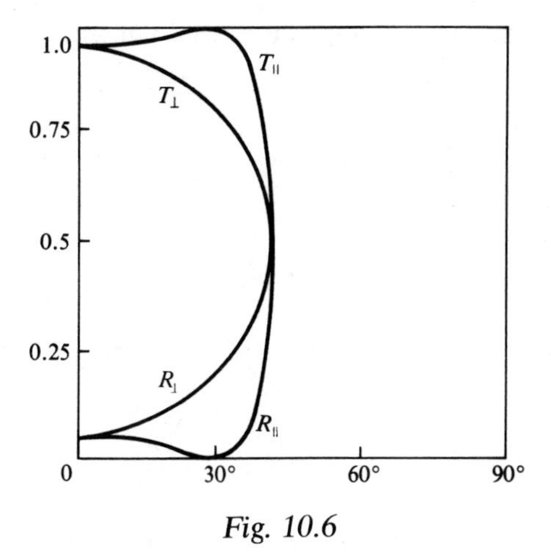

Fig. 10.6

§10.4 Reflection from a conductor: normal incidence

Let us now replace the dielectric (medium 2) by a uniform conductor and consider an electromagnetic wave incident normally on the interface $x=0$ (Fig. 10.7). For the incident and reflected wave we have, as in the dielectric case,

$$\mathbf{E}_0 = \hat{\mathbf{z}}E_{00}e^{i(kx-\omega t)}; \qquad \mathbf{H}_0 = \hat{\mathbf{y}}\sqrt{\frac{\varepsilon_1}{\mu_1}}\,E_{00}e^{i(kx-\omega t)} \tag{10.34}$$

$$\mathbf{E}_1 = -\hat{\mathbf{z}}E_{10}e^{-i(kx+\omega t)}; \qquad \mathbf{H}_1 = \hat{\mathbf{y}}\sqrt{\frac{\varepsilon_1}{\mu_1}}\,E_{10}e^{-i(kx-\omega t)} \tag{10.35}$$

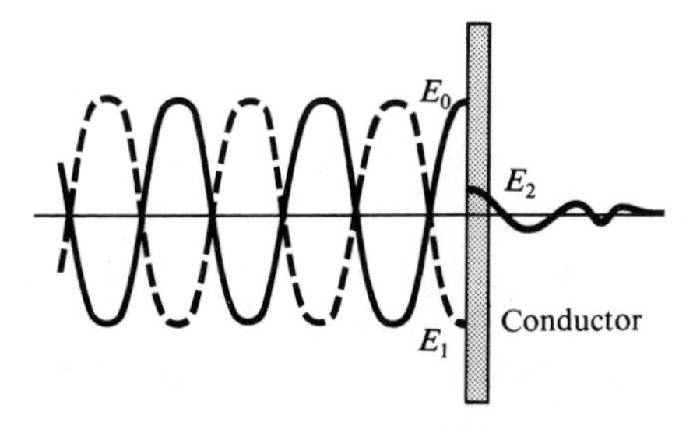

Fig. 10.7

However the wave in the conductor (medium 2) must now be expressed

$$\left.\begin{aligned} \mathbf{E}_2 &= \hat{\mathbf{z}} E_{20} e^{i(\kappa x - \omega t)} \\ \mathbf{H}_2 &= \hat{\mathbf{y}} \frac{\kappa E_{20}}{\omega \mu_2} e^{i(\kappa x - \omega t)} \end{aligned}\right\} \tag{10.36}$$

in which the propagation constant κ is now complex. From (9.55),

$$\kappa^2 = \omega^2(\varepsilon_2 \mu_2)\left[1 + \frac{i\sigma_2}{\varepsilon_2 \omega}\right] \tag{10.37}$$

Applying the boundary conditions as before gives

$$E_{00} - E_{10} = E_{20} \tag{10.38}$$

$$\sqrt{\frac{\varepsilon_1}{\mu_1}}(E_{00} + E_{10}) = \frac{\kappa}{\omega \mu_2} E_{20} \tag{10.39}$$

Since κ is complex it is apparent that E_{10} and E_{20} cannot both be real. Thus phase shifts other than 0 and π are to be expected for the reflected and transmitted waves. Solving (10.38), (10.39) gives

$$E_{10} = \frac{\dfrac{\kappa}{\omega \mu_2}\sqrt{\dfrac{\mu_1}{\varepsilon_1}} - 1}{\dfrac{\kappa}{\omega \mu_2}\sqrt{\dfrac{\mu_1}{\varepsilon_1}} + 1} E_{00}; \qquad E_{20} = \frac{2}{\dfrac{\kappa}{\omega \mu_2}\sqrt{\dfrac{\mu_1}{\varepsilon_1}} + 1} E_{00} \tag{10.40}$$

Substituting for κ from (10.37) we have

$$E_{10} = \frac{\sqrt{\dfrac{\varepsilon_2 \mu_1}{\varepsilon_1 \mu_2}\left(1 + \dfrac{i\sigma_2}{\varepsilon_2 \omega}\right)} - 1}{\sqrt{\dfrac{\varepsilon_2 \mu_1}{\varepsilon_1 \mu_2}\left(1 + \dfrac{i\sigma_2}{\varepsilon_2 \omega}\right)} + 1} E_{00}; \qquad E_{20} = \frac{2}{\sqrt{\dfrac{\varepsilon_2 \mu_1}{\varepsilon_1 \mu_2}\left(1 + \dfrac{i\sigma_2}{\varepsilon \omega}\right)} + 1} E_{00} \tag{10.41}$$

It is of interest to examine the case of a very good conductor in more detail. Then we may neglect displacement current compared with conduction current, i.e. $\sigma_2/\varepsilon_2\omega \gg 1$ (cf. §9.4) and

$$\kappa \simeq (1+i)\sqrt{\frac{\omega\sigma_2\mu_2}{2}}$$

so that (10.40) gives

$$E_{10} = \frac{(1+i)\sqrt{\dfrac{\sigma_2\mu_1}{2\omega\mu_2\varepsilon_1}}-1}{(1+i)\sqrt{\dfrac{\sigma_2\mu_1}{2\omega\mu_2\varepsilon_1}}+1}E_{00}; \qquad E_{20} = \frac{2}{(1+i)\sqrt{\dfrac{\sigma_2\mu_1}{2\omega\mu_2\varepsilon_1}}+1}E_{00} \tag{10.42a,b}$$

From (10.42a) we find, since $\sigma_2/\varepsilon_2\omega \gg 1$

$$E_{10} \simeq \left[1-(1-i)\sqrt{\frac{2\omega\mu_2\varepsilon_1}{\sigma_2\mu_1}}\right]E_{00} \tag{10.43}$$

The reflection coefficient is given by

$$R = \frac{|E_{10}|^2}{|E_{00}|^2} = 1-2\sqrt{\frac{2\omega\mu_2\varepsilon_1}{\sigma_2\mu_1}} \tag{10.44}$$

In the case of non-ferromagnetic metals (10.44) reduces to

$$\begin{aligned} R &= 1-2\sqrt{\frac{2\omega\varepsilon_1}{\sigma_2}} \\ &= 1-2\frac{\omega}{c}\delta n_1 \end{aligned} \tag{10.45}$$

These results are sometimes expressed in terms of a **surface impedance**.

§10.5 Guided electromagnetic waves

So far in this chapter we have examined boundary effects on the propagation of electromagnetic waves only in the sense of the waves being incident on a plane discontinuity, e.g. a dielectric–dielectric or a dielectric–metal interface. The dielectric or conducting material itself was assumed to be a semi-infinite slab. We now turn to a study of the behaviour of electromagnetic waves in the vicinity of conducting and dielectric boundaries when these boundaries are so configured as to guide the waves in specified directions. By a **guided wave** we mean one for which the direction of energy flow, as shown by the Poynting vector, follows the direction of the guiding system; in other words if the guide bends through some angle then the wave follows the geometry of the guide into a new direction of propagation.

Physically the importance of the boundaries is due to the fact that they determine the characteristics of a particular wave. Mathematically our aim is to find solutions of the wave equation which satisfy the physical boundary conditions imposed at walls of the guides, whether these be conductors or dielectrics. The subject of guided waves is of foremost importance in practical communication electronics, having evolved first for microwaves and, more recently, following the development of lasers, for light waves. Indeed a whole new subject, **optoelectronics**, has appeared, concerned with all aspects of optical communication. We shall discuss waveguides with conducting boundaries in §10.7 and dielectric waveguides in §10.9. In each instance we confine our attention to straight guides with uniform material and the discussion is restricted to some basic physical principles.

§10.6 Propagation of waves between conducting planes

As a simple extension of earlier work in this chapter let us consider electromagnetic wave propagation between parallel conducting planes (Fig. 10.8). Suppose that the space between the planes is a vacuum, i.e. $\varepsilon = \varepsilon_0$, $\mu = \mu_0$, $\sigma = 0$, $\rho = 0$ within this region. In §9.4 we showed that electromagnetic waves in free space are **transverse** with $(\mathbf{E}, \mathbf{B}, \mathbf{k})$ forming a right-handed orthogonal triad and we now label these TEM modes (**transverse electromagnetic**). There is no physical distinction between the x and z directions (since the conducting planes are infinite) and so there is no loss of generality in considering waves with wave-vector $\mathbf{k}_0$ in the plane Oyz. If we choose $\mathbf{E}$ to be in the x direction as in Fig. 10.8 then

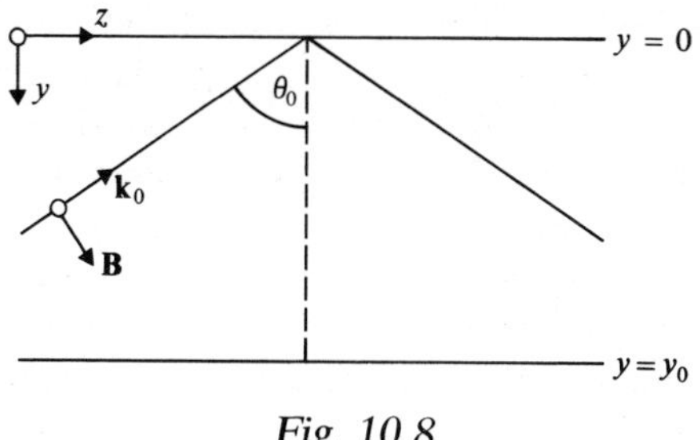

Fig. 10.8

B($\perp$**E**) lies in Oyz, i.e. $\mathbf{B}=B_y\hat{\mathbf{y}}+B_z\hat{\mathbf{z}}$ and so has a component in the direction to be followed by the guided wave, along Oz. A wave of this kind is known as a **transverse electric** or TE mode. Obviously if we choose the guided wave so that **B** is along Ox then **E** will have a longitudinal component, E_z, and such a wave is labelled **transverse magnetic** or TM. Waves propagating within **closed** conducting regions are either TE or TM modes – a TEM wave may not propagate (since it can have only the distributions of the corresponding two dimensional static problem and no electrostatic field can exist inside a source-free region completely enclosed by a conductor). For the wave in Fig. 10.8 we have

$$\mathbf{E}_0=\hat{\mathbf{x}}E_{00}e^{i(\mathbf{k}_0\cdot\mathbf{r}-\omega t)} \tag{10.46}$$

where

$$\mathbf{k}_0\cdot\mathbf{r}=k_0y\cos\theta_0+k_0z\sin\theta_0$$

and $k_0=\omega/c$. This wave will be reflected at the perfectly conducting boundary and, for the reflected wave,

$$\mathbf{E}_1=\hat{\mathbf{x}}E_{10}e^{-i(k_0y\cos\theta_0-k_0z\sin\theta_0-\omega t)} \tag{10.47}$$

For a perfect conductor $E_t=0$ at the boundary so that at $y=0$, $E_t=E_{00}+E_{10}=0$, i.e. the total electric field between the conducting planes is

$$\begin{aligned}\mathbf{E}&=\hat{\mathbf{x}}E_{00}(e^{ik_0y\cos\theta_0}-e^{-ik_0y\cos\theta_0})e^{i(k_0z\sin\theta_0-\omega t)}\\&=2i\hat{\mathbf{x}}E_{00}\sin(k_0y\cos\theta_0)e^{i(k_0z\sin\theta_0-\omega t)}\end{aligned} \tag{10.48}$$

Applying the boundary condition on **E** at $y=y_0$ gives

$$\sin(k_0y_0\cos\theta_0)=0$$

i.e.

$$k_0 y_0 \cos\theta_0 = n\pi, \ n \text{ integral,} \tag{10.49}$$

Defining an **effective wavelength** in the y direction by

$$\lambda_c = \frac{2\pi}{k_0 \cos\theta_0} = \frac{2\pi}{k_c} \tag{10.50}$$

then (10.49) becomes

$$\lambda_c = \frac{2y_0}{n} \tag{10.51}$$

We see that the **mode number** n and the separation of the conducting boundaries determine the values of the effective wavelength permitted by the boundary condition on **E**. In other words, for any given θ_0 only waves with frequencies $\{\omega_n\} = \{n\pi c/y_0 \cos\theta_0\}$ may propagate between the conducting planes.

We may also define a wavelength in the z direction by

$$\lambda_g = \frac{2\pi}{k_0 \sin\theta_0} = \frac{2\pi}{k_g} \tag{10.52}$$

From (10.50) and (10.52) it follows that

$$k_0^2 = k_c^2 + k_g^2 \tag{10.53}$$

Since

$$v_p = \frac{\omega}{k_g} = \frac{\omega}{k_0 \sin\theta_0} = \frac{c}{\sin\theta_0} \tag{10.54}$$

and

$$v_g = c \sin\theta_0$$

it follows that

$$v_p v_g = c^2 \tag{10.55}$$

From (10.54) we have

$$v_p = c\lambda_c(\lambda_c^2 - \lambda_0^2)^{-1/2} \tag{10.56}$$

which shows that for given λ_c (i.e. for some y_0 and a given mode number) as λ_0 increases, v_p increases and becomes infinite as $\lambda_0 \to \lambda_c$. At the same time $v_g \to 0$ and because of this limit to the propagation of the mode, λ_c is called the **cutoff wavelength**.

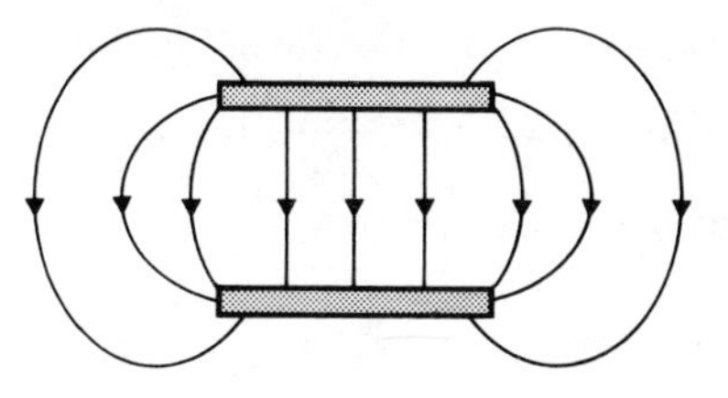

Fig. 10.9

From a practical standpoint parallel conducting planes could only be used to guide waves if their x dimension could be made small enough. In fact this may be done – such guides are known as **strip lines**. The electric field lines are sketched in Fig. 10.9 from which we see that the field is not entirely contained within the guide, leaking out at the edges. It is difficult to provide any simple mathematical discussion of a strip line, though, as with all other problems in wave propagation, we can solve the wave equation numerically.

§10.7 Waves in guides of rectangular cross-section

In this section we shall consider the propagation of waves inside guides of rectangular cross-section, the boundaries being taken to be perfect conductors. Our task is to solve the wave equations for the electric and magnetic fields within the guide subject to the appropriate boundary conditions. The waves travel as in §10.6 in the z direction and are confined in the x and y directions. We shall suppose that the guide is air-filled so that the wave equation to be solved is

$$\nabla^2\mathbf{E} = \frac{1}{c^2}\frac{\partial^2\mathbf{E}}{\partial t^2} \tag{10.57}$$

with an identical equation governing the magnetic field. The plane wave solutions are

$$\mathbf{E} = \mathbf{E}(x, y)\exp\left[i(k_g z - \omega t)\right] \tag{10.58}$$

where $\mathbf{E}(x, y)$ represents the field amplitude. Again, an identical expression may be written for the magnetic field. Defining a transverse Laplacian by $\nabla_T^2 = \nabla^2 - \dfrac{\partial^2}{\partial z^2}$, the wave equation in two dimensional form becomes

$$\left[\nabla_T^2 - k_g^2 + \frac{\omega^2}{c^2}\right]\mathbf{E}(x, y) = 0$$

i.e.

$$[\nabla_T^2 + k_c^2]\mathbf{E}(x, y) = 0 \tag{10.59}$$

The fields must satisfy Maxwell's equations:

$$\begin{aligned} \nabla \cdot \mathbf{E} &= 0 \qquad \nabla \times \mathbf{E} = i\omega \mathbf{B} \\ \nabla \cdot \mathbf{B} &= 0 \qquad \nabla \times \mathbf{B} = -\frac{i\omega}{c^2}\mathbf{E} \end{aligned} \tag{10.60}$$

If we now write $\mathbf{E} = \mathbf{E}_T + \mathbf{E}_z$, $\mathbf{B} = \mathbf{B}_T + \mathbf{B}_z$ it is a straightforward exercise to show that

$$\mathbf{E}_T = \frac{i}{k_c^2}[k_g \nabla_T E_z - k_0 c \hat{\mathbf{z}} \times \nabla_T B_z] \tag{10.61}$$

with a similarly structured expression for $\mathbf{B}_T$, thus showing that the transverse fields may be completely specified in terms of their z or longitudinal components. For $\mathbf{E}_T$ and $\mathbf{B}_T$ to remain finite for vanishing k_c, i.e. $k_g = k_0$, requires both E_z and B_z to be constant. We may set $E_z = 0 = B_z$ in this event, i.e. the wave is TEM and

$$\nabla_T^2 \mathbf{E} = 0; \qquad \nabla_T^2 \mathbf{B} = 0$$

Since each component of **E** satisfies Laplace's equation it follows that the surface of the guide is an equipotential; consequently **E** vanishes inside. It follows directly that a TEM mode cannot exist within a hollow guide with perfectly conducting walls. Note however that this result has **only** been established for a simply connected surface. With two or more (unconnected) surfaces as in a coaxial cable a TEM mode not only propagates but is in fact the dominant mode.

The boundary conditions to be satisfied at the surface of the guide are

$$E_t = 0; \qquad B_n = 0 \tag{10.62}$$

The wave equations for **E** and **B** together with these boundary conditions lead to eigenvalue problems. Thus for any prescribed ω only certain values of k_g will be consonant with the wave equation and the boundary conditions.

For TM modes $B_z = 0$ so that the boundary condition is $E_z = 0$. For TE modes $E_z = 0$ and the boundary condition, from (10.61), becomes $\dfrac{\partial B_z}{\partial n} = 0$. As an example let us consider TE modes in a guide of rectangular cross-section with dimensions $a, b\,(a > b)$. Then

$$\left[\frac{\partial^2}{\partial x^2} + \frac{\partial^2}{\partial y^2} + k_c^2\right] B_z(x, y) = 0 \tag{10.63}$$

subject to the boundary conditions

$$\left.\frac{\partial B_z}{\partial x}\right|_{x=0,a} = 0, \qquad \left.\frac{\partial B_z}{\partial y}\right|_{y=0,b} = 0 \tag{10.64}$$

From (10.63), (10.64) we find

$$B_z(x, y) = B_0 \cos\frac{m\pi x}{a} \cos\frac{n\pi y}{b} \tag{10.65}$$

The mode numbers m, n determine the cut-off frequency

$$\omega_{mn} = \pi c\left(\frac{m^2}{a^2} + \frac{n^2}{b^2}\right)^{1/2} \tag{10.66}$$

For $m = 0$, $n = 0$ (corresponding to the TEM mode) we get a trivial solution. Since $a > b$ the lowest cut-off frequency corresponds to $m = 1$, $n = 0$, i.e. $\omega_{10} = \dfrac{\pi c}{a}$. This is the **dominant** or principal TE mode, denoted by TE_{10}. For any prescribed frequency only modes for which $\omega_{mn} < \omega$ will propagate without attenuation. In practice one often wants only the principal mode to propagate since multimode operation is inefficient for many purposes. We see from (10.66) that this may be achieved by a judicious choice of guide dimensions.

The TE_{10} mode is important in engineering practice for a number of reasons. Since its cut-off frequency is independent of one of the dimensions of the guide this allows us – for any given ω – to make this dimension small enough for **only** the TE_{10} mode to propagate. TE_{10} field lines are plotted in Fig. 10.10a; electric field lines are continuous while magnetic field lines are represented as dashed.

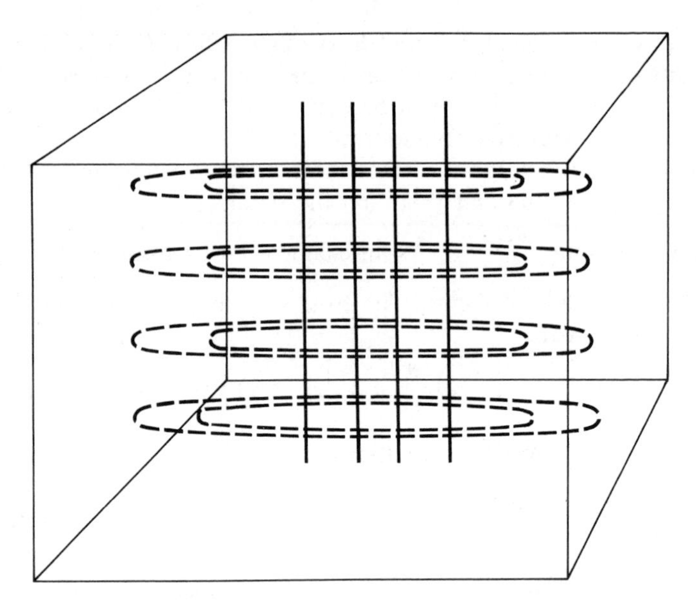

Fig. 10.10a

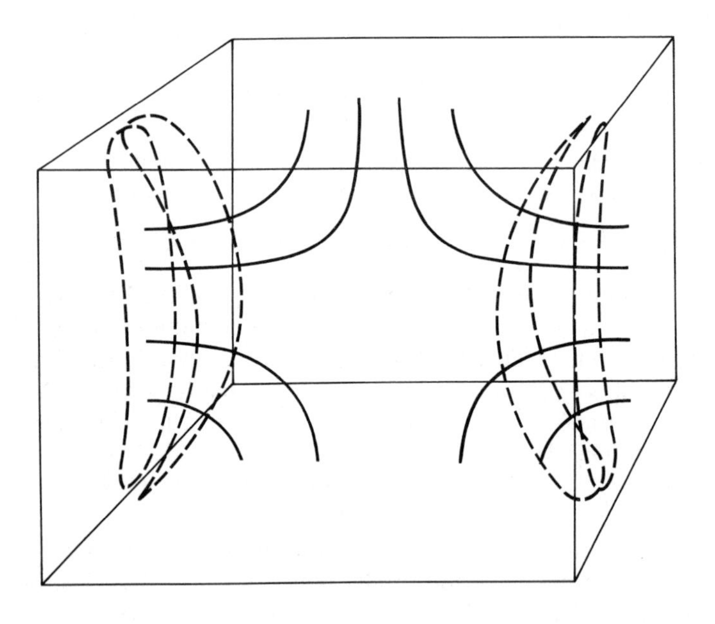

Fig. 10.10b

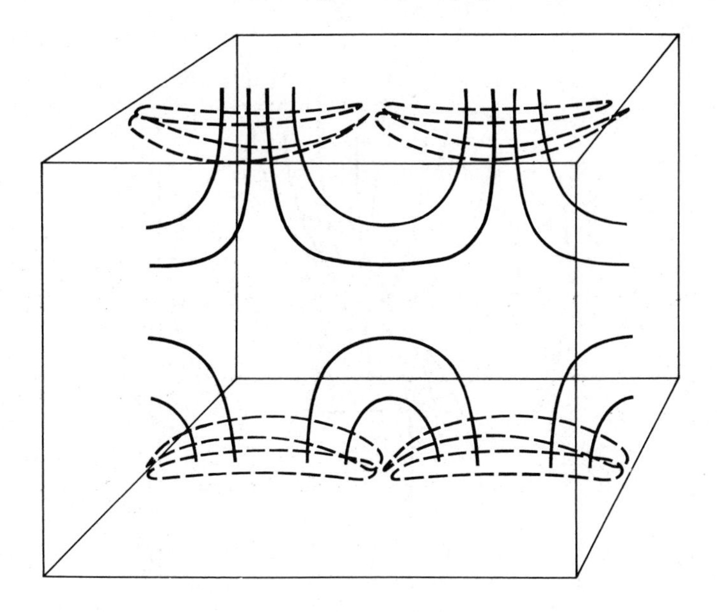

Fig. 10.10c

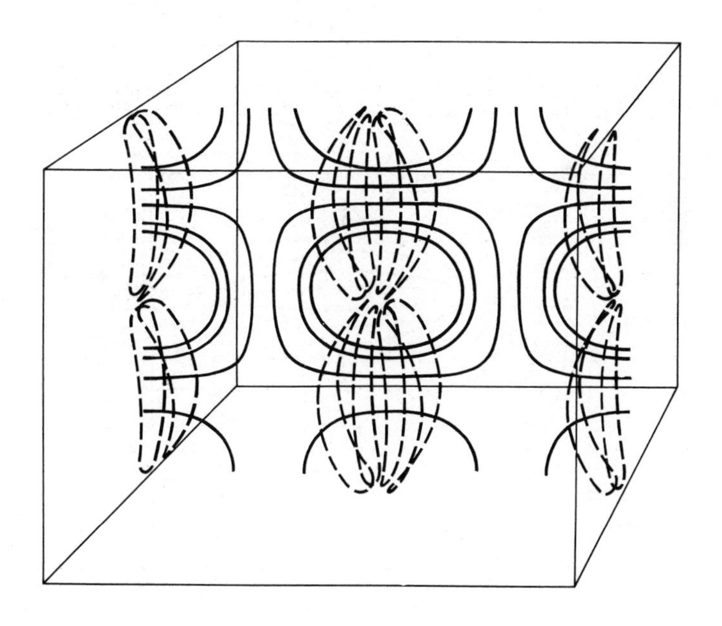

Fig. 10.10d

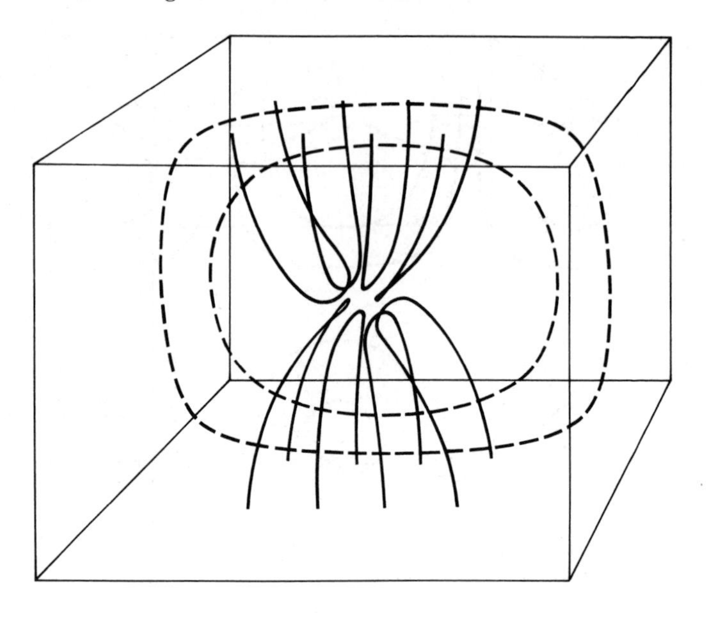

Fig. 10.11a

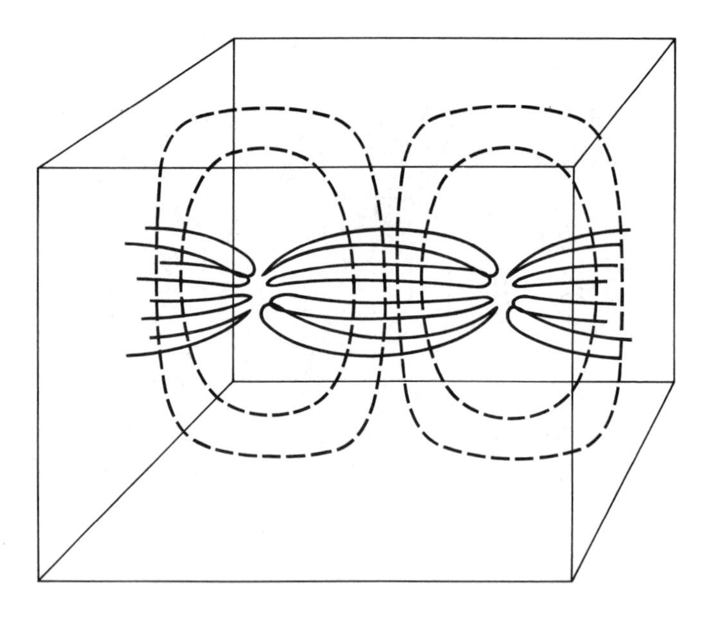

Fig. 10.11b

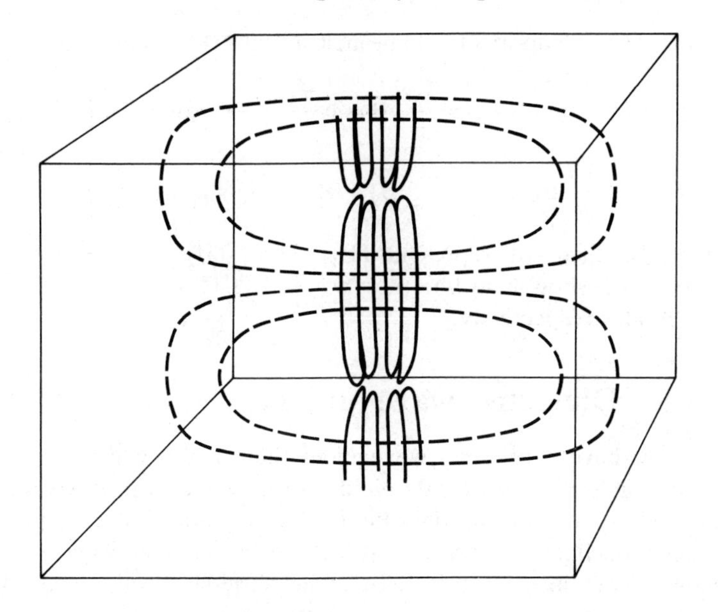

Fig. 10.11c

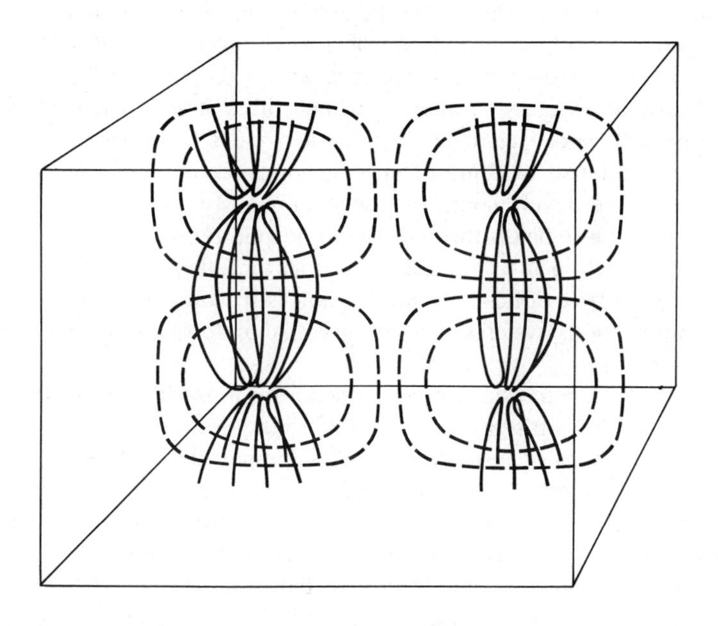

Fig. 10.11d

practice. The components of the field for the TE_{10} mode are

$$\left.\begin{aligned} E_x &= 0; \quad E_y = \frac{ikca}{\pi} B_0 \sin \frac{\pi x}{a} \quad E_z = 0; \\ B_x &= -\frac{ik_g a}{\pi} B_0 \sin \frac{\pi x}{a}; \quad B_y = 0; \quad B_z = B_0 \cos \frac{\pi x}{a} \end{aligned}\right\} \tag{10.67}$$

Figs. 10.10(b) – (d) represent TE_{11}, TE_{21}, TE_{22} modes. Figs. 10.11(a) – (d) show field line plots for TM_{11}, TM_{21}, TM_{12} and TM_{22} modes respectively.

§10.8 Dielectric waveguides

So far we have considered waveguides the walls of which are perfect conductors so that the fields are contained within the guide. These are not the only kinds of waveguide. For example, transmission lines also serve to direct electromagnetic waves. Again a wave may be guided by a dielectric slab which will have some similarities with the waveguides discussed in §§10.6–10.7 but also some differences on account of the different boundary conditions to be applied. The dielectric guide is a consequence of the property discussed in §10.3 of a wave propagating in a dielectric and incident on an interface with a second and less dense dielectric at an angle greater than the critical angle, when all the energy will be reflected. Dielectric waveguides have become of major interest in connection with optical communication following the development of lasers. Obviously one cannot simply transmit a laser beam through the atmosphere on account of the severe distortion it will suffer as well as the effects of diffraction. Consequently if optical communication is to compete with present communications systems it is essential to guide the light in some way.

One of the early approaches was to use a gas lens in which a pipe was filled with gas and a suitable temperature gradient established by means of heat sources. This temperature gradient caused a density gradient and consequently a radial gradient in the refractive index which provided a degree of guiding along the axis. This relatively cumbersome approach was superseded by using light "pipes" made of glass fibre. In these the guiding structure consists basically of a dielectric core of given refractive index surrounded by a cladding layer with a slightly lower refractive

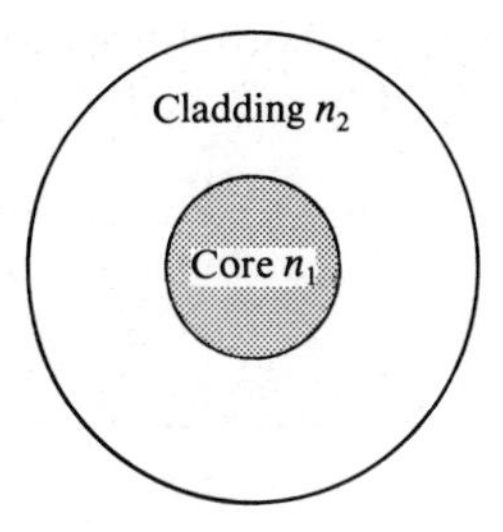

Fig. 10.12

index (Fig. 10.12). Such structures carry propagating optical modes, these modes being internally reflected at the core-cladding interface.

Optical fibres have features in common with the metallic waveguides discussed in §10.7. Both, for example, can support a finite number of guided modes at any given frequency. However, if the diameter of the core is much larger than the wavelength of the light then a large number of guided modes is possible. It is often desirable to limit the number of guided modes and, as mentioned already in §10.7, this may be achieved by a judicious choice of guide dimensions. The dimensions of the core which allow single mode operation depend critically on the ratio of refractive indices n_1/n_2. The larger this is, the smaller a_1 must be to ensure that only one guided mode can propagate. However, the inner core can be much larger (typically several microns) and still permit single mode operation if we choose n_1/n_2 to be close to unity.

In the following section we consider an example of dielectric waveguide propagation. To avoid mathematical complication we shall confine our attention to a plane slab guide rather than an optical fibre itself.

§10.9 Guided modes of an asymmetric slab waveguide

Consider an infinite, plane dielectric sheet surrounded by material of lower refractive index in the sandwich structure shown in Fig. 10.13. The outer layers which form the cladding are often referred to as the substrate (2) and cover or superstrate (3). Air is the

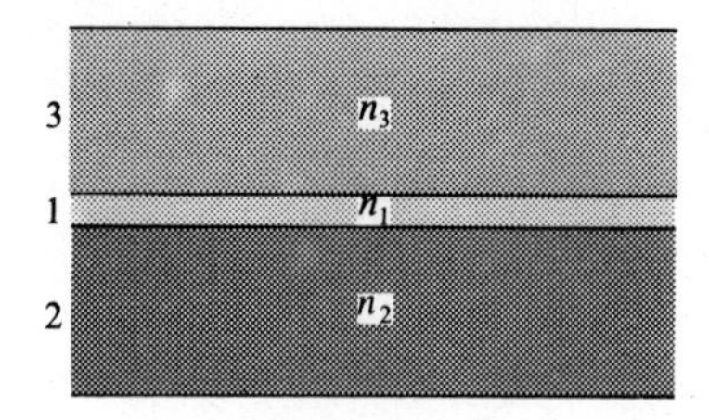

Fig. 10.13

most usual cover. The filling in the sandwich is a thin film (1) of thickness d which is of the order of the optical wavelength. All three layers should be weak absorbers of the propagating light, and to enable us to meet the condition for internal reflection, we must ensure that the dielectric constant of the film must be positive and larger than those of the cladding materials. The films used in practice are made from a wide variety of materials such as sputtered glasses, sputtered zinc oxide (ZnO), gallium arsenide (GaAs) as well as liquid crystals. Planar guides have also been formed by ion bombardment, which creates a thin layer of higher refractive index at the surface of a sample, so that the bulk sample then serves as the substrate.

We therefore assume that the guiding condition

$$n_1 > n_2 \geq n_3$$

is satisfied. If we assume a TE mode propagating along the z direction then $\mathbf{E} = (0, E_y e^{i(\omega t - \beta z)}, 0)$ and the wave equation may be written as

$$\frac{\partial^2 E_y}{\partial x^2} + (n^2 k^2 - \beta^2) E_y = 0 \qquad (10.68)$$

where β is the propagation constant in the z direction and is defined as

$$\beta = n_1 k \cos\theta \qquad (10.69)$$

For a guided mode, the fields must vanish at $x = \pm\infty$. A solution of (10.68) is

$$\left.\begin{aligned} E_y &= A e^{-\delta x} e^{i(\omega t - \beta z)} && x \geq 0 \\ &= (B \cos Kx + C \sin Kx)^{i(\omega t - \beta z)} && 0 \geq x \geq -d \\ &= D e^{\gamma x} e^{i(\omega t - \beta z)} && x \leq -d \end{aligned}\right\} \qquad (10.70)$$

where

$$\left.\begin{aligned}\delta^2 &= \beta^2 - n_3^2k^2 \ (>0)\\ K^2 &= n_1^2k^2 - \beta^2 \ (>0)\\ \gamma^2 &= \beta^2 - n_2^2k^2 \ (>0)\end{aligned}\right\} \tag{10.71}$$

Since

$$n_1 > n_2 \geq n_3$$

it follows that

$$n_3^2k^2 < \beta^2 < n_1^2k^2; \qquad n_3^2k^2 < n_1^2k^2 \cos^2\theta; \qquad n_3^2 < n_1^2 \cos^2\theta$$

which is the condition for total internal reflection at the upper boundary.

Snell's law requires β to be the same in all three media. Matching the tangential components of the electric field at $x=0$ and $x=-d$ gives

$$A = B \qquad (x=0)$$

$$B\cos Kd - C\sin Kd = De^{-\gamma d} \qquad (x=-d)$$

From Maxwell's equations we can now find the tangential component of the **H**-field, namely

$$\left.\begin{aligned}H_z &= \frac{-i\delta}{\omega\mu_0}[Ae^{-\delta x}e^{i(\omega t-\beta z)}] && x \geq 0\\ &= \frac{iK}{\omega\mu_0}[-A\sin Kx + C\cos Kx]e^{i(\omega t-\beta z)} && 0 \geq x \geq -d\\ &= \frac{i\gamma}{\omega\mu_0}[A\cos Kd - C\sin Kd]e^{\gamma(x+d)}e^{i(\omega t-\beta z)} && x \leq -d\end{aligned}\right\} \tag{10.72}$$

Applying the boundary condition on **H**, i.e. continuity of H_t across an interface, we have

$$-\delta A = KC \tag{10.73}$$

$$K[A\sin Kd + C\cos Kd] = \gamma[A\cos Kd - C\sin Kd] \tag{10.74}$$

For a non-trivial solution of these equations in A and C their system determinant must vanish, giving

$$\tan Kd = \frac{K(\gamma+\delta)}{K^2-\gamma\delta} \tag{10.75}$$

Equation (10.75) is the eigenvalue equation which can be solved either graphically or numerically [see program at the end of this section]. The equation determines β^2, which is equivalent to saying that we can have waves propagating in either the positive or negative z direction. From β, we can find θ, which is the mode angle at which energy must be coupled into the guide.

Cut-off in the guide is when $\gamma=0$, which corresponds to a loss of total internal reflection at the lower boundary. At this point γ becomes imaginary which results in radiation from the guide. The mode is now no longer guided (substrate radiation). The coefficients A and C can be found by relating them to the power in the guide. The time-averaged Poynting vector is $\frac{1}{2}(\mathbf{E}\times\mathbf{H}^*)$. The power, P, flowing through a plane perpendicular to the z axis is

$$P=\frac{1}{2}\, Re\int_{-\infty}^{\infty} \hat{\mathbf{k}}\cdot(\mathbf{E}\times\mathbf{H}^*)dx\, dy$$

$$=\frac{\beta}{2\omega\mu_0}\int_{-\infty}^{\infty} |E_y|^2 dx\, dy$$

($\hat{\mathbf{k}}$ is the unit vector in the z direction). It can be shown that

$$A^2=\frac{4K^2\omega\mu_0 P}{|\beta|\,[d+\gamma^{-1}+\delta^{-1}][K^2+\delta^2]}$$

In this case P is actually the power/unit length in the y direction.

A computer program which solves the eigenvalue equation (10.75) is given below. The refractive indices of the three media, wave number, mode number and guide thickness are required as input data. Typical values for these quantities may be found in the following section and in problems 21 and 22. After solving the eigenvalue equation, the program calculates the electric field distribution across the guide (see the example in §10.10).

```
      PROGRAM WAVGDS
      REAL NKSQ(3)
      REAL KAPPA
C--------------------------------------------------
C     STATEMENT FUNCTIONS
C     NOTE ALL QUANTITIES HAVE BEEN NORMALISED TO K(WAVNUM)
C     SO DELTA IS REALLY DELTA/K ETC.
      DELTA(BSQ)=SQRT(ABS(BSQ-NKSQ(3)))
      KAPPA(BSQ)=SQRT(ABS(NKSQ(1)-BSQ))
      GAMMA(BSQ)=SQRT(ABS(BSQ-NKSQ(2)))
      FUNC(DELTA,KAPPA,GAMMA)=ATAN2(DELTA,KAPPA)+ATAN2(GAMMA,KAPPA)
     + -DK*KAPPA+MODE*PI
      FDASH(DELTA,KAPPA,GAMMA)=0.5*(DK+1./GAMMA +1./DELTA)/KAPPA
C--------------------------------------------------
      DATA PI,NTTY/3.141592654,5/
C     WRITE INTRODUCTORY MESSAGE, AND GET REFRACTIVE INDICES
10    WRITE(NTTY,9000)
      DO 20 I=1,3
      WRITE(NTTY,9010)I
      READ(NTTY,9020)R
      NKSQ(I)=R*R
20    CONTINUE
C     TEST IF THEY OBEY THE ORDERING CONDITIONS
      IF (NKSQ(2).LE.NKSQ(1))GOTO 40
30    WRITE(NTTY,9030)
      GOTO 10
40    IF (NKSQ(3).GT.NKSQ(2))GOTO 30
C--------------------------------------------------
C     NOW INPUT THE OTHER DATA REQUIRED
      WRITE(NTTY,9040)
      READ(NTTY,9050)WAVNUM,MODE
50    WRITE(NTTY,9060)
      READ(NTTY,9070)D
C--------------------------------------------------
C     NOW TEST FOR THICKNESS BELOW CUTOFF
      D3=DELTA(NKSQ(2))
      D2=GAMMA(NKSQ(2))
      D1=KAPPA(NKSQ(2))
      DCUTOF=(ATAN2(D3,D1)+MODE*PI)/(D1*WAVNUM)
      IF (D.GT.DCUTOF)GOTO 100
C     BELOW CUTOFF- WRITE OUT A MESSAGE AND GET NEW DATA
      WRITE(NTTY,9080)DCUTOF
      GOTO 50
C--------------------------------------------------
100   CONTINUE
C     INITIAL GUESS IS MIDPOINT OF ADMISSIBLE VALUES
      DK=D*WAVNUM
      BSQOLD=(NKSQ(1)+NKSQ(2))*0.5
C     PERFORM 10 ITERATIONS USING NEWTON'S METHOD
      DO 110 IC=1,10
      D3=DELTA(BSQOLD)
      D1=KAPPA(BSQOLD)
      D2=GAMMA(BSQOLD)
      BSQNEW=BSQOLD-FUNC(D3,D1,D2)/FDASH(D3,D1,D2)
C     TEST IF NEW VALUE IS PERMISSIBLE-IF NOT PRINT OUT
C     NO CONVERGENCE MESSAGE
      IF (BSQNEW.GT.NKSQ(1))GOTO 130
      IF (BSQNEW.LT.NKSQ(2))GOTO 130
      IF (ABS(BSQNEW-BSQOLD).LT.1.0E-5)GOTO 140
110   BSQOLD=BSQNEW
C--------------------------------------------------
130   WRITE(NTTY,9090)
```

```
      STOP
C-------------------------------------------------
140   CONTINUE
C     ROOT HAS CONVERGED
      BETA=SQRT(BSQNEW)*WAVNUM
      DKAPPA=KAPPA(BSQNEW)
      DDELTA=DELTA(BSQNEW)
      DGAMMA=GAMMA(BSQNEW)
      DBETA=SQRT(BSQNEW)
      COSTHT=SQRT(BSQNEW/NKSQ(1))
      THETA=ACOS(COSTHT)*180./PI
      WRITE(NTTY,9100)(I,NKSQ(I),I=1,3),DCUTOF,D,WAVNUM,MODE
      WRITE(NTTY,9110)DK,BETA,DKAPPA,DDELTA,DGAMMA
      WRITE(NTTY,9120)DBETA,THETA
      XMIN=-2.*DK
      XMAX=2.*DK
C     ESTIMATE YMAX; THE MAX FIELD
C     USE EXPRESSION FOR H FIELD( DIFFERENTIAL OF E FIELD)
C     EQUATION 10.72=0 SOLVE FOR X AND USE THIS FOR EMAX.
C     ASSUMES MAX IS IN THE GUIDE SECTION.
      YMAX=ATAN2(DDELTA,DKAPPA)
      YMAX=COS(YMAX)-DDELTA/DKAPPA*SIN(YMAX)
      YMAX=AINT(ABS(YMAX)*2.+.9)
      YMIN=-YMAX
      WRITE(NTTY,9130)YMAX
C-------------------------------------------------
C     NOW INITIALISE AND SCALE THE SCREEN
      CALL INITAL
      CALL SCAL(XMIN,XMAX,YMIN,YMAX)
C-------------------------------------------------
C     NOW DRAW THE X-AXIS
      CALL POINT(XMIN,0.)
      CALL VECT(XMAX-XMIN,0.)
C     NOW DRAW THE Y-AXIS
      CALL POINT(0.,YMIN)
      CALL VECT(0.,YMAX-YMIN)
C     NOW A LINE AT X=-D I.E. TO REPRESENT THE BUFFER
      CALL POINT(-DK,YMIN)
      CALL VECT(0.,YMAX-YMIN)
C-------------------------------------------------
C     LABEL THE YAXIS
      CALL POINT(0.,.9*YMAX)
      CALL TTEXT(8,8HEY-FIELD)
C     LABEL THE REGIONS
      YMID=.7*YMAX
      CALL POINT((-DK+XMIN)/2.,YMID)
      CALL TTEXT(3,3HN2=)
      R=SQRT(NKSQ(2))
      CALL NUM(R)
      CALL POINT(-.5*DK,YMID)
      CALL TTEXT(3,3HN1=)
      R=SQRT(NKSQ(1))
      CALL NUM(R)
      CALL POINT(.5*XMAX,YMID)
      CALL TTEXT(3,3HN3=)
      R=SQRT(NKSQ(3))
      CALL NUM(R)
C-------------------------------------------------
C     DRAW THE RESULTING EY FIELD
C     USING THE APPROPRIATE FORMULAE IN EACH REGION
      XK=XMIN
      DXK=DK/20.
```

```
      Z=COS(DK*DKAPPA)+DDELTA/DKAPPA*SIN(DK*DKAPPA)
      EOBA=Z*EXP(DGAMMA*(XK+DK))
      CALL POINT(XK,EOBA)
      IMAX=INT((-DK-XMIN)/DXK)
      DO 230 I=1,IMAX
      XK=XK+DXK
      E=Z*EXP(DGAMMA*(XK+DK))
      CALL VECT(DXK,E-EOBA)
      EOBA=E
230   CONTINUE
C-------------------------------------------------
      IMAX=DK/DXK
      DO 240 I=1,IMAX
      XK=XK+DXK
      E=COS(DKAPPA*XK)-DDELTA/DKAPPA*SIN(DKAPPA*XK)
      CALL VECT(DXK,E-EOBA)
      EOBA=E
240   CONTINUE
C-------------------------------------------------
      IMAX=INT(XMAX/DXK)
      DO 250 I=1,IMAX
      XK=XK+DXK
      E=EXP(-DDELTA*XK)
      CALL VECT(DXK,E-EOBA)
      EOBA=E
250   CONTINUE
      CALL GREND
      STOP
C-------------------------------------------------
9000  FORMAT(1H ,35HINPUT REFRACTIVE INDICES, N1>N2>=N3)
9010  FORMAT(1H ,17HREFRACTIVE INDEX ,I1,5H IS: ,$)
9020  FORMAT(F6.4)
9030  FORMAT(1H ,30HREFRACTIVE INDICES INCORRECTLY,
     +21HORDERED PLEASE RETYPE)
9040  FORMAT(1H ,40HTYPE THE WAVE-NUMBER (IN RADIANS/MICRON),
     + 22H AND THE MODE NUMBER: ,$)
9050  FORMAT(F7.4,I2)
9060  FORMAT(1H ,47HWHAT IS THE THICKNESS OF THE GUIDE IN MICRONS: ,$)
9070  FORMAT(F7.3)
9080  FORMAT(1H,40HTHIS VALUE OF THICKNESS IS BELOW CUTOFF.,
     + /,1H ,29HINPUT A VALUE FOR THICKNESS >,F6.4)
9090  FORMAT(1X,39HROOT NON-CONVERGENT AFTER 10 ITERATIONS)
9100  FORMAT(1H0,25HRESULTS OF WAVGDS PROGRAM,/,
     + 1H ,32HREFRACTIVE INDICES SQUARED ARE:-,/,
     + 3(1H ,I2,10X,F6.4,/),
     + 1H ,18HCUTOFF THICKNESS :,F7.4,9H  MICRONS,/,
     + 1H ,18HGUIDE THICKNESS  :,F7.4,9H  MICRONS,/,
     + 1H ,18HWAVE NUMBER      :,F7.4,16H  RADIANS/MICRON,/,
     + 1H ,18HMODE NUMBER      :,I2)
9110  FORMAT(1H0,9HD*WAVNUM:,F8.4,/,1H ,9HBETA    :,F8.4,/,
     + 1H ,9HKAPPA/K :,F8.4,/,1H ,9HDELTA/K :,F8.4,/,
     + 1H ,9HGAMMA/K :,F8.4)
9120  FORMAT(1H ,34HBETA/K EFFECTIVE REFRACTIVE INDEX:,F7.4,/,
     + 1H ,9HTHETA   :,F9.2,8H DEGREES)
9130  FORMAT(1H ,9HEY MAX  :,F8.2)
C-------------------------------------------------
      END
```

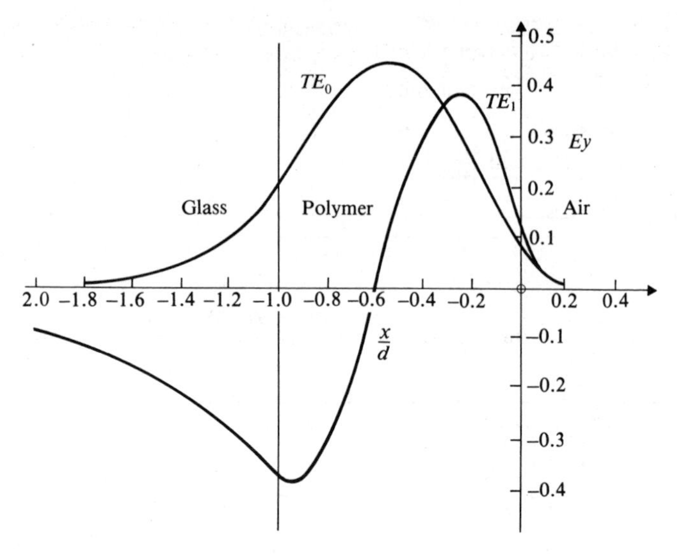

Fig. 10.14

§10.10 Example

Consider a polymer film ($n_1 = 1.588$) and of thickness 1.1 microns (μm) deposited on a glass substrate ($n_2 = 1.513$) with the top surface of the film exposed to air ($n_3) = 1.000$). Using the program in §10.9 we find that for light with wave number 9.29 rads/μm, this guide is capable of supporting TE_0 and TE_1 modes with mode angles 8.6° and 16.7° respectively.

Plots of E_y against x/d are shown in Fig. 10.14. Notice that a substantial part of the field penetrates into the substrate, particularly for the higher order mode.

§10.11 Problems

1. Verify (10.25) and (10.26).

2. Show that the reflection and transmission coefficients in the case in which **E** lies in the plane of incidence, are given by

$$R_{\|} = \frac{\tan^2(\theta_2 - \theta_0)}{\tan^2(\theta_2 + \theta_0)}; \qquad T_{\|} = \frac{\sin 2\theta_0 \sin 2\theta_2}{\sin^2(\theta_0 + \theta_2)\cos^2(\theta_0 - \theta_2)}$$

3. Sketch curves of $R_{\parallel}$, $R_{\perp}$, $T_{\parallel}$, $T_{\perp}$ against angle of incidence for (a) $n_1 = 1$, $n_2 = 1.5$; (b) $n_1 = 1$, $n_2 = 1.1$. Check from the curves that $R + T = 1$.

4. Calculate the Brewster angles from the data in problem 3 and compare them with the values from the curves.

5. Show that the reflection coefficient for light passing from glass to air at normal incidence is the same as from air to glass, but that the phase changes in the two cases are different.

6. A plane electromagnetic wave is incident normally on the plane interface between two dielectrics. If the refractive indices are n_1 and n_2 show that the reflected and transmitted energy fluxes are equal when $n_2/n_1 = (3 + \sqrt{2})$.

7. Light falls normally on a dielectric slab, bounded by two parallel faces. Show that the total fraction of energy reflected is $(n_2 - n_1)^2/(n_1^2 + n_2^2)$ and that transmitted is $2n_1n_2/(n_1^2 + n_2^2)$. [It is essential to take account of the multiple reflections which occur at each boundary.]

8. A light wave is incident on a dielectric slab bounded by two parallel faces. Assuming the wave is incident at the Brewster angle show that the refracted wave is incident on the back surface also at the Brewster angle.

9. Calculate the reflection coefficient for a plane electromagnetic wave incident on a plane quartz window 1 cm thick of refractive index 2.24 assuming free space on both sides of the window.

10. Consider a dielectric coating (2) of thickness l placed in a sandwich of two different dielectrics (1 and 3) such that $k_2 l = \pi/2$ (quarter-wave coating). Show that if $n_2^2 = n_1 n_3$ all reflection for light passing from the first dielectric to the third will be eliminated. [This forms the basis of a technique used to coat optical components with the aim of minimizing the reflected light.]

11. For **E** polarized perpendicular to the plane of incidence determine under what conditions the standing wave in x gives a minimum tangential electric field at the interface.

12. For the following dielectrics calculate the critical angle and the Brewster angle for an electromagnetic wave passing from each

dielectric into air:

Material	$\varepsilon/\varepsilon_0$
Water	81
Glass	9
Quartz	5

13. Show that a linearly polarized wave will in general be elliptically polarized after total reflection from the surface of a dielectric. Under what conditions will the polarization become circular?

14. A dielectric layer of permittivity ε_2 bounded by planes $z=0$, $z=d$, lies between dielectric materials with permittivities ε_1 and ε_3. An electromagnetic wave falls normally on the surface of the layer from the region $z<0$. Find the thickness of the layer corresponding to minimum reflection, and the ratio of ε_1, ε_2 and ε_3 for which there will be no reflection.

15. Consider the transport of energy in the case of total internal reflection and determine the energy flux along and normal to, the interface. Determine the Poynting flux **S**.

16. Find the reflection coefficient R of a plane-conducting surface for normally incident radiation in the limit of low conductivity.

17. Show that the ratio of the amplitudes of the reflected and incident waves in the case of reflection at the surface of a good conductor is given by $(1-2\pi\delta\cos\theta_0/\lambda_0)$ if the conductor has a refractive index of unity and is nonmagnetic.

18. A plane electromagnetic wave is incident normally on a slab of dielectric material, the back surface of which is in contact with a perfect conductor. What conditions must be satisfied to ensure that there will be no multiple reflections within the dielectric?

19. Calculate the rate of flow of energy along the axis for the TE_{01} mode of a rectangular waveguide.

20. Repeat the analysis of the dielectric waveguide in §10.9 for a TM mode and show that the corresponding eigenvalue equation may be written as

$$\cos Kd = (n_2^2 n_3^2 K^2 - n_1^4 \gamma\delta)/A$$

where

$$A = [(n_2^4 K^2 + n_1^4 \gamma^2)(n_3^4 K^2 + n_1^4 \delta^2)]^{1/2}$$

$$\sin Kd = n_1^2 K(n_3^2 \gamma + n_2^2 \delta)/A$$

where the symbols have the meaning attributed to them in §10.9. [Take $\mathbf{H} = (0, H_y, 0)$; $\mathbf{E} = (E_x, 0, E_z)$.]

21. Calculate the cut-off thickness for the fundamental TE mode in the dielectric slab waveguide discussed in §10.9 given $n_1 = 1.59$, $n_2 = 1.53$, $n_3 = 1$ and $k = 9.92$ rads/μm.

22. Given $n_1 = 1.01$, $n_2 = 1 = n_3$ and $k = 9.92$ rads/μm calculate the range of guide thickness which ensures that only one TE mode is present.

11

Electromagnetic radiation

§11.1 Electromagnetic potentials

In Chapters 9 and 10 we have examined electromagnetic waves propagating in a variety of conditions. So far no consideration has been given to the generation of electromagnetic waves which is a matter of quite fundamental importance in electronic engineering and physics. We shall examine the generation of radiation in this chapter dealing both with radiation from antennas and from charged particles. We shall find that radiation is only emitted by **accelerated** charges.

It is convenient to discuss the electrodynamics of radiation in terms of the potentials $\mathbf{A}$ and ϕ, the electric fields then being determined by

$$\mathbf{E}=-\boldsymbol{\nabla}\phi-\frac{\partial \mathbf{A}}{\partial t}; \qquad \mathbf{B}=\boldsymbol{\nabla}\times\mathbf{A} \tag{11.1}$$

Note that we may in fact add to $\mathbf{A}$ the gradient of an arbitrary scalar function Λ and leave $\mathbf{B}$ unchanged. Then clearly $\mathbf{E}$ will be unaffected if we replace ϕ by $\left(\phi-\frac{\partial \Lambda}{\partial t}\right)$. The invariance of (11.1) under the transformations

$$\mathbf{A}'=\mathbf{A}+\boldsymbol{\nabla}\Lambda \tag{11.2}$$

$$\phi'=\phi-\frac{\partial \Lambda}{\partial t} \tag{11.3}$$

is known as **gauge invariance**, and the transformation are called **gauge transformations**. Using (11.1) in the equations for the vacuum fields we find

$$\boldsymbol{\nabla}\cdot\mathbf{E}=-\nabla^2\phi-\frac{\partial}{\partial t}(\boldsymbol{\nabla}\cdot\mathbf{A})=\frac{\rho}{\varepsilon_0}$$

$$\begin{aligned}\boldsymbol{\nabla}\times\mathbf{B}&=\boldsymbol{\nabla}(\boldsymbol{\nabla}\cdot\mathbf{A})-\boldsymbol{\nabla}^2\mathbf{A}\\&=\mu_0\mathbf{j}-\frac{1}{c^2}\frac{\partial}{\partial t}\left(\boldsymbol{\nabla}\phi+\frac{\partial \mathbf{A}}{\partial t}\right)\end{aligned}$$

Rearranging these equations gives

$$\nabla^2\phi + \frac{\partial}{\partial t}(\nabla \cdot \mathbf{A}) = -\frac{\rho}{\varepsilon_0} \tag{11.4}$$

$$\nabla^2\mathbf{A} - \frac{1}{c^2}\frac{\partial^2\mathbf{A}}{\partial t^2} - \nabla\left(\nabla \cdot \mathbf{A} + \frac{1}{c^2}\frac{\partial\phi}{\partial t}\right) = -\mu_0\mathbf{j} \tag{11.5}$$

When we first introduced the vector potential in magnetostatics (§5.8) we remarked that having specified $\nabla \times \mathbf{A}$ we were free to assign to $\nabla \cdot \mathbf{A}$ any value we wish and, under static conditions, we chose $\nabla \cdot \mathbf{A} = 0$. In electrodynamics we make a different choice and write

$$\nabla \cdot \mathbf{A} + \frac{1}{c^2}\frac{\partial\phi}{\partial t} = 0 \tag{11.6}$$

This choice is known as the **Lorentz condition** and specifies the **Lorentz gauge** for the potentials under which (11.4) and (11.5) transform to

$$\nabla^2\phi - \frac{1}{c^2}\frac{\partial^2\phi}{\partial t^2} = -\frac{\rho}{\varepsilon_0} \tag{11.7}$$

$$\nabla^2\mathbf{A} - \frac{1}{c^2}\frac{\partial^2\mathbf{A}}{\partial t^2} = -\mu_0\mathbf{j} \tag{11.8}$$

If $\rho = 0$, $\mathbf{j} = 0$ we retrieve the homogeneous wave equation considered in Chapter 9 (cf. §9.5). Equations (11.7), (11.8) are known as **inhomogeneous wave equations**; the operator $\left[\nabla^2 - \frac{1}{c^2}\frac{\partial^2}{\partial t^2}\right]$, is identical with the homogeneous wave equation but **source terms**, ρ and $\mathbf{j}$ now appear on the right-hand side.

§11.2 Inhomogeneous wave equations

In Chapters 9 and 10 we were concerned with solutions of the **homogeneous** wave equation. We turn now to consider the **inhomogeneous** equations (11.7) and (11.8). In Chapter 7 we saw how a Green function approach could be used to solve the corresponding **static** equations. In what follows we introduce a time-dependent Green function to solve the inhomogeneous equations of electrodynamics. We require a Green function satisfying

$$\left[\nabla^2 - \frac{1}{c^2}\frac{\partial^2}{\partial t^2}\right] G(\mathbf{r}, t; \mathbf{r}', t') = -\delta(\mathbf{r}-\mathbf{r}')\delta(t-t') \tag{11.9}$$

The Green function $G(\mathbf{r}, t; \mathbf{r}', t')$ will also be required to satisfy certain boundary conditions. The right-hand side of (11.9) may be represented

$$\delta(\mathbf{r}-\mathbf{r}')\delta(t-t') = \frac{1}{(2\pi)^4}\int d\mathbf{k}\int d\omega e^{i\mathbf{k}\cdot(\mathbf{r}-\mathbf{r}')-i\omega(t-t')}$$

so that if we introduce the Fourier representation

$$G(\mathbf{r}, t; \mathbf{r}', t') = \frac{1}{(2\pi)^4}\int d\mathbf{k}\int d\omega G(\mathbf{k}, \omega) e^{i\mathbf{k}\cdot(\mathbf{r}-\mathbf{r}')-i\omega(t-t')} \tag{11.10}$$

we find, on Fourier transforming (11.9),

$$G(\mathbf{k}, \omega) = \frac{1}{k^2 - \dfrac{\omega^2}{c^2}} \tag{11.11}$$

Thus in determining $G(\mathbf{r}, t; \mathbf{r}', t')$ from (11.10), (11.11) we are faced with singularities in the integrand at $\omega = \pm kc$ and, without some prescription for dealing with these, the solution is of course meaningless. We may find the prescription needed if we consider (11.9) from a physical standpoint. G represents a wave created by a point source at $\mathbf{r}'$ which is pulsed at time t', and this wave propagates outwards from the source with velocity c. It is therefore reasonable to require that (i) $G=0$ over the whole domain for $t<t'$ and (ii) G should represent **outward** propagating waves for $t>t'$. In (11.10) consider the integration over ω as a Cauchy integral in the complex ω-plane. For $t<t'$ the integral along the real ω-axis is equivalent to a contour integral along C_1 (Fig. 11.1a) in the upper half plane, since the contribution from the semicircle at infinity vanishes. Similarly for $t>t'$. the contour C_2 (Fig. 11.1b) is closed in the lower half plane. Now our first

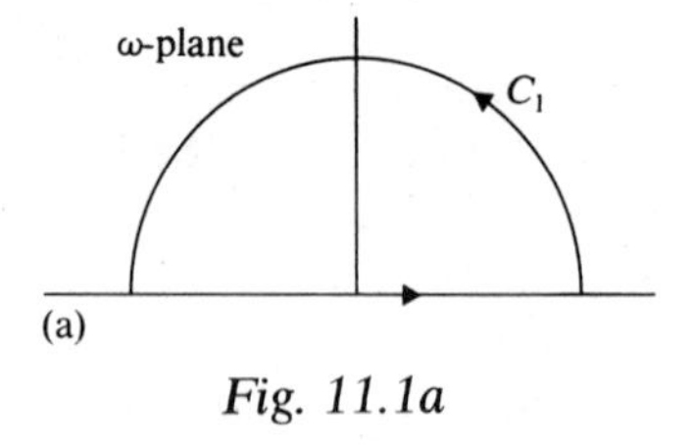

Fig. 11.1a

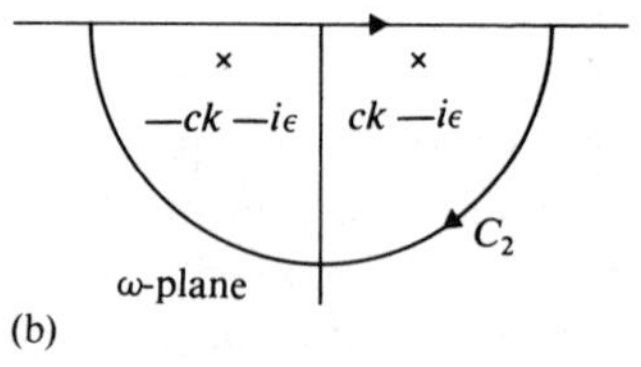

Fig. 11.1b

requirement – that $g=0$ over the whole domain for $t<t'$ – means that the poles at $\omega=\pm ck$ must be displaced **below** the $Re\omega$ axis (cf. **Fig 11.1b**) so that the integral around C_2 for $t>t'$ will **not** vanish. Thus by Cauchy's theorem the ω-integration for $t>t'$ gives

$$G(\mathbf{r}, t; \mathbf{r}', t') = \frac{c}{(2\pi)^3}\int e^{i\mathbf{k}\cdot(\mathbf{r}-\mathbf{r}')}\frac{\sin ck(t-t')}{k}\,d\mathbf{k} \tag{11.12}$$

$$\therefore G(\mathbf{r}, t; \mathbf{r}', t') = \frac{c}{(2\pi)^2}\int_0^\infty [\sin ck(t-t')]k\,dk \int_{-1}^{1} e^{ik|\mathbf{r}-\mathbf{r}'|\cos\theta}\,d(\cos\theta)$$

Writing $\xi = kc$ this gives

$$G(\mathbf{r}, t; \mathbf{r}', t') = \frac{1}{8\pi^2|\mathbf{r}-\mathbf{r}'|}\int_{-\infty}^{\infty}[e^{i(t-t'-|\mathbf{r}-\mathbf{r}'|/c)\xi} - e^{i(t-t'+|\mathbf{r}-\mathbf{r}'|/c)\xi}]\,d\xi$$

$$= \frac{1}{4\pi}\frac{\delta\left(t'+\dfrac{|\mathbf{r}-\mathbf{r}'|}{c}-t\right)}{|\mathbf{r}-\mathbf{r}'|} - \frac{1}{4\pi}\frac{\delta\left(t'-\dfrac{|\mathbf{r}-\mathbf{r}'|}{c}-t\right)}{|\mathbf{r}-\mathbf{r}'|} \tag{11.13}$$

The first term is the **retarded Green function** which expresses the fact that a signal observed at some field point $\mathbf{r}$ at time t originates from a source situated at $\mathbf{r}'$ emitting at an earlier time $t'=t-|\mathbf{r}-\mathbf{r}'|/c$. The second term in (11.13), the **advanced Green function**, is equally valid mathematically but leads to charge and current distributions determined by events at a time in the future $t'=t+|\mathbf{r}-\mathbf{r}'|/c$. These violate the principle of causality and are discarded in classical physics.

Knowing the Green function we may at once write down the solutions to the electrodynamic equations (11.7), (11.8):

$$\phi(\mathbf{r}, t) = \frac{1}{4\pi\varepsilon_0}\int\frac{[\rho(\mathbf{r}', t')]_{t'=t-|\mathbf{r}-\mathbf{r}'|/c}}{|\mathbf{r}-\mathbf{r}'|}\,d\mathbf{r}' \tag{11.14}$$

$$\mathbf{A}(\mathbf{r}, t) = \frac{\mu_0}{4\pi}\int\frac{[\mathbf{j}(\mathbf{r}', t')]_{t'=t-|\mathbf{r}-\mathbf{r}'|/c}}{|\mathbf{r}-\mathbf{r}'|}\,d\mathbf{r}' \tag{11.15}$$

§11.3 Oscillating electric dipole

Let us apply the solutions of the inhomogeneous wave equation to a simple electrodynamic configuration, namely an oscillating electric dipole – sometimes called a Hertzian oscillator – which is just an electric charge oscillating sinusoidally with time. We shall show that this system **radiates** electromagnetic waves.

Suppose we consider a unit positive charge fixed at the origin and an electron whose displacement from the positive charge is given by $\mathbf{d}(t) = Re(d_0\hat{\mathbf{z}}e^{i\omega t})$, i.e. we have a system with a time-dependent dipole moment $\mathbf{p}(t) = e\mathbf{d}(t)$, providing a current source over a region of scale length d_0. Let us also suppose that the wavelength of the radiation $\lambda = 2\pi c/\omega$ is large compared with the scale length of the radiating system, i.e.

$$d_0 \ll \lambda \tag{11.16}$$

Physically this means that $\dfrac{d_0}{c} \ll \dfrac{2\pi}{\omega} = T$, the period of the source, so that (11.16) may be interpreted as saying that the time d_0/c, taken for a signal to propagate across the radiating system or source, is very much less than that over which the source current changes appreciably.

Another ordering that may be imposed is to require

$$d_0 \ll r \tag{11.17}$$

where $\mathbf{r}$ denotes the position of the field point P at which an observer is sited; this is the usual dipole approximation. In terms of characteristic times it means simply that we neglect the time taken for a signal to propagate across the source compared with that for propagation from the source to the field point P.

Finally we may order λ relative to r and here there are two possibilities, $r \ll \lambda$ or $r \gg \lambda$. (We shall ignore cases for which $r \sim \lambda$.) When $r \ll \lambda$ we are in the **near zone** in which the electric field will be exactly like that calculated in §3.5 for the electrostatic dipole. However for $r \gg \lambda$ we shall find fields which vary as $1/r$ – the so-called radiation (as opposed to Coulomb) fields. This region of space is the far field or **radiation zone**, characterized by

$$d_0 \ll \lambda \ll r \tag{11.18}$$

Physically we may think of this region as one in which we may treat any radiation from the source as **plane waves** to a first approximation. Higher order corrections allow for the spherical character of the wavefront of the outgoing wave.

Using the approximation $d_0 \ll r$ in (11.15) means that $|\mathbf{r}'|$ may be neglected compared with $|\mathbf{r}|$

$$\mathbf{A}(\mathbf{r}, t) = \frac{\mu_0}{4\pi r} \int [\mathbf{j}(\mathbf{r}', t')]_{t'=t-(r/c)} \, d\mathbf{r}' \tag{11.19}$$

Since we are now dealing with a simple discrete system rather than a charge distribution we may put

$$\begin{aligned} \int \mathbf{j}(\mathbf{r}', t')\, d\mathbf{r}' &= \int \sum_\alpha e_\alpha \dot{\mathbf{r}}_\alpha(t')\delta(\mathbf{r}' - \mathbf{r}_\alpha(t'))\, d\mathbf{r}' \\ &= \frac{d}{dt'} \sum_\alpha e_\alpha \mathbf{r}_\alpha(t') = \dot{\mathbf{p}}(t') \end{aligned} \tag{11.20}$$

where $\mathbf{p}$ is the dipole moment of the system of charges. Then

$$\mathbf{A}(\mathbf{r}, t) = \frac{\mu_0}{4\pi r} [\dot{\mathbf{p}}(t')]_{t'=t-r/c} \tag{11.21}$$

Now

$$\mathbf{p}(t') = ed_0\hat{\mathbf{z}}e^{i\omega t'} = \mathbf{p}_0 e^{i\omega t'}$$

so that $A_x = 0 = A_y$ and A_z is given by

$$A_z(\mathbf{r}, t) = \frac{\mu_0}{4\pi r}\, i\omega p_0 e^{i\omega(t-(r/c))}$$

or since $\omega = kc$

$$A_z(\mathbf{r}, t) = \frac{\mu_0}{4\pi}\, i\omega p(t) \frac{e^{-ikr}}{r} \tag{11.22}$$

It is often more convenient to use the symmetry of the problem and switch to polar coordinates, giving

$$\left.\begin{aligned} A_r &= \frac{\mu_0}{4\pi}\, i\omega p \cos\theta \frac{e^{-ikr}}{r} \\ A_\theta &= \frac{\mu_0}{4\pi}\, i\omega p \sin\theta \frac{e^{-ikr}}{r} \\ A_\phi &= 0 \end{aligned}\right\} \tag{11.23}$$

Moreover ϕ may be written down directly by using the Lorentz gauge condition $\nabla \cdot \mathbf{A} + \frac{1}{c^2}\frac{\partial \phi}{\partial t} = 0$ (problem 5)

$$\phi = -\frac{1}{4\pi\varepsilon_0} p \cos\theta \frac{\partial}{\partial r}\left(\frac{e^{-ikr}}{r}\right) \tag{11.24}$$

From (11.23), (11.24) we may immediately determine **E** and **H**. We find (problem 6)

$$\left.\begin{aligned} H_r &= 0; \qquad H_\theta = 0 \\ H_\phi &= -\frac{cpk^2}{4\pi}\sin\theta\left[1 - \frac{i}{kr}\right]\frac{e^{-ikr}}{r} \end{aligned}\right\} \tag{11.25}$$

Since $\mathbf{E} = -\nabla\phi - \frac{\partial \mathbf{A}}{\partial t}$ it follows that (problem 7)

$$\left.\begin{aligned} E_r &= \frac{ipk}{2\pi\varepsilon_0 r}\cos\theta\left[1 - \frac{i}{kr}\right]\frac{e^{-ikr}}{r} \\ E_\theta &= -\frac{pk^2}{4\pi\varepsilon_0}\sin\theta\left[1 - \frac{i}{kr}\left(1 - \frac{i}{kr}\right)\right]\frac{e^{-ikr}}{r} \\ E_\phi &= 0 \end{aligned}\right\} \tag{11.26}$$

The set (11.25), (11.26) form **Hertz's relations** for the oscillating electric dipole and we shall examine these in the near-field ($kr \ll 1$) and far-field ($kr \gg 1$) approximations.

Near-zone: $kr \ll 1$.

$$E_r \sim \frac{p\cos\theta}{2\pi\varepsilon_0 r^3}; \qquad E_\theta \sim \frac{p\sin\theta}{4\pi\varepsilon_0 r^3}; \qquad H_\phi \sim \frac{i\omega p\sin\theta}{4\pi r^2} \tag{11.27}$$

i.e. the electric field is simply the dipole field in (3.10), specifically the **instantaneous** field since $p = p(t)$. These r^{-3} terms contribute to the stored field energy but not to the radiation field. The $r^{-2}(H_\phi)$ term likewise makes no contribution to the radiated power and is the source of magnetic induction field energy corresponding to the electrostatic capacitative energy from the electric field terms. Comparing the magnitude of magnetic to electric fields in the near zone we find

$$\frac{\omega\mu_0|\mathbf{H}|}{k|\mathbf{E}|} \sim \frac{\omega\mu_0}{k}\varepsilon_0\omega r = kr \ll 1$$

i.e. electric fields dominate in the near zone and, in the static limit $k \to 0$, the field is purely electric.

Radiation zone: $kr \gg 1$

In the far-field approximation the dominant terms are now

$$\left.\begin{aligned} E_\theta &\simeq -\frac{pk^2}{4\pi\varepsilon_0}\sin\theta\frac{e^{-ikr}}{r} \\ H_\phi &\simeq -\frac{cpk^2}{4\pi}\sin\theta\frac{e^{-ikr}}{r} \end{aligned}\right\} \tag{11.28}$$

with $H_r = 0 = H_\theta = E_\phi$ and E_r is $O(r^{-2})$. Thus at large distances from the source the electric and magnetic fields are each transverse and behave as r^{-1}. Moreover $|E_\theta|/|H_\phi| = (\mu_0/\varepsilon_0)^{1/2}$ and the electric and magnetic fields are in phase; in short, they show all the characteristics of radiation fields. (Note that the magnetic field from an oscillating electric dipole is transverse everywhere so that the radiation is in general TM.) The electric field lines may be determined (problem 8) and are plotted in Fig. 11.2.

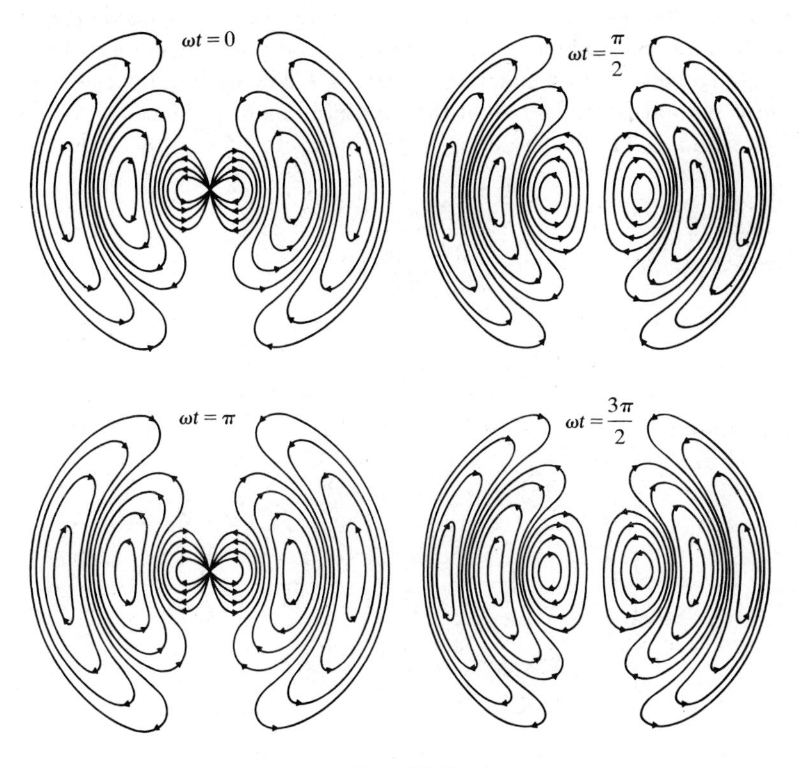

Fig. 11.2

From (9.48) the Poynting vector $\langle \mathbf{S} \rangle$ is

$$\langle \mathbf{S} \rangle = \tfrac{1}{2} Re(\mathbf{E} \times \mathbf{H}^*)$$
$$= \frac{c p_0^2 k^4}{32\pi^2 \varepsilon_0} \frac{\sin^2 \theta}{r^2} \hat{\mathbf{n}} \tag{11.29}$$

and the total radiated power is therefore

$$\langle P \rangle = \int_{-1}^{+1} \int_0^{2\pi} \langle \mathbf{S} \rangle \cdot \hat{\mathbf{n}} r^2 \, d(\cos \theta) \, d\phi$$
$$= \frac{c p_0^2}{12\pi\varepsilon_0} \frac{(2\pi)^4}{\lambda^4} \tag{11.30}$$

where λ is the wavelength. The total average radiated power thus varies as the square of the amplitude of the electric dipole and inversely as the wavelength to the fourth power.

Note from the average Poynting flux that since the energy flow varies as $\sin^2 \theta$ it vanishes along the dipole axis and is maximum in the equatorial plane as represented in the polar diagram in Fig. 11.3. To electrical engineers the current I flowing in an aerial is a more natural parameter than the dipole moment p. Since $I = i\omega p/d_0$, in (11.30)

$$\langle P \rangle = \frac{1}{12\pi\varepsilon_0} \frac{I_0^2}{c} (kd_0)^2$$
$$= \frac{2\pi}{3} \sqrt{\frac{\mu_0}{\varepsilon_0}} \left(\frac{d_0}{\lambda}\right)^2 \frac{I_0^2}{2} \tag{11.31}$$

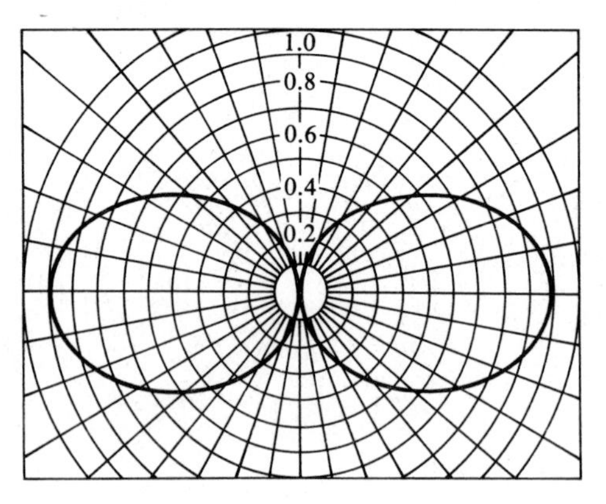

Fig. 11.3

A resistance R carrying a current $I = I_0 e^{i\omega t}$ dissipates energy at an average rate $RI_0^2/2$ and so comparing this with (11.31) we define the **radiation resistance** of a dipole by

$$R_r = \frac{2\pi}{3}\sqrt{\frac{\mu_0}{\varepsilon_0}}\left(\frac{d_0}{\lambda}\right)^2$$

$$= 789\left(\frac{d_0}{\lambda}\right)^2 \text{ ohms} \tag{11.32}$$

The radiation resistance is simply that resistance which would dissipate as heat the same power as that radiated by the electric dipole if it carried the same current. Remember that (11.32) is only valid provided $d_0 \ll \lambda$. If we have an antenna for which $d_0/\lambda \sim 10^{-2}$ then the radiation resistance is approximately 0.08 ohm. The ohmic resistance of the antenna could be appreciably larger and this would result in most of the input power being dissipated as heat rather than being radiated as electromagnetic energy. A short dipole antenna is therefore in general a rather inefficient radiator. To get appreciable radiation would require $d_0 \sim \lambda$ but in this situation the dipole approximation is no longer valid.

§11.4 Radiation from a linear antenna

A more realistic antenna is a centre driven linear antenna shown in Fig. 11.4. We suppose that the input signal and hence the current density vary harmonically with time. The current density is also taken to vary harmonically **along** the antenna; it must of course vanish at the ends, $z = \pm d_0/2$. Thus

$$\mathbf{j}(\mathbf{r}', t') = \hat{\mathbf{z}} I_0 e^{i\omega t'} \sin\left(\frac{kd_0}{2} - k\,|z|\right) \delta(x)\,\delta(y) \tag{11.33}$$

and the input signal from the coaxial line is

$$I(t') = I_0 e^{i\omega t'} \sin\frac{kd_0}{2}$$

As with the electric dipole considered in §11.3 we may determine the properties of the radiation field from (11.15), now substituting

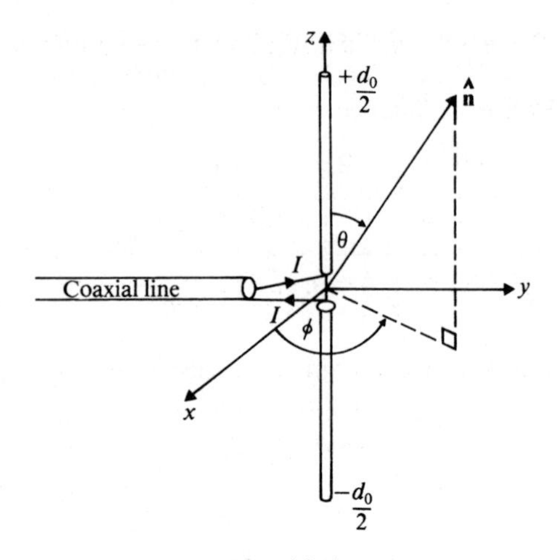

Fig. 11.4

(11.33) for $\mathbf{j}(\mathbf{r}', t')$, giving after integrating over x, y,

$$\mathbf{A}(\mathbf{r}, t) = \frac{\mu_0 \hat{\mathbf{z}}}{4\pi} I_0 \int_{-d_0/2}^{d_0/2} \frac{e^{i\omega(t-|\mathbf{r}-\mathbf{z}'|/c} \sin\left(\frac{kd_0}{2} - k|z'|\right)}{|\mathbf{r}-\mathbf{z}'|} dz' \tag{11.34}$$

In the far field approximation we may replace the denominator in (11.34) by r and in the exponent we may write

$$|\mathbf{r}-\mathbf{z}'| \simeq r - \hat{\mathbf{n}} \cdot \mathbf{z}' = r - z' \cos\theta$$

so that

$$\mathbf{A}(\mathbf{r}, t) = \frac{\mu_0 \hat{\mathbf{z}}}{4\pi} I_0 e^{i\omega t} \frac{e^{-ikr}}{r} \int_{-d_0/2}^{d_0/2} \sin\left(\frac{kd_0}{2} - k|z'|\right) e^{ikz'\cos\theta} dz'$$

The integration is straightforward (problem 15) and we find

$$\mathbf{A}(\mathbf{r}, t) = \frac{\mu_0 \hat{\mathbf{z}}}{2\pi} I_0 e^{i\omega t} \frac{e^{-ikr}}{kr} \left[\frac{\cos\left(\frac{kd_0}{2}\cos\theta\right) - \cos\left(\frac{kd_0}{2}\right)}{\sin^2\theta}\right] \tag{11.35}$$

From this we may calculate the magnetic field in the radiation zone. Since $\mathbf{B}=\nabla\times\mathbf{A}$, $\mathbf{B}=ik\hat{\mathbf{n}}\times\mathbf{A}$ and so $\mathbf{H}=\dfrac{ik}{\mu}\hat{\mathbf{n}}\times\mathbf{A}$. Forming the Poynting vector

$$\langle\mathbf{S}\rangle=\tfrac{1}{2}Re(\mathbf{E}\times\mathbf{H}^*)$$

$$=\frac{1}{2}\sqrt{\frac{\mu_0}{\varepsilon_0}}\,\hat{\mathbf{n}}\,\frac{I_0^2}{4\pi^2}\frac{1}{r^2}\left[\frac{\cos\left(\dfrac{kd_0}{2}\cos\theta\right)-\cos\dfrac{kd_0}{2}}{\sin\theta}\right]^2 \tag{11.36}$$

The time-averaged power radiated into unit solid angle is

$$\left\langle\frac{dP}{d\Omega}\right\rangle=r^2\langle\mathbf{S}\rangle\cdot\hat{\mathbf{n}}$$

i.e.

$$\left\langle\frac{dP}{d\Omega}\right\rangle=\sqrt{\frac{\mu_0}{\varepsilon_0}}\frac{I_0^2}{8\pi^2}\left[\frac{\cos\left(\dfrac{kd_0}{2}\cos\theta\right)-\cos\dfrac{kd_0}{2}}{\sin\theta}\right]^2 \tag{11.37}$$

The radiation is polarized in the plane containing the antenna and $\hat{\mathbf{n}}$. The angular distribution of the radiation depends on the value of kd_0. First let us retrieve the dipole result from (11.37) by considering the long wavelength limit $kd_0\ll 1$. In this limit

$$I(t')=I_0e^{i\omega t'}\left(\frac{kd_0}{2}-k\,|z'|\right) \tag{11.38}$$

whereas for the electric dipole

$$I(t')=I_{0D}e^{i\omega t'} \tag{11.39}$$

To make a proper comparison between (11.37) and its counterpart in the case of the electric dipole we must identify (11.38) with (11.39). While this clearly cannot be done in general we may satisfy a relation involving **average** currents, i.e.

$$I_{0D}=I_0\frac{1}{d_0}\int_{-d_0/2}^{+d_0/2}\left(\frac{kd_0}{2}-k\,|z'|\right)dz'=\frac{kd_0}{4}I_0$$

This gives, from (11.37),

$$\left\langle\frac{dP}{d\Omega}\right\rangle_D=\sqrt{\frac{\mu_0}{\varepsilon_0}}\left(\frac{4I_{0D}}{kd_0}\right)^2\frac{1}{8\pi^2}\frac{k^4d_0^4}{64}\frac{(1-\cos^2\theta)^2}{\sin^2\theta}$$

i.e.

$$\left\langle\frac{dP}{d\Omega}\right\rangle_D=\frac{1}{8}\sqrt{\frac{\mu_0}{\varepsilon_0}}\left(\frac{d_0}{\lambda}\right)^2 I_{0D}^2\sin^2\theta \tag{11.40}$$

$$\therefore\quad \langle P\rangle_D=\frac{2\pi}{3}\sqrt{\frac{\mu_0}{\varepsilon_0}}\left(\frac{d_0}{\lambda}\right)^2\frac{I_{0D}^2}{2} \tag{11.41}$$

which is identical with the dipole result (11.31). The total power radiated by the linear antenna in free space is obtained from (11.37) by integrating over the solid angle Ω:

$$\langle P\rangle=\sqrt{\frac{\mu_0}{\varepsilon_0}}\frac{I_0^2}{4\pi}\int_0^\pi\left[\frac{\cos\left(\dfrac{kd_0}{2}\cos\theta\right)-\cos\dfrac{kd_0}{2}}{\sin\theta}\right]^2\sin\theta\,d\theta$$

Since the current varies along the antenna the radiation resistance will depend on the current used to define it. If we use I_0 then

$$R_r=\frac{2\langle P\rangle}{I_0^2}=\frac{1}{2\pi}\sqrt{\frac{\mu_0}{\varepsilon_0}}\int_0^\pi\frac{\left[\cos\left(\dfrac{kd_0}{2}\cos\theta\right)-\cos\dfrac{kd_0}{2}\right]^2}{\sin\theta}\,d\theta$$

which may be evaluated in terms of sine and cosine integrals $Si(x)$, $Ci(x)$ (problem 16).

§11.5 Half-wave antenna

An important special case of the long centre-fed linear antenna is the half-wave antenna in which $d_0=\lambda/2$, i.e. $kd_0=\pi$, so that (11.37) becomes

$$\left\langle\frac{dP}{d\Omega}\right\rangle=\sqrt{\frac{\mu_0}{\varepsilon_0}}\frac{I_0^2}{8\pi^2}\left[\frac{\cos\left(\dfrac{\pi}{2}\cos\theta\right)}{\sin\theta}\right]^2 \tag{11.42}$$

while the radiation resistance now reduces to (problem 17)

$$\begin{aligned}R_r&=\frac{1}{2\pi}\sqrt{\frac{\mu_0}{\varepsilon_0}}\left[0.289+\ln\pi-\frac{1}{2}\ln\frac{\pi}{2}-\frac{1}{2}Ci2\pi\right]\\&=73.1\text{ ohms}\end{aligned} \tag{11.43}$$

Comparing this with (11.32) the equivalent expression for an electric dipole, with $d_0/\lambda\sim10^{-2}$, we see that the half-wave antenna is a much more efficient radiator. The full-wave antenna is more efficient still.

A polar plot of the power density is shown in Fig. 11.5.

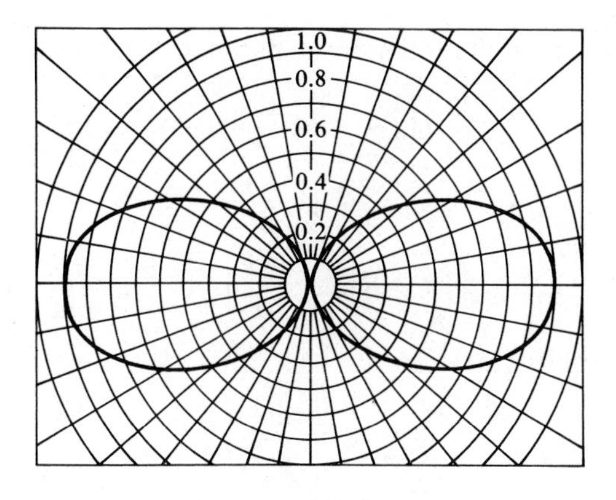

Fig. 11.5

§11.6 Dipole radiation: generalization

When obtaining the dipole fields in §11.3 we assumed a harmonic time dependence, writing $\mathbf{p}(t)=\mathbf{p}_0e^{i\omega t}$. We may generalize our expressions for the fields by retaining

$$\mathbf{A}(\mathbf{r},t)=\frac{\mu_0}{4\pi r}[\dot{\mathbf{p}}(t')]_{t'=t-r/c} \tag{11.21}$$

so that in place of (11.23) we have

$$\left.\begin{aligned} A_r &= \frac{\mu_0}{4\pi r}[\dot{p}]\cos\theta \\ A_\theta &= -\frac{\mu_0}{4\pi r}[\dot{p}]\sin\theta \\ A_\phi &= 0 \end{aligned}\right\} \tag{11.44}$$

where the brackets [ ] imply that the quantity contained is to be evaluated **at the retarded time,** $t-r/c$. Since $\mu\mathbf{H}=\nabla\times\mathbf{A}$ it follows that $H_r=0=H_\theta$ and

$$H_\phi=\frac{1}{\mu_0 r}\left\{\frac{\partial}{\partial r}(rA_\theta)-\frac{\partial A_r}{\partial\theta}\right\}$$

i.e.

$$H_\phi=\frac{\sin\theta}{4\pi r}\left\{-\frac{\partial}{\partial r}[\dot{p}]+\frac{[\dot{p}]}{r}\right\}$$

Carrying out the differentiation in the first term gives us

$$\frac{\partial}{\partial r}\dot{p}\left(t-\frac{r}{c}\right)=-\frac{1}{c}\ddot{p}\left(t-\frac{r}{c}\right)$$

so that

$$H_\phi=\frac{\sin\theta}{4\pi}\left[\frac{\ddot{p}}{cr}+\frac{\dot{p}}{r^2}\right] \qquad (11.45)$$

Similarly (problem 21)

$$E_\theta=\frac{\sin\theta}{4\pi\varepsilon_0 c}\left[\frac{\ddot{p}}{cr}+\frac{\dot{p}}{r^2}+\frac{cp}{r^3}\right] \qquad (11.46)$$

$$E_r=\frac{\cos\theta}{2\pi\varepsilon_0 c}\left[\frac{\dot{p}}{r^2}+\frac{cp}{r^3}\right] \qquad (11.47)$$

Again we must remember that **all** the fields are evaluated at the retarded time.

The radiation fields are simply

$$H_\phi=\frac{\sin\theta}{4\pi cr}[\ddot{p}]; \qquad E_\theta=\frac{\sin\theta}{4\pi\varepsilon_0 c^2 r}[\ddot{p}]$$

or in vector notation

$$\mathbf{H}_{\text{rad}}=\frac{1}{4\pi cr}[\ddot{\mathbf{p}}]\times\hat{\mathbf{n}}$$
$$\mathbf{E}_{\text{rad}}=\frac{1}{4\pi\varepsilon_0 c^2 r}([\ddot{\mathbf{p}}]\times\hat{\mathbf{n}})\times\hat{\mathbf{n}} \qquad (11.48)$$

The Poynting vector for the radiation field is

$$\mathbf{S}_{\text{rad}}=\mathbf{E}_{\text{rad}}\times\mathbf{H}_{\text{rad}}$$
$$=\frac{1}{\varepsilon_0 c^3(4\pi r)^2}|[\ddot{\mathbf{p}}\times\hat{\mathbf{n}}]|^2\,\hat{\mathbf{n}}$$

$$\therefore\ \mathbf{S}_{\text{rad}}=\frac{1}{\varepsilon_0 c^3(4\pi r)^2}[\ddot{p}]^2\sin^2\theta\,\hat{\mathbf{n}} \qquad (11.49)$$

The power radiated per unit solid angle, $\frac{dP}{d\Omega}=(\mathbf{S}_{\text{rad}}\cdot\hat{\mathbf{n}})r^2$ is therefore

$$\frac{dP}{d\Omega}=\frac{[\ddot{p}]^2\sin^2\theta}{16\pi^2\varepsilon_0 c^3} \qquad (11.50)$$

The power radiated in all directions is obtained by integrating (11.50) over the solid angle, i.e.

$$P = \frac{[\ddot{p}]^2}{6\pi\varepsilon_0 c^3} \tag{11.51}$$

a result known as **Larmor's formula**, having been first derived by J. J. Larmor in 1897.

In the dipole approximation we see that the radiation is governed by the second derivative of the dipole moment $\mathbf{p}$. Since $\mathbf{p} = \sum e\mathbf{d}$, $\ddot{\mathbf{p}} = \sum e\ddot{\mathbf{d}} = \sum e\dot{\mathbf{v}}$ so that uniformly moving charges do not radiate.

The Larmor formula, (11.51), together with other results obtained thus far in this chapter, is simply a non-relativistic approximation to the exact electrodynamic expression. This is to be expected since the dipole approximation is equivalent to assuming that velocities of charges in the system are small compared with that of light. In §11.3 we assumed that $d_0 \ll \lambda$, i.e. that the dimensions of the radiating system were small compared with the wavelength of the radiation. If we take v as a velocity characteristic of the charges, and denote by T the order of magnitude of the time over which the charge distribution changes significantly, then $T \sim d_0 v$. The frequency of the radiation ω will be of order T^{-1}, i.e. its wavelength $\lambda \sim cT \sim cd_0/v$. So the ordering $d_0 \ll \lambda$ implies that

$$v \ll c \tag{11.52}$$

Thus all of the results in §§11.3–11.6 are **only** valid in this limit.

§11.7 Scattering of radiation

In the previous section we obtained expressions for the radiation fields without specifying the source of the time dependence of $\mathbf{p}(t)$. If we consider an electromagnetic wave interacting with a system of charges, these charges are set in motion to the tune of the wave. The motion of these charges in turn is the source of radiation. We describe this process as **scattering** of the original radiation incident on the charge system.

The simplicity consider a plane monochromatic linearly polarized wave incident on a charge e, at rest, and assume that the subsequent velocity of this charge v is such that $v \ll c$ so that our treatment – as in the preceding sections – will be non-relativistic. The motion of the charge will be described by the Lorentz equation which, in non-relativistic form, is

$$m\ddot{\mathbf{r}} = e(\mathbf{E} + \mathbf{v} \times \mathbf{B})$$

and since $vB/E = v/c$ (cf. §9.10) we can ignore the second term and write

$$m\ddot{\mathbf{r}} = e\mathbf{E} = e\mathbf{E}_0 \exp i(\mathbf{k}_0 \cdot \mathbf{r} - \omega_0 t) \qquad (11.53)$$

If the charge oscillates about the origin we may assume that the field acting on the charge at all times is the same as that at $\mathbf{r} = 0$ so that (11.53) gives

$$m\ddot{\mathbf{r}} = e\mathbf{E}_0 e^{-i\omega_0 t}$$

Writing $\mathbf{p}(t) = e\mathbf{r}(t)$, this becomes

$$\ddot{\mathbf{p}} = \frac{e^2\mathbf{E}_0}{m} e^{-i\omega_0 t} \qquad (11.54)$$

Substituting (11.54) in (11.50) and time averaging gives for the power **re-radiated** per unit solid angle

$$\left\langle \frac{dP}{d\Omega} \right\rangle = \frac{e^4 E_0^2}{16\pi^2 \varepsilon_0 m^2 c^3} \sin^2\theta$$

If we now define a differential **scattering cross-section** $d\sigma/d\Omega$ as the energy re-radiated per unit time per unit solid angle divided by the incident energy flux we have

$$\frac{d\sigma}{d\Omega} = \left\langle \frac{dP}{d\Omega} \right\rangle \Big/ \left(\frac{\varepsilon_0}{\mu_0}\right)^{1/2} E^2$$

$$= \left(\frac{e^2}{4\pi\varepsilon_0 mc^2}\right)^2 \sin^2\theta \qquad (11.55)$$

The quantity within the brackets is the **classical electron radius**, r_e, i.e.

$$r_e = \frac{e^2}{4\pi\varepsilon_0 mc^2} \qquad (11.56)$$

(cf. §3.4) so that the differential scattering cross-section is now

$$\frac{d\sigma}{d\Omega} = r_e^2 \sin^2\theta \qquad (11.57)$$

In (11.57), θ is the angle between the direction of re-radiation or scattering $\hat{\mathbf{n}}$ and the direction of the electric field of the incident radiation $\mathbf{E}$. For a linearly polarized wave we may take $\psi = 0$ in Fig. 11.6 and so the total cross-section σ is simply

$$\sigma_T = \frac{8\pi}{3} r_e^2 \tag{11.58}$$

The scattering of radiation by a classical charged particle is known as **Thomson scattering** and (11.58) as the Thomson scattering cross-section. For scattering by a free electron

$$\sigma_T = 0.665 \times 10^{-28}\,\text{m}^2$$

In general if we want to describe scattering from incident radiation that is unpolarized then following the geometry in Fig. 11.6 we have

$$\cos\theta = \sin\xi \cos(\phi - \psi)$$

and

$$\frac{d\sigma}{d\Omega} = r_e^2 [1 - \sin^2\xi \cos^2(\phi - \psi)]$$

For unpolarized incident radiation we average over ψ to get

$$\frac{d\sigma}{d\Omega} = \frac{r_e^2}{2}(1 + \cos^2\xi)$$

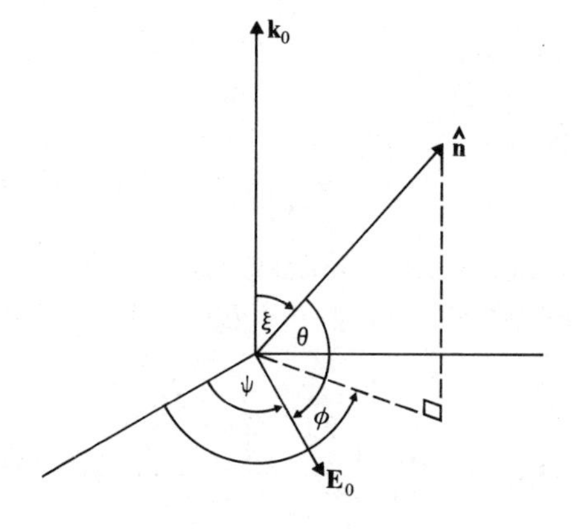

Fig. 11.6

where ξ, the angle between the directions of incident and scattered radiation, is known as the **scattering angle**.

Thomson scattering results in a force driving the electron in the direction of the incident electromagnetic wave. The electron absorbs energy from the wave at an average rate $\langle S\rangle\sigma_T$. If we take the **momentum** of the radiation field as its energy divided by the velocity of light it follows that the electron gains momentum from the wave at a rate $\langle S\rangle\sigma_T/c$. The average force acting on the electron is equal to this average rate of gain of momentum, i.e.

$$
\begin{aligned}
F &= \frac{\langle S\rangle\sigma_T}{c}\hat{\mathbf{n}} \\
&= \frac{\hat{\mathbf{n}}}{12\pi\varepsilon_0}\left(\frac{e^2E_0}{mc^2}\right)^2
\end{aligned}
\tag{11.59}
$$

This force acts in the direction of the incident wave. The force per unit area corresponds to the **radiation pressure** and provides a small correction to the Lorentz equation (problem 23).

§11.8 Liénard–Wiechert potentials

Let us now return to the expressions for the retarded potentials given in §11.2 and consider a rather special source, namely a single charged particle moving **quite arbitrarily** with velocity $\dot{\mathbf{r}}_0(t)$ at the point $\mathbf{r}_0(t)$. Then

$$\mathbf{j}(\mathbf{r}, t) = e\dot{\mathbf{r}}_0(t)\delta(\mathbf{r}-\mathbf{r}_0(t)) \tag{11.60}$$

Using this as a source we find

$$\mathbf{A}(\mathbf{r}, t) = \frac{\mu_0 e}{4\pi}\iint \frac{\dot{\mathbf{r}}_0(t')\delta\left(t'+\dfrac{|\mathbf{r}-\mathbf{r}'|}{c}-t\right)}{|\mathbf{r}-\mathbf{r}'|}\delta(\mathbf{r}'-\mathbf{r}_0(t'))\,d\mathbf{r}'\,dt'$$

where the dot implies differentiation with respect to the argument. The delta function in (11.60) allows us to integrate over $\mathbf{r}'$ to give

$$\mathbf{A}(\mathbf{r}, t) = \frac{\mu_0 e}{4\pi}\int \frac{\dot{\mathbf{r}}_0(t')\delta\left(t'+\dfrac{R(t')}{c}-t\right)}{R(t')}\,dt' \tag{11.61}$$

where $R(t') = |\mathbf{r}-\mathbf{r}_0(t')|$, with the geometry shown in Fig. 11.7.

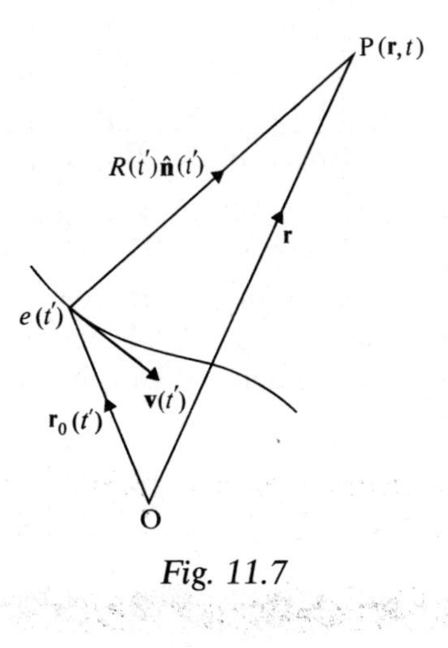

Fig. 11.7

Integrating over t' is not quite as straightforward as the $\mathbf{r}'$ integration because $R(t')$ appears in the argument of the delta function. However, using the result (problem 24)

$$\int_{-\infty}^{\infty} \delta\{g(x)\}\gamma(x)\,dx = \sum_i \gamma(x_i)\bigg/\left|\frac{dg(x_i)}{dx}\right|$$

where the x_i are the zeros of $g(x)=0$, the integration over t' gives

$$\mathbf{A}(\mathbf{r},t)=\frac{\mu_0 e}{4\pi}\left[\frac{\dot{\mathbf{r}}_0(t')}{R\left(1+\dfrac{1}{c}\dfrac{dR}{dt'}\right)}\right]_{t'=t-(R(t')/c)} \tag{11.62}$$

Now

$$\frac{dR}{dt'}=\frac{(\mathbf{r}_0(t')-\mathbf{r})\cdot\dot{\mathbf{r}}_0(t')}{|\mathbf{r}-\mathbf{r}_0(t')|}=-\hat{\mathbf{n}}(t')\cdot\mathbf{v}(t')$$

where $\hat{\mathbf{n}}(t')$ is the unit vector from the source (at the retarded position) to the field point P (Fig. 11.7) and $\mathbf{v}(t')=\dot{\mathbf{r}}_0(t')$. Thus

$$\mathbf{A}(\mathbf{r},t)=\frac{\mu_0}{4\pi}\left[\frac{e\mathbf{v}c}{cR-\mathbf{v}\cdot\mathbf{R}}\right]_{t'=t-R(t')/c} \tag{11.63}$$

and, since $\rho(\mathbf{r},t)=e\delta(\mathbf{r}-\mathbf{r}_0(t))$, it follows by a similar argument that

$$\phi(\mathbf{r}, t) = \frac{1}{4\pi\varepsilon_0}\left[\frac{ec}{cR - \mathbf{v}\cdot\mathbf{R}}\right]_{t'=t-R(t')/c} \qquad (11.64)$$

The potentials in (11.63), (11.64) are known as the **Liénard–Wiechert potentials**.

We have not had to make any assumption about the magnitude of $\mathbf{v}$ relative to c and in fact (11.63), (11.64) are relativistically correct as we shall see in Chapter 12. In the non-relativistic limit we find the familiar results

$$\phi \to \frac{1}{4\pi\varepsilon_0}\left[\frac{e}{R}\right]; \qquad \mathbf{A} \to \frac{\mu_0}{4\pi}\left[\frac{\mathbf{j}}{R}\right]$$

where the brackets imply that the quantities within are to be evaluated at the retarded times.

§11.9 Potentials for charge in uniform motion

As a prelude to relativistic electrodynamics considered in the next chapter, we shall apply the Liénard–Wiechert potentials derived in §11.8 to the special case of a charge moving with uniform velocity in a straight line. If we consider an observer moving with the charge – i.e. in the **rest frame** of the charge – then that observer will see the situation as electrostatic and assert that the scalar potential is given by the usual

$$\phi(x, y, z) = \frac{e}{4\pi\varepsilon_0}\frac{1}{(x^2+y^2+z^2)^{1/2}}$$

We can use (11.64) to answer the question: what is the scalar potential appropriate to an observer **at rest**? In the next chapter we will see that the answer to questions like this is provided by applying a **Lorentz transformation** (see §12.2). This transformation – and indeed the theory of special relativity itself as formulated by Einstein – are founded in electrodynamics. The following is essentially the argument used by Lorentz in arriving at the transformation named after him. Suppose the charge e moves uniformly along Ox with velocity $v\hat{\mathbf{x}}$ so that in the notation of Fig. 11.8

$$\begin{aligned} R(t') &= |\mathbf{r} - \mathbf{r}_0(t')| \\ &= |\mathbf{r} - vt'\hat{\mathbf{x}}| \end{aligned}$$

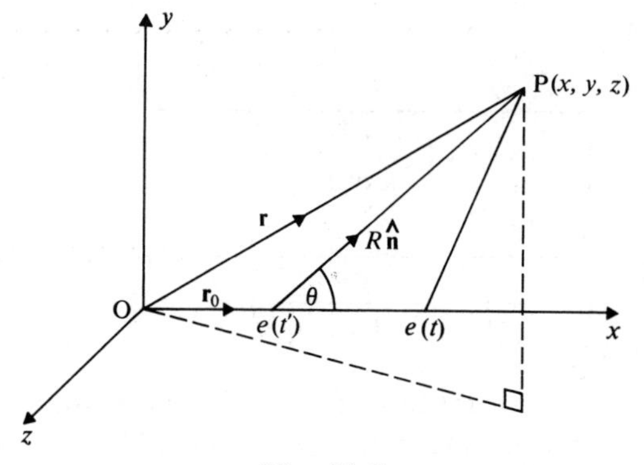

Fig. 11.8

Also

$$R(t') = ((x - vt')^2 + y^2 + z^2)^{1/2}$$

and since $t' = t - R(t')/c$ it follows that

$$c^2(t - t')^2 = (x - vt')^2 + y^2 + z^2$$

i.e.

$$(c^2 - v^2)t'^2 - 2(c^2t - xv)t' - (r^2 - c^2t^2) = 0$$

Solving for t' gives

$$(c^2 - v^2)t' = (c^2t - xv) - \{(c^2t - xv)^2 + (c^2 - v^2)(r^2 - c^2t^2)\}^{1/2}$$

$$\therefore \left(1 - \frac{v^2}{c^2}\right)t' = \left(t - \frac{xv}{c^2}\right) - \frac{1}{c}\left\{(x - vt)^2 + \left(1 - \frac{v^2}{c^2}\right)(y^2 + z^2)\right\}^{1/2} \tag{11.65}$$

Further

$$\mathbf{v} \cdot \mathbf{R} = vR \cos\theta = v(x - vt')$$

i.e.

$$R - \mathbf{v} \cdot \mathbf{R} = c(t - t') - \frac{v}{c}(x - vt')$$

$$= \left\{(x - vt)^2 + \left(1 - \frac{v^2}{c^2}\right)(y^2 + z^2)\right\}^{1/2} \tag{11.66}$$

from (11.65). Substituting in the Liénard–Wiechert expression for

$\phi(\mathbf{r}, t)$ gives

$$\phi(\mathbf{r}, t) = \frac{e}{4\pi\varepsilon_0} \frac{1}{\left\{(x - vt)^2 + \left(1 - \frac{v^2}{c^2}\right)(y^2 + z^2)\right\}^{1/2}}$$

$$= \frac{e}{4\pi\varepsilon_0} \frac{1}{\left(1 - \frac{v^2}{c^2}\right)^{1/2}} \frac{1}{\left\{\left(\frac{x - vt}{\sqrt{1 - v^2/c^2}}\right)^2 + y^2 + z^2\right\}^{1/2}} \tag{11.67}$$

In the limit $v \to 0$ we retrieve the familiar electrostatic result. This suggests that in a moving frame of reference the coordinates should transform as

$$x \to \frac{x - vt}{\sqrt{1 - v^2/c^2}}, \qquad y \to y, \qquad z \to z \tag{11.68}$$

This is in fact **part** of the Lorentz transformation. But it is only part; as we shall see in Chapter 12, time itself is relative. The full Lorentz transformation will be derived in §12.2.

§11.10 Field of an accelerated point charge

Returning to the arbitrarily moving charge considered in §11.8 let us obtain an expression for the electric field **E**. We may either tackle the problem directly by using (11.63), (11.64) in (11.1) but in fact it is rather more straightforward to return to $\mathbf{A}(\mathbf{r}, t)$ given by (11.61) and the corresponding expression for $\phi(\mathbf{r}, t)$. Then

$$\mathbf{E}(\mathbf{r}, t) = -\frac{e\hat{\mathbf{n}}}{4\pi\varepsilon_0} \frac{\partial}{\partial R} \int \frac{\delta\left(t' + \frac{R(t')}{c} - t\right)}{R(t')} dt'$$

$$- \frac{\mu_0 e}{4\pi} \frac{\partial}{\partial t} \int \frac{\mathbf{v}(t')\delta\left(t' + \frac{R(t')}{c} - t\right)}{R(t')} dt'$$

$$= \frac{e}{4\pi\varepsilon_0} \int \left[\frac{\hat{\mathbf{n}}}{R^2} \delta\left(t' + \frac{R(t')}{c} - t\right) + \frac{(\boldsymbol{\beta} - \hat{\mathbf{n}})}{cR} \delta'\left(t' + \frac{R(t')}{c} - t\right)\right] dt'$$

where $\boldsymbol{\beta}(t') = \mathbf{v}(t')/c$ and the expression δ' implies differentiation with respect to the argument of the delta function. Integrating the second term by parts (problem 25) we find

$$\mathbf{E}(\mathbf{r}, t) = \frac{e}{4\pi\varepsilon_0}\left[\frac{\hat{\mathbf{n}}}{gR^2} + \frac{1}{cg}\frac{d}{dt'}\left(\frac{\hat{\mathbf{n}} - \boldsymbol{\beta}}{gR}\right)\right]_{\text{ret}} \tag{11.69}$$

in which $g = 1 - \hat{\mathbf{n}} \cdot \boldsymbol{\beta}$ and the subscript means that the expression within the brackets is to be evaluated at the retarded time

$$t' = t - R(t')/c$$

It is instructive to rewrite (11.69) in a form originally given by Feynman. Using the result

$$\frac{1}{c}\frac{d\hat{\mathbf{n}}}{dt'} = \frac{\hat{\mathbf{n}} \times (\hat{\mathbf{n}} \times \boldsymbol{\beta})}{R}$$

it follows on rearranging that

$$\frac{\boldsymbol{\beta}}{R} = \frac{(\hat{\mathbf{n}} \cdot \boldsymbol{\beta})\hat{\mathbf{n}}}{R} - \frac{1}{c}\frac{d\hat{\mathbf{n}}}{dt'} \tag{11.70}$$

Substituting (11.70) in (11.69) and using $dt/dt' = g$, we find

$$\mathbf{E}(\mathbf{r}, t) = \frac{e}{4\pi\varepsilon_0}\left[\frac{\hat{\mathbf{n}}}{R^2} + \frac{(\hat{\mathbf{n}} \cdot \boldsymbol{\beta})\hat{\mathbf{n}}}{gR^2} + \frac{1}{c}\frac{d}{dt}\left(\frac{\hat{\mathbf{n}}}{gR}\right) - \frac{1}{c}\frac{d}{dt}\left(\frac{(\hat{\mathbf{n}} \cdot \boldsymbol{\beta})\hat{\mathbf{n}}}{gR}\right) + \frac{1}{c^2}\frac{d^2\hat{\mathbf{n}}}{dt^2}\right]_{\text{ret}} \tag{11.71}$$

It is straightforward (problem 26) to combine the three middle terms in (11.71) to give

$$\mathbf{E}(\mathbf{r}, t) = \frac{e}{4\pi\varepsilon_0}\left[\frac{\hat{\mathbf{n}}}{R^2} + \frac{R}{c}\frac{d}{dt}\left(\frac{\hat{\mathbf{n}}}{R^2}\right) + \frac{1}{c^2}\frac{d^2\hat{\mathbf{n}}}{dt^2}\right]_{\text{ret}} \tag{11.72}$$

The first term in (11.72) represents the Coulomb field of the charge e at its retarded position. The second term is a correction to this retarded Coulomb field, being the product of the rate of change of this field and the retardation delay time R/c. Thus the first and second terms together correspond to computing the retarded Coulomb field and extrapolating it forward in time by R/c, i.e. to the observer's time. In other words for fields varying slowly enough these two terms represent the **instantaneous** Coulomb field of the charge. The final term, the second time derivative of the unit vector from the retarded position of the charge to the observer, is vital to the study of radiation, the first two terms being Coulombic.

It is again straightforward (problem 27) to show that the **radiation electric field** contained in (11.72) is

$$\mathbf{E}^{\text{rad}}(\mathbf{r}, t) = \frac{e}{4\pi\varepsilon_0}\left[\frac{\hat{\mathbf{n}}\times\{(\hat{\mathbf{n}}-\boldsymbol{\beta})\times\dot{\boldsymbol{\beta}}\}}{cg^3R}\right]_{\text{ret}} \tag{11.73}$$

Finally the total electric field is

$$\mathbf{E}(\mathbf{r}, t) = \frac{e}{4\pi\varepsilon_0}\left[\frac{(1-\beta^2)(\hat{\mathbf{n}}-\boldsymbol{\beta})}{g^3R^2}\right]_{\text{ret}} + \mathbf{E}^{\text{rad}}(\mathbf{r}, t) \tag{11.74}$$

Similarly (problem 28) the magnetic field is

$$\mathbf{B}(\mathbf{r}, t) = \frac{e}{4\pi\varepsilon_0 c}\left[\frac{(1-\beta^2)(\boldsymbol{\beta}\times\hat{\mathbf{n}})}{g^3R^2} + \frac{\hat{\mathbf{n}}\times(\hat{\mathbf{n}}\times\{(\hat{\mathbf{n}}-\boldsymbol{\beta})\times\dot{\boldsymbol{\beta}}\})}{cg^3R}\right]_{\text{ret}} \tag{11.75}$$

so that

$$c\mathbf{B} = \hat{\mathbf{n}}\times\mathbf{E}$$

§11.11 Radiation from an accelerated charge

Our expressions for the electric field contain one part which is identical with the field from a particle moving with uniform velocity and a part which involves the acceleration, the radiation field. Corresponding to this radiation field we have an energy flux determined by the Poynting vector

$$\begin{aligned}\mathbf{S} &= \mathbf{E}^{\text{rad}}\times\mathbf{H}^{\text{rad}}\\ &= (c\mu_0)^{-1}\mathbf{E}^{\text{rad}}\times(\hat{\mathbf{n}}\times\mathbf{E}^{\text{rad}})\\ &= (c\mu_0)^{-1}\,|\mathbf{E}^{\text{rad}}|^2\,\hat{\mathbf{n}}\end{aligned} \tag{11.76}$$

since $\hat{\mathbf{n}}\cdot\mathbf{E}^{\text{rad}}\equiv 0$. It follows directly that the instantaneous power radiated per unit solid angle, $dP/d\Omega$, is

$$\frac{dP(t)}{d\Omega} = (\mathbf{S}\cdot\hat{\mathbf{n}})R^2 = (c\mu_0)^{-1}\,|R\mathbf{E}^{\text{rad}}|^2 \tag{11.77}$$

Now $P(t)$ is the energy per unit time measured by an observer at time t due to radiation by the charge at the earlier time t'. It is more convenient to work with $P(t')$, i.e.

$$\frac{dP(t')}{d\Omega} = (\mathbf{S}\cdot\hat{\mathbf{n}})R^2\frac{dt}{dt'} = g(c\mu_0)^{-1}\,|R\mathbf{E}^{\text{rad}}|^2 \tag{11.78}$$

Then from (11.73) and (11.78)

$$\frac{dP(t')}{d\Omega}=\left(\frac{e}{4\pi\varepsilon_0}\right)^2\frac{1}{c^3\mu_0}\frac{|\hat{\mathbf{n}}\times\{(\hat{\mathbf{n}}-\boldsymbol{\beta})\times\dot{\boldsymbol{\beta}}\}|^2}{g^5} \tag{11.79}$$

For $\beta \ll 1$ (i.e. the non-relativistic limit) this reduces to

$$\frac{dP}{d\Omega}=\frac{e^2\dot{v}^2}{16\pi^2\varepsilon_0 c^3}\sin^2\theta \tag{11.80}$$

where θ is the angle between $\dot{\mathbf{v}}$ and $\hat{\mathbf{n}}$. We see that this is identical with (11.50). The full expression (11.79) is quite generally correct and contains relativistic effects both in the numerator and denominator. Clearly as $\beta \to 1$ (the ultra-relativistic limit) the effect of the denominator will be dominant. To see this it is simplest to consider the particular example in which $\boldsymbol{\beta}$ and $\dot{\boldsymbol{\beta}}$ are **collinear**. Then (11.79) reduces to

$$\frac{dP(t')}{d\Omega}=\left(\frac{e^2}{16\pi^2\varepsilon_0 c}\right)\frac{\dot{\beta}^2\sin^2\theta}{(1-\beta\cos\theta)^5} \tag{11.81}$$

As $\beta \to 1$ the denominator determines the radiation pattern. The characteristic dipole distribution deforms with the lobes switching towards the forward direction and extending as shown in Fig. 11.9. The angle at which the peak power is radiated is given by

$$\cos\theta_{\text{max}}=\frac{1}{3\beta}[(1+15\beta^2)^{1/2}-1] \tag{11.82}$$

For the case in which $\dot{\boldsymbol{\beta}} \perp \boldsymbol{\beta}$ and the particle executes instantaneously circular motion see problem 29.

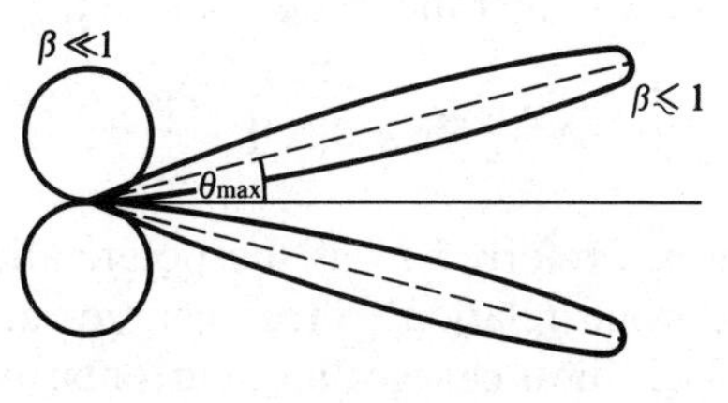

Fig. 11.9

§11.12 Problems

1. Obtain (11.12).

2. Verify by direct substitution that the retarded potentials (11.14), (11.15) satisfy the inhomogeneous wave equations (11.7), (11.8) and the Lorentz condition (11.6).

3. What equations are satisfied by $\mathbf{A}$, ϕ if in place of the Lorentz gauge we were to adopt the Coulomb gauge condition, $\nabla \cdot \mathbf{A} = 0$?

4. Show that in the wave-zone (where the far field approximation is valid) the potentials ϕ and $\mathbf{A}$ are related by $\phi = c\hat{\mathbf{n}} \cdot \mathbf{A}$ where $\hat{\mathbf{n}}$ denotes the direction of wave propagation.

5. Derive (11.24).

6. Derive (11.25).

7. Determine $\mathbf{E}$ in (11.26) from $\mathbf{E} = -\nabla\phi - \dfrac{\partial \mathbf{A}}{\partial t}$

8. Using the field components from (11.26) in the differential equation for the electric field lines show that the equation determining the electric field is (in polar coordinates)

$$\frac{\sin^2\theta}{r}[\cos(kr - \omega t) + kr \sin(kr - \omega t)] = \text{constant}$$

Show that the magnetic field lines are circles perpendicular to the axis of the dipole, with their centres on this axis.

9. The **Hertz vector Z** combines the potentials $\mathbf{A}$ and ϕ to satisfy the Lorentz condition and is defined by

$$\phi = -\nabla \cdot \mathbf{Z}; \qquad \mathbf{A} = \frac{1}{c^2}\frac{\partial \mathbf{Z}}{\partial t}$$

Similarly the source of the radiation fields ρ and $\mathbf{j}$ may be represented by a single vector function $\mathbf{p}$ such that

$$\rho = -\nabla \cdot \mathbf{p}; \qquad \mathbf{j} = \frac{\partial \mathbf{p}}{\partial t}$$

where $\mathbf{p}$ is sometimes referred to as the polarization vector because it bears the same relation to true charges and currents as does $\mathbf{P}$ to the polarization charges and currents; it is only a mathematical parallel.

Show that the Hertz potential satisfies

$$\left(\nabla^2-\frac{1}{c^2}\frac{\partial^2}{\partial t^2}\right)\mathbf{Z}=-\frac{\mathbf{p}}{\varepsilon_0}$$

and solve for $\mathbf{Z}$. Show also that

$$\mathbf{E}=\nabla\times(\nabla\times\mathbf{Z});\qquad \mathbf{B}=\frac{1}{c^2}\frac{\partial}{\partial t}(\nabla\times\mathbf{Z})$$

10. Two electric dipoles each of amplitude $\mathbf{p}_0$ are aligned along a given straight line and oscillate exactly out of phase so that there is zero net dipole moment. The separation between their centres is $d/2$ and we shall suppose $d \ll \lambda$. Determine the electromagnetic field at distances $r \gg d$ and show that the average power radiated is given by

$$\left\langle\frac{dP}{d\Omega}\right\rangle=\frac{\omega^6 p_0^2 d^2}{128\pi^2\varepsilon_0 c^5}\sin^2\theta\,\sin^2\left(\frac{\pi}{2}\cos\theta\right)$$

At what angle is the power radiated a maximum? Sketch the radiation pattern.

11. In the previous question investigate the limit as $d\to 0$ in such a way that $p_0 d$ remains constant. This limit is known as a **point quadrupole**.

12. Consider two electric dipoles oscillating out of phase in the z direction and centred at the points $\left(-\frac{\lambda}{4},0,0\right)$, $\left(+\frac{\lambda}{4},0,0\right)$. Show that in this case

$$\left\langle\frac{dP}{d\Omega}\right\rangle=\frac{\omega^4 p_0^2}{8\pi^2\varepsilon_0 c^3}\sin^2\theta\,\sin^2\left(\frac{\pi}{2}\sin\theta\cos\phi\right)$$

13. Sketch the radiation pattern obtained in the previous question.

14. Determine the radiation pattern from an electric dipole rotating in a plane with constant angular velocity ω.

15. Obtain (11.35).

16. Show that the radiation resistance of the linear antenna considered in §11.4 is given by

$$R_r = \frac{1}{\sqrt{2}\pi}\left(\frac{\mu_0}{\varepsilon_0}\right)^{1/2}\{0.577 + \ln kd_0 - Ci(kd_0)$$
$$+\tfrac{1}{2}\sin kd_0[Si(2kd_0) - 2Si(kd_0)]$$
$$+\tfrac{1}{2}\cos kd_0\left[0.577 + \ln\frac{kd_0}{2} + Ci(2kd_0) - 2Ci(kd_0)\right]\Big\}\text{ ohms}$$

where $Si(x)$ and $Ci(x)$ are the sine and cosine integrals

$$Si(x) = \int_0^x \frac{\sin x}{x}\,dx, \qquad Ci(x) = -\int_x^\infty \frac{\cos x}{x}\,dx$$

17. Obtain (11.43).

18. A travelling current wave $I = I_0 e^{i(kz-\omega t)}$ propagates in a linear antenna of length d_0 where z is the coordinate of a point on the antenna. Assuming that no reflected waves arise at the ends of the antenna determine the average power radiated and the radiation resistance R_r.

19. Calculate the electric field intensity in millivolts per metre at a distance of 1 km in the equatorial plane of a half-wave antenna radiating 1 kW of power.

20. Consider a system of charged particles all having the same charge-to-mass ratio e/m and such that the centre of mass of the system is in uniform motion. Show that this system cannot emit dipole radiation.

21. Establish (11.46) and (11.47).

22. Show that if electromagnetic radiation is incident on electrons moving with velocity $\mathbf{v}_0$, the Thomson scattered radiation is Doppler shifted by an amount $\omega = (\mathbf{k}_s - \mathbf{k}_0)\cdot\mathbf{v}_0$ where $\mathbf{k}_0, \mathbf{k}_s$ are the wave-vectors of the incident and scattered radiation.

23. The non-relativistic equation of motion of a charge e in an electromagnetic wave is

$$m\dot{\mathbf{v}} = e\mathbf{E} + e\mathbf{v}\times\mathbf{B} + \frac{e^2}{6\pi\varepsilon_0 c^3}\ddot{\mathbf{v}}$$

where the last term is a small correction to the Lorentz equation due to radiation emission; it is known as the **radiation reaction** and represents a damping force. Show that the time average of this damping force is $\dfrac{1}{12\pi\varepsilon_0}\left(\dfrac{e^2E_0}{mc^2}\right)^2$.

24. Show that

$$\int_{-\infty}^{\infty} \delta\{g(x)\}\gamma(x)\,dx = \sum_i \gamma(x_i)\Big/\left|\frac{dg(x_i)}{dx}\right|$$

in which the x_i are zeros of $g(x)=0$. This follows most easily by changing variable from x to g.

25. Establish (11.69).

26. By combining the three middle terms of (11.71) obtain Feynman's expression for the electric field (11.72).

27. Show that the contribution from $\dfrac{e^2}{4\pi\varepsilon_0 c^2}\left[\dfrac{d^2\hat{\mathbf{n}}}{dt^2}\right]_{\text{ret}}$ to the radiation field is

$$\mathbf{E}^{\text{rad}}(\mathbf{r}, t) = \frac{e}{4\pi\varepsilon_0}\left[\frac{\hat{\mathbf{n}}\times\{(\hat{\mathbf{n}}-\boldsymbol{\beta})\times\dot{\boldsymbol{\beta}}\}}{cg^3R}\right]_{\text{ret}}$$

28. Obtain (11.75).

29. Consider a charged particle executing instantaneously circular motion. Show that

$$\frac{dP}{d\Omega} = \frac{e^2\dot{\beta}^2}{16\pi^2\varepsilon_0 c}\frac{1}{(1-\beta\cos\theta)^3}\left[1-\frac{(1-\beta^2)\sin^2\theta\cos^2\phi}{(1-\beta\cos\theta)^2}\right]$$

Compare this angular distribution with that (11.81). Note that in this case too the peak power is radiated in the forward direction.

[Choose a coordinate system with $\boldsymbol{\beta}$ instantaneously along Oz, $\dot{\boldsymbol{\beta}}$ along Ox.]

30. Write down expressions for the **advanced potentials** analogous to (11.14) and (11.15). Show that if we were to choose solutions to the wave equations of the form

$$\phi = \tfrac{1}{2}(\phi_{\text{ret.}} + \phi_{\text{adv.}}); \qquad \mathbf{A} = \tfrac{1}{2}(\mathbf{A}_{\text{ret.}} + \mathbf{A}_{\text{adv.}})$$

the total energy radiated by an oscillating dipole vanishes.

12

Relativistic electrodynamics

§12.1 The principle of relativity

The theory of special relativity put forward by Einstein in 1905 has its origins in a dilemma arising out of the development of electrodynamics. The idea of relativity itself was not new; Newtonian mechanics recognized that absolute motion was a meaningless concept. Events in one frame of reference are related to those in a second frame moving uniformly with respect to the first, by a Galilean transformation. The laws of classical mechanics are invariant under Galilean transformations. The dilemma appears because this cannot be true for the laws of classical electrodynamics, which contain a fundamental limiting velocity, namely the velocity of light. It is straightforward to demonstrate that under a Galilean transformation the wave equation (9.24) is not invariant (problem 1).

As a way round this dilemma it was suggested that the laws of electrodynamics were valid only in a special frame of reference, known as the **aether frame**. If that were true it should then be possible to detect motion with respect to the aether frame. One of the great experiments of physics, the Michelson–Morley experiment, was designed to look for this effect. The idea was to detect the motion of the earth itself relative to the aether. The experiment, carried out first by Michelson in 1881 and repeated by Michelson and Morley in 1887, was a masterpiece of experimental ingenuity and showed a null result. There was no detectable effect of the motion of the earth relative to the aether. In 1892 FitzGerald proposed an ingenious explanation of the failure to detect motion through the aether, based on the hypothesis that bodies change their dimensions when in motion. The same explanation occurred to Lorentz and is known as **FitzGerald–Lorentz contraction**. The Michelson–Morley experiment has been repeated many times since, with yet greater refinement but the original conclusion is unchanged. That being so, the concept of the aether frame becomes untenable. Maxwell's equations must be universally valid. If that is the case then classical mechanics cannot be universally valid.

Following FitzGerald's hypothesis various attempts were made by Larmor, Lorentz and others to incorporate this idea in electrodynamic theory. In 1900 Larmor produced a transformation in which time as well as space was transformed, and in 1904 Lorentz showed that Maxwell's equations were invariant under this transformation. But it took the genius of Einstein to see that the dilemma had its roots in the classical concept of **absolute time**. The absolute nature of time is inherent in the Galilean transformation, under which Newton's laws are invariant. Einstein, however, postulated that statements in physical theory involving time should be on the same basis as those involving space, i.e. time should have only a **relative** meaning. Statements about time depend on the observer making them and will be different for observers in motion with respect to one another.

Einstein based the theory of special relativity on two axioms:

(i) the velocity of light is the same when viewed from any frame of reference;

(ii) it is impossible to detect uniform motion since the laws of physics are identical in all **inertial frames** (an inertial frame being one in which a body on which no forces act suffers no acceleration).

Classical electrodynamics is already consistent with Einstein's theory. In this chapter we shall examine the formulation of electrodynamics from the standpoint of relativity theory [11].

§12.2 The Lorentz transformation

The transformation equations relating space and time in frames of reference in uniform relative motion are generally known as the **Lorentz transformation**. They might more appropriately have been labelled the Larmor–Lorentz transformation, since as remarked in §12.1 it was Larmor who, in 1900, first obtained the formulae and, in so doing, recognized the need for a transformation in time as well as those in the space coordinates [5].

In deriving the transformation we shall follow essentially arguments first given by Poincaré based on the two postulates of relativity. The first postulate requires that for two frames of reference S and S' moving relatively to one another with uniform velocity $\mathbf{v}$, if

$$|\mathbf{r}_2 - \mathbf{r}_1|^2 = c^2 t^2$$

then

$$|\mathbf{r}_2' - \mathbf{r}_1'|^2 = c^2 t'^2$$

and vice versa. By choosing the origins of the coordinate systems in S, S' suitably we may set

$$\mathbf{r}_2 - \mathbf{r}_1 = \mathbf{r}; \qquad \mathbf{r}_2' - \mathbf{r}_1' = \mathbf{r}'$$

so that

$$r^2 - c^2 t^2 = K(r'^2 - c^2 t'^2) \tag{12.1}$$

Our object is to find the most general transformation which satisfies (12.1) **and** the second postulate of relativity. The need to ensure that motion which is uniform in S be also uniform in S' requires that K in (12.1), which in general could be $K(\mathbf{r}, \mathbf{v}, t)$, should **not** in fact depend on $\mathbf{r}$ or t. The fact that space is **isotropic** means that K cannot be a function of $\mathbf{v}$ and so at most (12.1) reduces to

$$r^2 - c^2 t^2 = K(v)(r'^2 - c^2 t'^2)$$

Moreover if we consider the inverse transformation it follows that

$$\begin{aligned} r'^2 - c^2 t'^2 &= K(-v)(r^2 - c^2 t^2) \\ &= K(v)(r^2 - c^2 t^2) \end{aligned}$$

since we assert that K cannot be a function of $\mathbf{v}$. Thus

$$K^2 = 1$$

Moreover in the limit as $v \to 0$ clearly S and S' become identical, and consequently

$$K = +1$$

i.e.

$$r^2 - c^2 t^2 = r'^2 - c^2 t'^2 \tag{12.2}$$

Now suppose that the inertial frame S' moves relative to S along Ox. The most general transformation between $(\mathbf{r}', t')$ and $(\mathbf{r}, t)$ may then be expressed by

$$\begin{aligned} x' &= Ax + Bt \qquad & z' &= Dz \\ y' &= Cy & t' &= Ex + Ft \end{aligned} \tag{12.3}$$

If we assume that at $t=0$, $x=0$, $z=0$ then (12.2) requires $y^2=y'^2$ so that $C=\pm1$; similarly $D=\pm1$. We choose the positive sign; the negative sign merely corresponds to reflection of the coordinates. Then, from (12.2),

$$x^2-c^2t^2=A^2x^2+2ABxt+B^2t^2-c^2(E^2x^2+2EFxt+F^2t^2)$$

i.e.

$$A^2-c^2E^2=1; \qquad B^2-c^2F^2=-c^2$$

$$AB=c^2EF=0$$

Moreover if we let S' move along Ox in S in the positive direction with velocity v then

$$\frac{B}{A}=-v$$

From these four conditions it follows that

$$A=\gamma; \qquad B=-v\gamma; \qquad E=-\frac{v\gamma}{c^2}; \qquad F=\gamma$$

where $\gamma=(1-\beta^2)^{-1/2}$, $\beta=v/c$. Thus the transformation (12.3) may be written

$$\begin{aligned} x'&=\gamma(x-vt) \qquad & z'&=z \\ y'&=y & t'&=\gamma\left(t-\frac{vx}{c^2}\right) \end{aligned} \tag{12.4}$$

The corresponding inverse transformation is

$$\begin{aligned} x&=\gamma(x'+vt') \qquad & z&=z' \\ y&=y' & t&=\gamma\left(t'+\frac{vx'}{c^2}\right) \end{aligned} \tag{12.5}$$

Note that in the limit as $c\rightarrow\infty$, $\beta\rightarrow0$, $\gamma\rightarrow1$ and the transformation reduces to a Galilean transformation. In this limit Newtonian theory, including the concept of time as an absolute quantity, is a valid approximation.

It is a straightforward matter to obtain the transformation laws for velocities. Defining velocity in S, S' by $\mathbf{u}$, $\mathbf{u}'$ we have

$$u_x=dx/dt,\ u_y=dy/dt,\ u_z=dz/dt \tag{12.6}$$

$$u'_x=dx'/dt',\ u'_y=dy'/dt',\ u'_z=dz'/dt' \tag{12.7}$$

Since the transformation relations (12.4) are of the form $x' = x'(x, t)$ and $t' = t'(x, t)$ it follows that

$$\frac{dx'}{dt'} = \frac{\dfrac{\partial x'}{\partial x}\dfrac{dx}{dt} + \dfrac{\partial x'}{\partial t}}{\dfrac{\partial t'}{\partial x}\dfrac{dx}{dt} + \dfrac{\partial t'}{\partial t}}$$

i.e.

$$u'_x = \frac{u_x - v}{1 - \dfrac{vu_x}{c^2}} \tag{12.8}$$

The corresponding inverse transformation is

$$u_x = \frac{u'_x + v}{1 + \dfrac{vu'_x}{c^2}} \tag{12.9}$$

We see at once that the sum of any two velocities is less than c unless one or other of them is the velocity of light itself, when the transformation shows that the transformed velocity is also c. For the components u'_y, u'_z we have

$$u'_y = \frac{u_y}{\gamma\left(1 - \dfrac{vu_x}{c^2}\right)}; \qquad u'_z = \frac{u_z}{\gamma\left(1 - \dfrac{vu_x}{c^2}\right)} \tag{12.10}$$

with inverse transformations

$$u_y = \frac{u'_y}{\gamma\left(1 + \dfrac{vu_x}{c^2}\right)}; \qquad u_z = \frac{u'_z}{\gamma\left(1 + \dfrac{vu_x}{c^2}\right)} \tag{12.11}$$

§12.3 The Minkowski diagram

In the previous section we saw how the principle of relativity enabled us to express a relation between the three space coordinates and (relative) time. In 1908 Minkowski discovered an important geometrical interpretation of this relation by introducing a four-dimensional manifold, with every point of which are associated three space coordinates and a time coordinate. A point in this four-space is known as a **world point**; its locus is a **world line**. The motion of a point in Minkowski space is therefore represented by world lines. The slope of a world line with respect to the time axis represents the velocity of the point as seen by an observer. A straight world line corresponds to uniform motion.

Considering one space dimension x and time, let us represent the transformation (12.4) in the following form. Since $\gamma^2(1-\beta^2)=1$ we may introduce α such that if

$$\gamma = \cosh \alpha, \text{ then } \beta\gamma = \sinh \alpha$$

and (12.4) may be expressed as

$$x' = x \cosh \alpha - ct \sinh \alpha \tag{12.12a}$$

$$ct' = -x \sinh \alpha + ct \cosh \alpha \tag{12.12b}$$

Thus if we set $t'=0$, which holds along the $0x'$ axis, then $ct = x \tanh \alpha = \beta x$ so that, in Fig. 12.1, $\tan \phi_1 = \beta$. Similarly by setting $x'=0$, $x=\beta ct$ and consequently $\tan \phi_2 = \beta$, i.e. the diagram is symmetrical. In the (x, ct) plane, position is determined by the usual Cartesian grid. In the (x', ct') plane lines of constant x' are parallel to the axis Oct' and those of constant t' are parallel to Ox'. The Minkowski metric is determined by (12.2), i.e.

$$x^2 - c^2t^2 = x'^2 - c^2t'^2 = \text{constant}$$

which is **hyperbolic** and is sketched in Fig. 12.1.

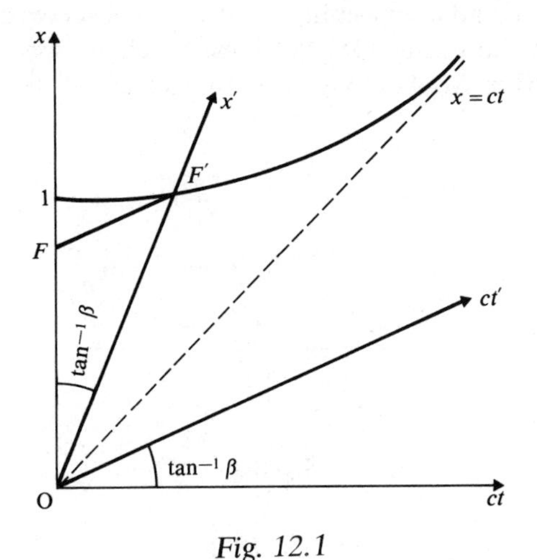

Fig. 12.1

The paths of light rays (photons) passing through the origin are the lines $x = \pm ct$, which divide the plane into four regions (Fig. 12.2). representing, for an observer at rest at O, **future**, **past** and **elsewhere** (so labelled because any event in these elsewhere regions can never be known to the observer since c is the maximum velocity for transmitting information). The accessible parts of the diagram are shaded and referred to as the **light cone**, since the lines $x = \pm ct$ generate a cone in the four-dimensional space. The origin represents the present instant. Figure 12.2 is known as a **Minkowski diagram**. A Galilean transformation, corresponding to $c \to \infty$, would be represented by the disappearance of the elsewhere region.

Expressing the invariance of equation (12.2) in differential form we may write

$$ds^2 = c^2 dt^2 - dx^2 - dy^2 - dz^2 \tag{12.13}$$

where ds denotes an **interval** of space-time. We shall define intervals as **time-like** if $ds^2 > 0$ and **space-like** if $ds^2 < 0$. Thus in Fig. 12.2 the past and future are time-like zones and elsewhere is space-like.

We can use a Minkowski diagram to illustrate the FitzGerald contraction. The frame S' moves along Ox in the frame S with velocity v. Consider a measuring rod of unit length at rest in S' with one end at O, the other at F' (Fig. 12.1). The world line of the first end is the ct' axis while that of the other is the line FF', parallel to Oct' and intersecting Ox at F. An observer in S will see the length of the rod as OF. We must be circumspect when comparing lengths in Minkowski space on account of the hyperbolic

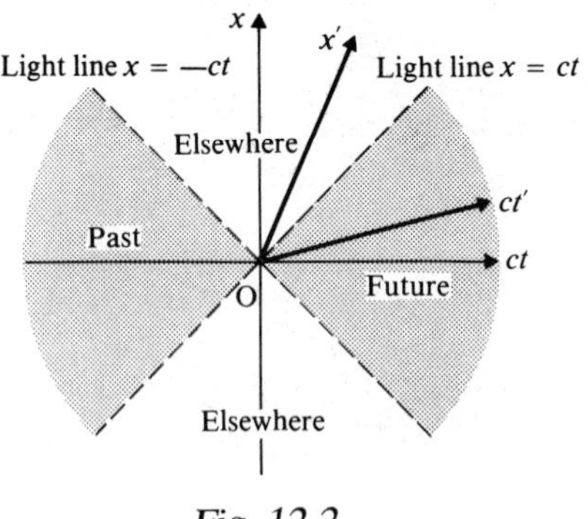

Fig. 12.2

metric. Marking unit length as shown along the Ox axis we see that O$F<1$, i.e. to an observer in the frame S the rod of unit length to the observer in S' appears to have contracted. To measure the length OF requires measuring the distance between the two world lines of the ends of the rod at a time when they are observed simultaneously in S. The world line of F' intersects Ox at F; for this line $ct=0$, $ct'=-\beta x'$. Thus from the invariance of $x^2-c^2t^2$ it follows that

$$x^2=x'^2-c^2t'^2=1-\beta^2$$

In other words to an observer in S the rod is seen to have contracted by a factor $(1-\beta^2)^{1/2}$.

Using a parallel argument we may represent the **dilation of time** on a Minkowski diagram (problem 5).

§12.4 Covariance

In §12.2 we used the principle of relativity to establish the space–time coordinate transformations. We found too the transformation formulae for components of velocity. We would now like to go further and find how other physical quantities transform; in particular we should like to establish how potentials and fields transform. For example, if we consider a charge at rest at the origin of an inertial frame S then of course we may immediately write down the corresponding electric field which is simply the Coulomb field. However, viewed from a frame S' which moves uniformly along Ox in S the charge will be seen to move uniformly. In other words a current in the Ox direction will be seen, together with the magnetic field associated with it. Thus just as the Lorentz transformation of the space–time coordinates mixes together space and time when relating events in different frames, so it appears that the transformations describing the electromagnetic field should do likewise for the component electric and magnetic fields.

In general we may say that if some physical law in S is expressed by the statement

$$C_l(q)=0$$

where q is a generalized coordinate (e.g. $\mathbf{r}$, t, $\mathbf{j}$, ρ), then viewed from the S' frame in which the corresponding coordinate is q', the same law will be expressed by

$$C'_m(q')=0$$

where not only is $q \neq q'$ but C_l and C'_m are also distinct, as in the example. We now wish to exploit a powerful concept which will enable us to extend the use of the relativity principle to obtain transformation equations of general physical variables. The concept is that of **covariance** which may be expressed by the statement:

$$C_l(q)=0 \Leftrightarrow C'_m(q')=0 \tag{12.14}$$

This is equivalent to saying that the quantities C are related to the quantities C' by the transformation (summation convention implied)

$$C'_m = Q_{ml} C_l, \quad |Q_{ml}| \neq 0 \tag{12.15}$$

In particular if this holds for a Lorentz transformation then the relation $C_l(q)=0$ is said to be **Lorentz covariant**.

In terms of the four-dimensional Minkowski space introduced in §12.2 let us now write the coordinates (x, y, z, ct) as (x^1, x^2, x^3, x^4) and express the Lorentz transformation (12.4) as

$$x^{\mu\prime} = Q^\mu_\nu x^\nu \tag{12.16}$$

where again – and throughout the rest of this chapter – the summation convention is implied. The transformation matrix of which Q^μ_ν is the element in row μ, column ν, is

$$Q^\mu_\nu = \begin{bmatrix} \gamma & 0 & 0 & -\beta\gamma \\ 0 & 1 & 0 & 0 \\ 0 & 0 & 1 & 0 \\ -\beta\gamma & 0 & 0 & \gamma \end{bmatrix} \tag{12.17}$$

We define any set of quantities C^μ which transform in this way, i.e. as

$$C^{\mu\prime} = Q^\mu_\nu C^\nu \tag{12.18}$$

to be a **contravariant four-vector**.

The inverse transformation to (12.16) expresses the x^ν in terms of the $x^{\mu\prime}$ as

$$x^\nu = (Q^\mu_\nu)^{-1} x^{\mu\prime} \tag{12.19}$$

where $(Q^{\mu}_{\nu})^{-1}$ is the inverse matrix of Q^{μ}_{ν} and is given by

$$(Q^{\mu}_{\nu})^{-1} = \begin{bmatrix} \gamma & 0 & 0 & +\beta\gamma \\ 0 & 1 & 0 & 0 \\ 0 & 0 & 1 & 0 \\ +\beta\gamma & 0 & 0 & \gamma \end{bmatrix} \quad (12.20)$$

with

$$(Q^{\mu}_{\nu})^{-1} Q^{\sigma}_{\nu} = \delta^{\sigma}_{\mu}$$

where δ^{σ}_{μ} is the unit matrix. A four-vector with components D_{ν} which transforms as the inverse transformation of the x^{ν} in (12.19), i.e. as

$$D'_{\mu} = (Q^{\mu}_{\nu})^{-1} D_{\nu} \quad (12.21)$$

is called a **covariant four-vector.**

Covariant and contravariant four-vectors are tensors of rank one. The set of sixteen quantities $C^{\mu}D^{\nu}$ which transform according to

$$(C^{\mu}D^{\nu})' = Q^{\mu}_{\sigma} Q^{\nu}_{\rho} C^{\sigma} D^{\rho} \quad (12.22)$$

form a contravariant tensor of rank two. Its covariant counterpart transforms as the product of two covariant four-vectors, i.e.

$$(C_{\mu}D_{\nu})' = (Q^{\mu}_{\sigma})^{-1} (Q^{\nu}_{\rho})^{-1} C_{\sigma} D_{\rho} \quad (12.23)$$

A mixed tensor of rank two is one which transforms as the product of a covariant and a contravariant four-vector, i.e. like

$$(C_{\mu}D^{\nu})' = (Q^{\mu}_{\sigma})^{-1} Q^{\nu}_{\rho} C_{\sigma} D^{\rho} \quad (12.24)$$

The extension to higher ranks is obvious. The tensor of rank zero, a scalar, is invariant with respect to any coordinate transformations. The tensor $T_{\mu\nu}$ is **symmetric** if $R_{\mu\nu} = T_{\nu\mu}$ and **anti-symmetric** or **skew-symmetric** if $T_{\mu\nu} = -T_{\nu\mu}$, implying $T_{\mu\mu} = 0$, and similarly for $T^{\mu\nu}$.

If a tensor vanishes in one frame, it will vanish in any other. Any tensor relation, if valid in one frame, holds in any other. We have seen in (12.22)–(12.24) that the **outer product** of two tensors of rank one is a tensor of rank two and this is true generally; e.g. $T_{\mu\nu}C^{\rho} = S^{\rho}_{\mu\nu}$. The operation of **contraction**, which is applicable to any mixed tensor, consists of setting one of the lower indices equal to one of the upper and summing (in the case of tensors in

Minkowski space from $n=1$ to $n=4$). Thus a contraction of the tensor $S^{\rho}_{\mu\nu}$ gives $S^{\nu}_{\mu\nu}=S_{\mu}$, a four-vector.

The **inner product** is the outer product contracted once or more. For example the inner product of C_{μ}, D^{ν} is $C_{\mu}D^{\mu}$ which is a scalar. Similarly the inner product of $T_{\mu\nu}S^{\sigma\rho}$ gives $T_{\mu\nu}S^{\sigma\nu}=C^{\sigma}_{\mu}$ (a mixed tensor) and after a second contraction, $T_{\mu\nu}S^{\mu\nu}=C^{\mu}_{\mu}=$ a scalar. If we apply two successive coordinate transformations then $x^{\sigma''}=Q^{\sigma}_{\mu}x^{\mu'}=Q^{\sigma}_{\mu}Q^{\mu}_{\nu}x^{\nu}=Q^{\sigma}_{\nu}x^{\nu}$ on contracting. Thus the application of two transformations is equivalent to another transformation (see problem 7).

If we consider the inner product of x_{μ} with x^{ν} then

$$x_{\mu}x^{\mu}=\text{invariant}$$

which is simply (12.2) expressed in the language of four-vectors, i.e.

$$\begin{aligned} x_{\mu}x^{\mu} &\equiv x^2+y^2+z^2-c^2t^2 \\ &= \text{invariant} \end{aligned}$$

Since the contravariant vector (x^1, x^2, x^3, x^4) was taken to be (x, y, z, ct) it follows that the covariant vector (x_1, x_2, x_3, x_4) is $(x, y, z, -ct)$.

§12.5 Transformation of electrodynamic variables

We shall now use the concepts introduced in the previous section to express Maxwell's equations in covariant form and so determine how the electrodynamic variables transform under a Lorentz transformation. Maxwell's equations in the absence of any dielectric or magnetic materials take the form

$$\nabla\cdot\mathbf{E}=\rho/\varepsilon_0 \nabla\cdot\mathbf{B}=0 \qquad (12.25\text{a,b})$$

$$\nabla\times\mathbf{E}=-\frac{\partial\mathbf{B}}{\partial t} \qquad \nabla\times\mathbf{H}=\mathbf{j}+\varepsilon_0\frac{\partial\mathbf{E}}{\partial t} \qquad (12.25\text{c,d})$$

with $\mathbf{H}=\mathbf{B}/\mu_0$. Moreover we know that the current and charge densities in (12.25) are related by the charge conservation

$$\frac{\partial\rho}{\partial t}+\nabla\cdot\mathbf{j}=0$$

Since this law is contained in the field equations it follows that like them it must be covariant under a Lorentz transformation. The transformation properties of ρ and $\mathbf{j}$ follow at once from this. Introducing the operator, ∂_μ, defined by

$$\partial_\mu = \frac{\partial}{\partial x_\mu} \equiv \left(\boldsymbol{\nabla}, \frac{\partial}{\partial(ct)}\right)$$

This operator transforms according to (12.21) and consequently is covariant. Thus we need a contravariant four-vector $j^\mu \equiv (\mathbf{j}, c\rho)$ to ensure that

$$\partial_\mu j^\mu = 0 \tag{12.26}$$

is covariant. (The use of the word **covariant** in two different senses in relativity is unfortunate but universal; when we say (12.26) is covariant we mean that its form is invariant under a Lorentz transformation. When we say a four-vector is covariant we mean that it transforms as in (12.21).) Thus j^μ transforms according as (12.18), i.e.

$$\begin{aligned} j^{1\prime} &= \gamma(j^1 - \beta j^4) \qquad & j^{3\prime} &= j^3 \\ j^{2\prime} &= j^2 & j^{4\prime} &= \gamma(j^4 - \beta j^1) \end{aligned} \tag{12.27}$$

In §11.1 we introduced the Lorentz gauge for the electrodynamic potentials by requiring that (see (11.6)),

$$\boldsymbol{\nabla} \cdot \mathbf{A} + \frac{1}{c^2}\frac{\partial \phi}{\partial t} = 0$$

The covariance of this gauge condition demands that $A^\mu = \left(\mathbf{A}, \frac{\phi}{c}\right)$ be a contravariant four-vector. The gauge condition is then simply

$$\partial_\mu A^\mu = 0$$

From this it is an easy step to express (11.7), (11.8) in covariant form:

$$\partial_\nu \partial^\nu A^\mu = -\mu_0 j^\mu \tag{12.28}$$

From these results we may proceed to the transformation of the fields themselves since

$$\mathbf{E} = -\boldsymbol{\nabla}\phi - \frac{\partial \mathbf{A}}{\partial t}; \qquad \mathbf{B} = \boldsymbol{\nabla} \times \mathbf{A}$$

It follows from the four-vector nature of A^μ that the field components form a tensor of rank two, $F^{\mu\nu}$, which must be antisymmetric since there are in all just six field quantities, namely the three components of **E** and **B**. Thus

$$F^{\mu\nu} = \partial^\mu A^\nu - \partial^\nu A^\mu \tag{12.29}$$

and $F^{\mu\nu} = -F^{\nu\mu}$. Then, in terms of the field components,

$$\begin{aligned}
&F^{\nu\nu} = 0 && F^{14} = -F^{41} = \frac{\partial}{\partial x}\left(\frac{\phi}{c}\right) + \frac{\partial A_x}{\partial(ct)} \\
&F^{12} = -F^{21} = B_z && \qquad\qquad = -E_x/c \\
&F^{13} = -F^{31} = -B_y && F^{24} = -F^{42} = -E_y/c \\
&F^{23} = -F^{32} = B_x && F^{34} = -F^{43} = -E_z/c
\end{aligned} \tag{12.30}$$

The **contravariant field tensor** $F^{\mu\nu}$ is therefore

$$F^{\mu\nu} = \begin{array}{c} \nu \rightarrow \\ \mu \downarrow \begin{bmatrix} 0 & B_z & -B_y & -E_x/c \\ -B_z & 0 & B_x & -E_y/c \\ B_y & -B_x & 0 & -E_z/c \\ \dfrac{E_x}{c} & \dfrac{E_y}{c} & \dfrac{E_z}{c} & 0 \end{bmatrix} \end{array} \tag{12.31}$$

The corresponding **covariant field tensor** $F_{\mu\nu}$ has the form

$$F_{\mu\nu} = \begin{array}{c} \nu \rightarrow \\ \mu \downarrow \begin{bmatrix} 0 & B_z & -B_y & E_x/c \\ -B_z & 0 & B_x & E_y/c \\ B_y & -B_x & 0 & E_z/c \\ \dfrac{-E_x}{c} & \dfrac{-E_y}{c} & \dfrac{-E_z}{c} & 0 \end{bmatrix} \end{array} \tag{12.32}$$

Maxwell's equations may now be expressed in terms of the field tensor. Consider first the two inhomogeneous equations (12.25a,d). From (12.29), forming the covariant derivative of the field tensor and applying the gauge condition, we get

$$\begin{aligned}
\partial_\nu F^{\mu\nu} &= \partial_\nu \partial^\mu A^\nu - \partial_\nu \partial^\nu A^\mu \\
&= -\partial_\nu \partial^\nu A^\mu
\end{aligned}$$

which, using (12.28) gives

$$\partial_\nu F^{\mu\nu} = \mu_0 j^\mu \tag{12.33a}$$

Both of the inhomogeneous Maxwell equations are contained in (12.33a). For example, with $\mu = 1$ we have

$$\partial_2 F^{12} + \partial_3 F^{13} + \partial_4 F^{14} = \mu_0 j^1$$

i.e.

$$\frac{\partial B_z}{\partial y} - \frac{\partial B_y}{\partial z} - \frac{1}{c}\frac{\partial E_x}{\partial (ct)} = \mu_0 j_x$$

which is simply the x component of

$$\nabla \times \left(\frac{\mathbf{B}}{\mu_0}\right) = \mathbf{j} + \varepsilon_0 \frac{\partial \mathbf{E}}{\partial t} \tag{12.25d}$$

Similarly with $\mu = 4$ we find

$$\partial_1 F^{41} + \partial_2 F^{42} + \partial_3 F^{43} = \mu_0 j^4$$

i.e.

$$\frac{1}{c}\left(\frac{\partial E_x}{\partial x} + \frac{\partial E_y}{\partial y} + \frac{\partial E_y}{\partial z}\right) = \mu_0 c\rho$$

or

$$\nabla \cdot \mathbf{E} = \frac{\rho}{\varepsilon_0} \tag{12.25a}$$

The two homogeneous Maxwell equations (12.25b,c) may be shown to correspond to

$$\partial_\lambda F_{\mu\nu} + \partial_\mu F_{\nu\lambda} + \partial_\nu F_{\lambda\mu} = 0 \tag{12.33b}$$

Choosing (λ, μ, ν) to be $(1, 2, 3)$ gives simply

$$\nabla \cdot \mathbf{B} = 0 \tag{12.25b}$$

while $(4, 2, 3)$ gives

$$\frac{1}{c}\frac{\partial B_x}{\partial t} + \frac{1}{c}\frac{\partial E_z}{\partial y} - \frac{1}{c}\frac{\partial E_y}{\partial z} = 0$$

which is the x component of

$$\nabla \times \mathbf{E} = -\frac{\partial \mathbf{B}}{\partial t} \tag{12.25c}$$

Thus in covariant form Maxwell's equations become

$$\partial_\nu F^{\mu\nu} = \mu_0 j^\mu \tag{12.33a}$$

$$\partial_\lambda F_{\mu\nu} + \partial_\mu F_{\nu\lambda} + \partial_\nu F_{\lambda\mu} = 0 \tag{12.33b}$$

Note too that the gauge transformations (11.2), (11.3) are covariant:

$$A'_\mu = A_\mu + \partial_\mu \Lambda \tag{12.34}$$

From the field tensor we may determine how the fields themselves transform under a Lorentz transformation. Using (12.22)

$$F^{\mu\nu\prime} = Q^\mu_\sigma Q^\nu_\rho F^{\sigma\rho} \tag{12.35}$$

we may show (problem 11) that

$$\begin{aligned}
\mathbf{E}'_\parallel &= \mathbf{E}_\parallel \\
\mathbf{B}'_\parallel &= \mathbf{B}_\parallel \\
\mathbf{E}'_\perp &= \gamma(\mathbf{E} + \mathbf{v}\times\mathbf{B})_\perp \\
\mathbf{B}'_\perp &= \gamma\left(\mathbf{B} - \frac{\mathbf{v}\times\mathbf{E}}{c^2}\right)_\perp
\end{aligned} \tag{12.36}$$

From these transformation formulae we see that electric and magnetic fields are relative. Thus it is possible for either **E** or **B** to be zero in one frame S and to be present in another frame S'. The inverse transformations are

$$\begin{aligned}
\mathbf{E}_\parallel &= \mathbf{E}'_\parallel & \mathbf{B}_\parallel &= \mathbf{B}'_\parallel \\
\mathbf{E}_\perp &= \gamma(\mathbf{E}' - \mathbf{v}\times\mathbf{B}')_\perp & \mathbf{B}_\perp &= \gamma\left(\mathbf{B}' + \frac{\mathbf{v}\times\mathbf{E}'}{c^2}\right)_\perp
\end{aligned} \tag{12.37}$$

§12.6 Invariants of the electromagnetic field

Quantities which remain invariant when transformations are made from one inertial frame to another are particularly important. The invariants of the electromagnetic field are easily obtained. First let us define the **dual** of a tensor $F_{\mu\nu}$ by the relation

$$G^{\alpha\beta} = \varepsilon^{\alpha\beta\mu\nu} F_{\mu\nu} \tag{12.38}$$

where $\varepsilon^{\alpha\beta\mu\nu}$ is the completely antisymmetric unit tensor of fourth rank. This unit tensor is itself defined so that its components are zero unless all four indices are distinct and the non-vanishing components are ± 1 depending on whether the indices $\alpha\beta\mu\nu$ form an even or odd permutation of 1234. In fact $\varepsilon^{\alpha\beta\mu\nu}$ is a **pseudo-tensor**; while it shows true tensor transformation properties under translations and rotations of coordinates it does not behave like a tensor under reflections of coordinates.

Making use of $\varepsilon^{\alpha\beta\mu\nu}$ it is at once clear that the product $\varepsilon^{\alpha\beta\mu\nu}F_{\alpha\beta}F_{\mu\nu}=F_{\alpha\beta}G^{\alpha\beta}$ is a scalar (strictly $F_{\alpha\beta}G^{\alpha\beta}$ is a pseudo-scalar since $G^{\alpha\beta}$ is itself a pseudo-tensor). Thus we have

$$\begin{aligned} F_{\mu\nu}F^{\mu\nu} &= \text{invariant} \\ \varepsilon^{\alpha\beta\mu\nu}F_{\alpha\beta}F_{\mu\nu} &= \text{invariant} \end{aligned} \tag{12.39}$$

Using the field tensor components we may show (problem 12) that (12.39) reduces to

$$E^2 - c^2B^2 = \text{invariant} \tag{12.40}$$

$$\mathbf{E}\cdot\mathbf{B} = \text{invariant} \tag{12.41}$$

From the invariance of these two quantities it is easy to establish the following theorems (problems 13–16):

(i) If in a reference frame S the electric and magnetic fields are mutually orthogonal, then they will remain so for **all** other inertial frames.

(ii) If $\mathbf{E}\cdot\mathbf{B}=0$ we can always find a frame in which $\mathbf{E}=0$ or $\mathbf{B}=0$ (according as $E^2-c^2B^2<0$ or >0. Conversely, if in any frame $\mathbf{E}=0$ or $\mathbf{B}=0$, then they are mutually orthogonal in **all** other inertial frames.

(iii) If $|\mathbf{E}|>|c\mathbf{B}|$ in some frame S, then $|\mathbf{E}|>|c\mathbf{B}|$ in all other inertial frames and vice-versa.

(iv) If $\theta=\cos^{-1}(\mathbf{E}\cdot\mathbf{B}/|\mathbf{E}|\,|\mathbf{B}|)$ is acute (obtuse) in some frame S, then θ will be acute (obtuse) in all other inertial frames.

§12.7 Field of a point charge in uniform motion

We turn next to a problem first considered in Chapter 11 where we saw that (§11.10) the electric field of a uniformly moving charge e was given by (see (11.74))

$$\mathbf{E}(\mathbf{r},t)=\frac{e}{4\pi\varepsilon_0}\left[\frac{(1-\beta^2)(\hat{\mathbf{n}}-\boldsymbol{\beta})}{g^3R^2}\right]_{\text{ret}} \tag{12.42}$$

Let us consider the charge e moving with uniform velocity $\mathbf{v}$ along the positive x-axis in the frame of reference S. Then in the frame S', which moves with the charged particle, the charge will be at rest so that (Fig. 12.3)

$$\mathbf{E}' = \frac{e\mathbf{r}'}{4\pi\varepsilon_0 r'^3}; \qquad \mathbf{B}' = 0 \tag{12.43}$$

For convenience, the origin O′ in S' has been chosen to coincide with the position of the charge. To determine the fields in the laboratory frame S we have only to apply (12.37), i.e.

$$E_x = E'_x = \frac{ex'}{4\pi\varepsilon_0 r'^3} \tag{12.44}$$

$$B_x = B'_x = 0 \tag{12.45}$$

$$\mathbf{E}_\perp = \gamma \mathbf{E}'_\perp = \frac{e\gamma \mathbf{r}'_\perp}{4\pi\varepsilon_0 r'^3} \tag{12.46}$$

$$\mathbf{B}_\perp = \frac{\gamma}{c^2}(\mathbf{v}\times\mathbf{E}') = \frac{e\gamma(\mathbf{v}\times\mathbf{r}')}{4\pi\varepsilon_0 r'^3 c^2} \tag{12.47}$$

Defining vectors

$$\gamma \mathbf{R}^* = \mathbf{r}' = \{\gamma(x - vt), y, z\}$$

and $\mathbf{R}$, the position vector of the charge relative to the point of observation at time t by

$$\mathbf{R} = \{(x - vt), y, z\}$$

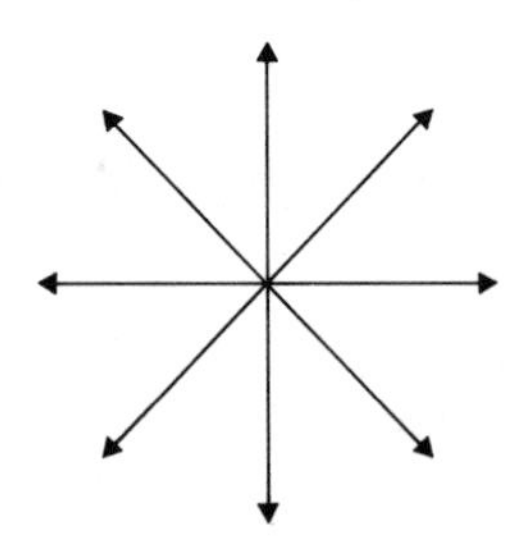

Fig. 12.3

it follows that the electric field **E** may be expressed as

$$\mathbf{E} = \frac{e}{4\pi\varepsilon_0} \frac{\mathbf{R}}{\gamma^2 R^{*3}} \tag{12.48}$$

and

$$\mathbf{B} = \frac{e\mu_0}{4\pi} \frac{\mathbf{v} \times \mathbf{R}}{\gamma^2 R^{*3}} \tag{12.49}$$

In terms of $\theta = \cos^{-1}(\mathbf{v} \cdot \mathbf{R}/|\mathbf{v}|\,|\mathbf{R}|)$ (12.48) may be written

$$\mathbf{E} = \frac{e\mathbf{R}(1-\beta^2)}{4\pi\varepsilon_0 R^3(1-\beta^2 \sin^2\theta)^{3/2}} \tag{12.50}$$

In this form it is immediately clear that the electric field is no longer spherically symmetric. Along the line of motion ($\theta = 0$) the field is reduced from the Coulomb value $e\mathbf{R}/4\pi\varepsilon_0 R^3$ by a factor $(1-\beta^2)$. Perpendicular to the line of motion ($\theta = \pi/2$) the electric field is enhanced over the Coulomb field by $(1-\beta^2)^{-1/2}$. In the limit, as $\beta \to 1$ the spherically symmetric Coulomb field is squashed in the direction of motion as illustrated in Fig. 12.4.

It is not at once obvious that (12.50) is equivalent to (12.42), obtained from the Liénard–Wiechert potentials. To establish their equivalence we have to evaluate (12.42) so that the field **E** is expressed in the **present**, rather than the **retarded**, position of the charge (see problem 19).

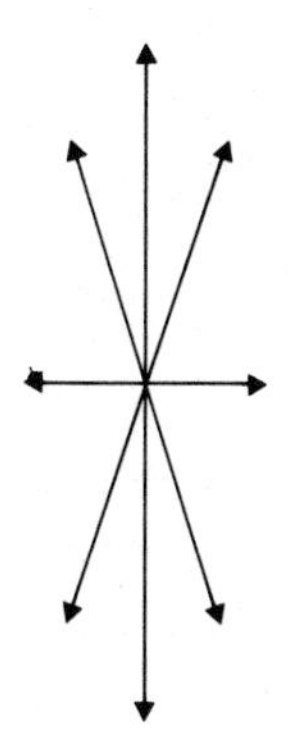

Fig. 12.4

§12.8 The Lorentz equation

To complete this introduction to relativistic electrodynamics we wish to establish the equations of motion of a charge in a prescribed electromagnetic field. We shall obtain these equations by postulating a Lagrangian in four-space and using this function in the principle of least action. Before we can do this we need to introduce a four-velocity. As they stand the velocity components considered in §12.2 do not form part of a four-vector.

Let us define a differential four-vector by $(dx, dy, dz, c\,dt)$. We know from (12.13) that the square of the length of this four-vector is an invariant. In other words the quantity $d\tau$ given by

$$c^2 d\tau^2 = c^2 dt^2 - dx^2 - dy^2 - dz^2 \tag{12.51}$$

is a differential invariant. From (12.51)

$$\begin{aligned} d\tau^2 &= dt^2\left[1 - \frac{1}{c^2}\left(\frac{dx}{dt}\right)^2 - \frac{1}{c^2}\left(\frac{dy}{dt}\right)^2 - \frac{1}{c^2}\left(\frac{dz}{dt}\right)^2\right] \\ &= dt^2\left(1 - \frac{v^2}{c^2}\right) \end{aligned}$$

i.e.

$$d\tau = \frac{dt}{\gamma} \tag{12.52}$$

The invariant time τ is known as the **proper time**. If we divide dx^μ by $d\tau$ we will obtain a new four-vector $dx^\mu/d\tau$ given by

$$\begin{aligned} \frac{dx^\mu}{d\tau} &= \left(\frac{dx}{d\tau}, \frac{dy}{d\tau}, \frac{dz}{d\tau}, \frac{d(ct)}{d\tau}\right) \\ &= \left(\gamma\frac{dx}{dt}, \gamma\frac{dy}{dt}, \gamma\frac{dz}{dt}, \gamma c\right) \end{aligned} \tag{12.53}$$

This four-vector is the required **four-velocity** v^μ,

$$v^\mu = (\gamma \mathbf{v}, \gamma c) \tag{12.54}$$

Note that this four-velocity satisfies the identity

$$v_\mu v^\mu = -c^2 \tag{12.55}$$

Moreover

$$\left.\begin{aligned} v^{1'} &= \gamma(v^1 - \beta v^4) \qquad & v^{3'} &= v^3 \\ v^{2'} &= v^2 & v^{4'} &= \gamma(v^4 - \beta v^1) \end{aligned}\right\} \tag{12.56}$$

where $\beta = u/c$. Writing out the components using (12.54) we get from (12.56)

$$\frac{v'_x}{\left(1-\frac{v'^2}{c^2}\right)^{1/2}} = \frac{v_x - u}{\left(1-\frac{u^2}{c^2}\right)^{1/2}\left(1-\frac{v^2}{c^2}\right)^{1/2}}$$

i.e.

$$\gamma(v')v'_x = \gamma(u)\gamma(v)(v_x - u) \tag{12.57a}$$

Also

$$\gamma(v')v'_y = \gamma(v)v_y \tag{12.57b}$$

$$\gamma(v')v'_z = \gamma(v)v_z \tag{12.57c}$$

$$\gamma(v') = \gamma(u)\gamma(v)\left(1-\frac{uv_x}{c^2}\right) \tag{12.57d}$$

From (12.57d) and (12.57a) we find that

$$v'_x = \frac{v_x - u}{1-\frac{uv_x}{c^2}} \tag{12.58}$$

which reproduces (12.8) (apart from a notational interchange of v with u).

We now wish to use this new four-vector in constructing a four-space Lagrangian to use in the principle of least action

$$\delta\int_{t_1}^{t_2} L_4(x^\mu, v^\mu)\, dt = 0$$

i.e.

$$\delta\int_{\tau_1}^{\tau_2} \gamma L_4(x^\mu, v^\mu)\, d\tau = 0 \tag{12.59}$$

where τ is the proper time and the variation takes place between fixed world points.

We require the four-Lagrangian to be a scalar function and since the equation of motion is linear it cannot contain terms in v^μ higher than second order. For a charged particle in the absence of electromagnetic fields we choose

$$L_4 = \frac{m_0 c}{\gamma}(-v_\mu v^\mu)^{1/2} \tag{12.60}$$

the coefficient being chosen to give the correct classical expression in the limit $c \to \infty$. The scalar term representing the interaction of the particle with the electromagnetic field described by a four-potential $A^\mu = \left(\mathbf{A}, \frac{\phi}{c}\right)$ is $\frac{e}{\gamma} v_\mu A^\mu$. Thus the complete Lagrangian may be written

$$L_4 = \frac{m_0 c}{\gamma} (-v_\mu v^\mu)^{1/2} + \frac{e}{\gamma} v_\mu A^\mu \tag{12.61}$$

To obtain the equations of motion for the particle in covariant form we use (12.61) in (12.59)

$$\begin{aligned}\delta \int_1^2 \gamma L_4(x^\mu, v^\mu)\, d\tau &= \int_1^2 [m_0 c \delta(-v_\mu v^\mu)^{1/2} + e\delta(v_\mu A^\mu)]\, d\tau \\ &= \int_1^2 \left[m_0 v_\mu \delta v^\mu + e\left(A_\mu \delta v^\mu + \frac{\partial A_\mu}{\partial x^\nu} \delta x^\nu v^\mu\right)\right] d\tau\end{aligned} \tag{12.62}$$

Now $\delta v^\mu = \frac{d}{d\tau}(\delta x^\mu)$ and $\frac{\partial A_\mu}{\partial \tau} = \frac{\partial A_\mu}{\partial x^\nu} v^\nu$ so that, on integrating by parts

$$\delta \int_1^2 \gamma L_4(x^\mu, v^\mu)\, d\tau = \int_1^2 \left[-m_0 \frac{dv_\nu}{d\tau} + e\left(\frac{\partial A_\mu}{\partial x^\nu} - \frac{\partial A_\nu}{\partial x^\mu}\right) v^\mu\right] \delta x^\nu\, d\tau \tag{12.63}$$

Since δx^ν is an arbitrary variation it follows, using (12.29), that

$$m_0 \frac{dv_\nu}{d\tau} = eF_{\nu\mu} v^\mu \tag{12.64}$$

which is the covariant equation of motion of the particle of charge e and rest mass m_0.

Switching back now to three-vector notation we find from (12.64) for $\nu = 1, 2, 3$

$$m_0 \frac{d}{d\tau}(\gamma \mathbf{v}) = e\gamma[\mathbf{E} + \mathbf{v} \times \mathbf{B}]$$

i.e.

$$\frac{d\mathbf{p}}{dt} = e[\mathbf{E} + \mathbf{v} \times \mathbf{B}] \tag{12.65}$$

where we have introduced the momentum $\mathbf{p}=m_0\gamma\mathbf{v}=m(v)\mathbf{v}$, where the mass

$$m(v)=\gamma m_0=\frac{m_0}{\left(1-\frac{v^2}{c^2}\right)^{1/2}} \tag{12.66}$$

Thus the space part of (12.64) is simply the generalization of the **Lorentz equation**, previously introduced in Chapter 5 for velocities small compared with c. We see that the form of the equation is preserved with the momentum $\mathbf{p}$ now being given by $\mathbf{p}=m_0\gamma\mathbf{v}$.

There remains the time-like component of (12.64). This tells us that

$$m_0\frac{d}{d\tau}(\gamma c)=-e\frac{\gamma}{c}\mathbf{E}\cdot\mathbf{v}$$

or

$$\frac{d}{dt}(m_0\gamma c^2)=-e\mathbf{E}\cdot\mathbf{v} \tag{12.67}$$

How are we to interpret the expression on the left-hand side? The right-hand side of (12.67) simply expresses the rate at which work is done on the particle by the electric field. We would expect this by analogy with non-relativistic mechanics to be equal to the rate of change of the kinetic energy T with time. Making this identification and integrating with respect to time gives

$$\int_{t_1}^{t_2}\frac{dT}{dt}\,dt=[m_0\gamma c^2]_{t_1}^{t_2}$$

Suppose that the particle is at rest at $t=t_1$, then

$$\begin{aligned}T&=m_0c^2(\gamma-1)\\&=mc^2-m_0c^2\end{aligned} \tag{12.68}$$

Thus the kinetic energy T is simply the difference between the quantity mc^2 and the **rest energy**, m_0c^2. Put otherwise, we may express W, the total energy, as

$$W=T+m_0c^2 \tag{12.69}$$

and (12.67) becomes

$$\frac{dW}{dt}=-e\mathbf{E}\cdot\mathbf{v} \tag{12.70}$$

Note that from (12.69) $W = m_0\gamma c^2$, while in (12.65) we wrote the momentum $\mathbf{p} = m_0\gamma\mathbf{v}$. Thus

$$\frac{W^2}{c^2} = m_0^2\gamma^2c^2 \text{ and } p^2 = m_0^2\gamma^2v^2$$

so that

$$\frac{W^2}{c^2} = p^2 + m_0^2\gamma^2(c^2 - v^2)$$

i.e.

$$W = [m_0^2c^4 + p^2c^2]^{1/2} \tag{12.71}$$

We shall now consider a number of examples of particle motion in a variety of fields and solve the appropriate Lorentz equation in each case.

§12.9 Charged particle in a constant, uniform E field

The simplest problem is that of a charged particle moving in a constant uniform electric field. Generally the best approach is to start from (12.64) in four-vector form. We may of course use the equivalent three-vector Lorentz equation (12.65). To show both approaches we shall adopt (12.65) here and use (12.64) for the more complicated problem in the following section. We have then

$$\frac{d\mathbf{p}}{dt} = \frac{d}{dt}(m_0\gamma\mathbf{v}) = m_0\gamma\frac{d\mathbf{v}}{dt} + m_0\gamma^3\mathbf{v}\frac{v}{c^2}\frac{dv}{dt}$$

$$= m_0\gamma^3\left[\frac{1}{\gamma^2}\frac{d\mathbf{v}}{dt} + \frac{v\mathbf{v}}{c^2}\frac{dv}{dt}\right] \tag{12.72}$$

For collinear $\mathbf{v}$, $\dfrac{d\mathbf{v}}{dt}$ it follows that

$$\frac{d\mathbf{p}_L}{dt} = m_0\gamma^3\frac{d\mathbf{v}}{dt} \tag{12.73}$$

If on the other hand $\dfrac{d\mathbf{v}}{dt}$ is orthogonal to $\mathbf{v}$, then

$$\frac{d\mathbf{p}_T}{dt} = m_0\gamma\frac{d\mathbf{v}}{dt} \tag{12.74}$$

Suppose the electric field is directed along Oz *and that there is* a component of velocity at right angles to the field, i.e.

$$\mathbf{E} = E\hat{\mathbf{z}}, \qquad \mathbf{v} = v_x\hat{\mathbf{x}} + v_z\hat{\mathbf{z}}$$

Then from (12.72)

$$e\mathbf{E} = m_0\gamma^3\left[\frac{1}{\gamma^2}\frac{d\mathbf{v}}{dt} + \frac{v\mathbf{v}}{c^2}\frac{dv}{dt}\right] \tag{12.75}$$

and in component form

$$0 = \left[\left(1 - \frac{v^2}{c^2}\right)\frac{dv_x}{dt} + \frac{vv_x}{c^2}\frac{dv}{dt}\right] \tag{12.76a}$$

$$eE = m_0\gamma^3\left[\left(1 - \frac{v^2}{c^2}\right)\frac{dv_z}{dt} + \frac{vv_z}{c^2}\frac{dv}{dt}\right] \tag{12.76b}$$

The first equation is separable and may be integrated directly to give

$$\frac{v_x}{v_0} = \left(\frac{c^2 - v^2}{c^2 - v_0^2}\right)^{1/2} \tag{12.77}$$

where v_0 is the initial velocity taken to be in the x direction. It follows from (12.77) that

$$v_z^2 \equiv v^2 - v_x^2 = v^2 - \frac{v_0^2(c^2 - v^2)}{c^2 - v_0^2} = \frac{c^2(v^2 - v_0^2)}{c^2 - v_0^2}$$

which, when substituted into (12.76b), gives

$$\frac{eE}{m_0c^2(c^2 - v_0^2)^{1/2}} = \frac{v}{(v^2 - v_0^2)^{1/2}(c^2 - v^2)^{3/2}}\frac{dv}{dt} \tag{12.78}$$

On integrating we find

$$v(t) = \left(\frac{v_0^2 + c^2\omega_0^2t^2}{1 + \omega_0^2t^2}\right)^{1/2} \tag{12.79}$$

with

$$\omega_0 = \frac{eE}{m_0c^2}(c^2 - v_0^2)^{1/2} = \frac{eE}{mc} \tag{12.80}$$

We then have

$$v_x = \frac{v_0}{(1 + \omega_0^2t^2)^{1/2}}; \qquad v_z = \frac{c\omega_0 t}{(1 + \omega_0^2t^2)^{1/2}} \tag{12.81}$$

Integrating once more with respect to time and taking $x(0)=0$, $z(0)=0$, gives

$$x=\frac{v_0}{\omega_0}\sinh^{-1}\omega_0 t$$
$$z=\frac{c}{\omega_0}[(1+\omega_0^2t^2)^{1/2}-1] \tag{12.82}$$

The particle trajectory is therefore

$$\frac{\omega_0 z}{c}=\cosh\left(\frac{\omega_0 x}{v_0}\right)-1 \tag{12.83}$$

which reduces to the familiar classical result at low velocities.

§12.10 Motion of a charged particle in static, uniform **E** and **B** fields

In Fig. 5.14 (and problem 20, §5.14) we considered the non-relativistic motion of a charged particle in static and uniform electric and magnetic fields. Here we shall determine the motion for a particle of relativistic energy. Writing the Lorentz equation (12.64) we find, on taking the z-axis to be in the direction of the magnetic field,

$$m_0\frac{d^2x}{d\tau^2}=eE_x\frac{dt}{d\tau}+eB\frac{dy}{d\tau}$$
$$m_0\frac{d^2y}{d\tau^2}=eE_y\frac{dt}{d\tau}-eB\frac{dx}{d\tau}$$
$$m_0\frac{d^2z}{d\tau^2}=eE_z\frac{dt}{d\tau} \tag{12.84}$$
$$m_0c^2\frac{d^2t}{d\tau^2}=e\left[E_x\frac{dx}{d\tau}+E_y\frac{dy}{d\tau}+E_z\frac{dz}{d\tau}\right]$$

in which m_0 denotes the rest mass of the particle.

In the case when $\mathbf{E}=0$ the solution to (12.84) will be

$$\left.\begin{aligned} x=\frac{v_\perp}{\Omega}\sin(\Omega_0\tau+\alpha); \qquad y=\frac{v_\perp}{\Omega}\cos(\Omega_0\tau+\alpha) \\ v^2=v_\perp^2+v_\parallel^2=\text{constant} \end{aligned}\right\} \tag{12.85}$$

with $\Omega_0 = |e|B/m_0$. In terms of t rather than the proper time τ, we see that the Larmor frequency for a relativistic particle is given by

$$\Omega = \Omega_0(1-\beta^2)^{1/2} \tag{12.86}$$

The other special case, i.e. when $\mathbf{B}=0$, has been considered in §12.9. When considering the non-relativistic motion in fields **E** and **B** the analysis was simplified by using a reference frame moving with the particle drift velocity. It is possible to follow the same procedure in the relativistic case.

We established in §12.6 that the electromagnetic field possesses two invariants

$$E^2 - c^2B^2 = I_1; \qquad \mathbf{E}\cdot\mathbf{B} = I_2$$

This result allows us to distinguish fields in which the first invariant vanishes and those in which both I_1 and I_2 vanish. In the first case **E** and **B** can be made parallel to one another by an appropriate choice of reference frame. In the second case, **E** and **B** are mutually orthogonal in all frames.

In the case in which $I_1 = 0$, (12.84) decouples into two sets of equations:

$$\begin{aligned} m_0\frac{d^2x}{d\tau^2} &= eB\frac{dy}{d\tau} \qquad & m_0\frac{d^2z}{d\tau^2} &= eE\frac{dt}{d\tau} \\ m_0\frac{d^2y}{d\tau^2} &= -eB\frac{dx}{d\tau} \qquad & m_0c^2\frac{d^2t}{d\tau^2} &= eE\frac{dz}{d\tau} \end{aligned} \tag{12.87}$$

These equations have solutions

$$\begin{aligned} x &= \frac{A}{\Omega_0}\sin\Omega_0\tau; \qquad & y &= \frac{A}{\Omega_0}\cos\Omega_0\tau \\ z &= \frac{cB}{E}\frac{(1+A^2)^{1/2}}{\Omega_0}\cosh\frac{\Omega_0E\tau}{cB}; \qquad & t &= \frac{B}{E}\frac{(1+A^2)^{1/2}}{\Omega_0}\sinh\frac{\Omega_0E\tau}{cB} \end{aligned} \tag{12.88}$$

The trajectory is plotted in Fig. 12.5.

In the second case when $I_1 = 0 = I_2$ then $E = cB$ and (12.84) becomes

$$\begin{aligned} \frac{d^2x}{d\tau^2} &= \Omega_0\frac{dy}{d\tau} \qquad & \frac{d^2z}{d\tau^2} &= 0 \\ \frac{d^2y}{d\tau^2} &= \Omega_0\left(c\frac{dt}{d\tau} - \frac{dx}{d\tau}\right) \qquad & c\frac{d^2t}{d\tau^2} &= \Omega_0\frac{dy}{d\tau} \end{aligned} \tag{12.89}$$

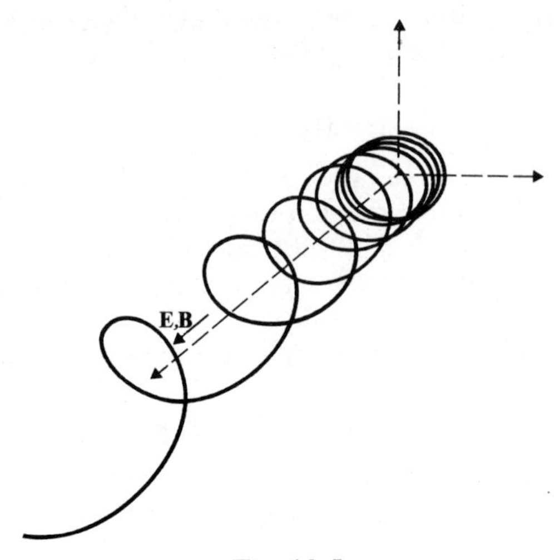

Fig. 12.5

The solution to (12.89) may be written (problem 22)

$$\begin{aligned}
x &= \frac{A}{6}(\Omega_0\tau)^3 + \frac{C^2 - A^2 + c^2\Omega_0^{-2}}{2A}\Omega_0\tau \\
y &= \frac{A}{2}(\Omega_0\tau)^2 \\
z &= C\Omega_0\tau \\
ct &= \frac{A}{6}(\Omega_0\tau)^3 + \frac{C^2 + A^2 + c^2\Omega_0^{-2}}{2A}\Omega_0\tau
\end{aligned} \tag{12.90}$$

in which A and C are arbitrary constants.

§12.11 Motion of a charged particle in a monochromatic plane wave

Another problem in which the relativistic Lorentz equation may be integrated is that of the motion of an electron (or proton) in a monochromatic plane wave. We saw in §11.7 that the velocity of the electron is proportional to the amplitude of the wave. Thus if we have radiation of sufficient intensity – as in the most powerful lasers – relativistic particle velocities may be attained. The analysis presented here reveals a **drift** in the direction of propagation of the wave.

For a plane polarized wave propagating in the x direction, $\mathbf{E}=(0, E, 0)$ and $\mathbf{B}=(0, 0, B)$ (there is no background magnetic field in this problem). If we describe the fields by their vector potential, $\mathbf{A}=(0, a(\tau)\cos\omega\tau, 0)$ where a is the wave amplitude, ω the frequency and τ the proper time, then $\mathbf{E}$ and $\mathbf{B}$ are given by

$$\mathbf{E}=\frac{\partial \mathbf{A}}{\partial t}, \qquad \mathbf{B}=\nabla\times\mathbf{A} \tag{12.91}$$

Note that we have written the amplitude as $a(\tau)$. For a truly monochromatic wave a is constant. In practice, however, a can never be constant since the amplitude must grow from zero when the wave is switched on. If we assume constant a we are led to solutions which depend critically on the initial phase and so predict drifts in arbitrary directions. For an almost monochromatic wave $a(\tau)$ must be a slowly varying function and the only significant effect of including its variation is to ensure that the fields are initially zero; dependence on the initial phase does not then appear. From (12.91)

$$E=cB=-\frac{dA}{d\tau}$$

The relativistic Lorentz equation, (12.65), (12.67) then reads

$$\frac{d}{dt}(\gamma\dot{x})=-\frac{e}{m_0c}\,\dot{y}\,\frac{dA}{d\tau} \tag{12.92}$$

$$\frac{d}{dt}(\gamma\dot{y})=-\frac{e}{m_0c}(c-\dot{x})\frac{dA}{d\tau} \tag{12.93}$$

$$\frac{d}{dt}(\gamma\dot{z})=0 \tag{12.94}$$

$$\frac{d}{dt}(\gamma c)=-\frac{e}{m_0c}\,\dot{y}\,\frac{dA}{d\tau} \tag{12.95}$$

Substracting (12.92) from (12.95) gives, on integrating,

$$\gamma(c-\dot{x})=c \tag{12.96}$$

assuming that the electron is initially at rest at the origin. Since

$$\frac{d\tau}{dt}=1-\frac{\dot{x}}{c} \tag{12.97}$$

(12.93) may be integrated directly to give

$$\gamma\dot{y} = -\frac{eA}{m_0} \tag{12.98}$$

Substituting for $\dot{y}$ from (12.98) and using (12.96), (12.97), (12.92) becomes

$$\frac{d}{dt}\left(\frac{\gamma\dot{x}}{c}\right) = \left(\frac{e}{m_0 c}\right)^2 A \frac{dA}{d\tau}\frac{d\tau}{dt}$$

so that on integrating

$$\frac{\gamma\dot{x}}{c} = \frac{1}{2}\left(\frac{eA}{m_0 c}\right)^2$$

Finally, using (12.96), (12.97) again

$$\dot{x} = \frac{c}{2}\left(\frac{eA}{m_0 c}\right)^2 \frac{d\tau}{dt} = \frac{c}{2}\left(\frac{ea}{m_0 c}\right)^2 \frac{d\tau}{dt}\cos^2\omega\tau \tag{12.99}$$

$$\dot{y} = -c\left(\frac{eA}{m_0 c}\right)\frac{d\tau}{dt} = -c\left(\frac{ea}{m_0 c}\right)\frac{d\tau}{dt}\cos\omega\tau \tag{12.100}$$

and, from (12.94)

$$\dot{z} = 0 \tag{12.101}$$

Averaging $\dot{x}$ over a period $T = 2\pi/\omega$, $a(\tau)$ may be taken as constant, so that

$$\langle\dot{x}\rangle = \frac{\int \dot{x}\,dt}{\int dt} = \frac{\frac{c}{2}\left(\frac{ea}{m_0 c}\right)^2 \int \cos^2\omega\tau\,d\tau}{\int\left(d\tau + \frac{dx}{c}\right)}$$

$$= \frac{\frac{c}{2}\left(\frac{ea}{m_0 c}\right)^2 \int \cos^2\omega\tau\,d\tau}{\int d\tau\left[1 + \frac{1}{2}\left(\frac{ea}{m_0 c}\right)^2 \cos^2\omega\tau\right]}$$

$$= \frac{c\left(\frac{ea}{2m_0 c}\right)^2}{1 + \left(\frac{ea}{2m_0 c}\right)^2} \tag{12.102}$$

and

$$\dot{y} = 0$$

Thus the electron drifts in the direction of propagation of the wave. If $\frac{ea}{m_0 c} < 1$ the motion is mainly in the y direction (i.e. oscillation due to $\mathbf{E}$). The drift velocity is given approximately by

$$\mathbf{v}_D = \frac{1}{2}\left(\frac{e}{m_0 c\omega}\right)^2 \langle \mathbf{E} \times \mathbf{B} \rangle$$

An interesting effect of this drift is to predict a Doppler shift in the frequency of light scattered by free electrons.

§12.12 Radiation from relativistic particles

We saw in §11.6 that the power radiated by an accelerated particle in the non-relativistic limit is given by Larmor's formula

$$P = \frac{e^2[\dot{v}]^2}{6\pi\varepsilon_0 c^3} \tag{12.103}$$

We may generalize this result to relativistic energies by first re-writing it as

$$P = \frac{e^2}{6\pi\varepsilon_0 m_0^2 c^3}\left[\frac{d\mathbf{p}}{dt} \cdot \frac{d\mathbf{p}}{dt}\right]$$

and then expressing this in covariant form, i.e.

$$P = \frac{e^2}{6\pi\varepsilon_0 m_0^2 c^3}\left[\frac{dp_\mu}{d\tau}\frac{dp^\mu}{d\tau}\right] \tag{12.104}$$

Evaluating this we find

$$\begin{aligned} P &= \frac{e^2\gamma^6}{6\pi\varepsilon_0 c^3}\left[\frac{\dot{v}^2}{\gamma^4} + \frac{2(\mathbf{v}\cdot\dot{\mathbf{v}})^2}{c^2\gamma^2} + \frac{(\mathbf{v}\cdot\dot{\mathbf{v}})^2 v^2}{c^4} - \frac{(\mathbf{v}\cdot\dot{\mathbf{v}})^2}{c^2}\right] \\ &= \frac{e^2\gamma^4}{6\pi\varepsilon_0 c}[(1-\beta^2)\dot{\beta}^2 + 2(1-\beta^2)(\boldsymbol{\beta}\cdot\dot{\boldsymbol{\beta}})^2 - (1-\beta^2)(\boldsymbol{\beta}\cdot\dot{\boldsymbol{\beta}})^2] \end{aligned}$$

Thus

$$P = \frac{e^2}{6\pi\varepsilon_0 c}\frac{[(\dot{\boldsymbol{\beta}})^2 - (\boldsymbol{\beta}\times\dot{\boldsymbol{\beta}})^2]}{(1-\beta^2)^3} \tag{12.105}$$

which is the relativistic generalization of Larmor's formula.

It follows that the total power radiated in all directions for $\boldsymbol{\beta}$, $\dot{\boldsymbol{\beta}}$ collinear is

$$P = \frac{e^2}{6\pi\varepsilon_0 c} \frac{\dot{\beta}^2}{(1-\beta^2)^3}$$

while for $\boldsymbol{\beta} \perp \dot{\boldsymbol{\beta}}$

$$P = \frac{e^2}{6\pi\varepsilon_0 c} \frac{\dot{\beta}^2}{(1-\beta^2)^2}$$

§12.13 Problems

1. Show that the wave equation $\nabla^2 \mathbf{E} = \dfrac{1}{c^2}\dfrac{\partial^2 \mathbf{E}}{\partial t^2}$ is not invariant under a Galilean transformation.

2. Show that the Lorentz transformation may be written in vector form as

$$\mathbf{r}' = \gamma\{(\hat{\mathbf{n}} \cdot \mathbf{r})\hat{\mathbf{n}} - \mathbf{v}t\} + \hat{\mathbf{n}} \times (\mathbf{r} \times \hat{\mathbf{n}})$$

$$t' = \gamma\left(t - \frac{\mathbf{v} \cdot \mathbf{r}}{c^2}\right)$$

where $\hat{\mathbf{n}} = \mathbf{v}/|\mathbf{v}|$.

3. Show that in vector form the velocity addition result may be expressed as

$$\mathbf{u} = \frac{\mathbf{u}' + \mathbf{v} + (\gamma - 1)[(\mathbf{u}' \cdot \hat{\mathbf{n}}) + v]\hat{\mathbf{n}}}{\gamma\left(1 + \dfrac{\mathbf{u}' \cdot \mathbf{v}}{c^2}\right)}$$

where $\mathbf{v} = v\hat{\mathbf{n}}$.

4. Show that the Lorentz transformations form a group, known as the **Lorentz group**, G_L, by establishing the group properties i.e.
 (i) for any a, $b \in G_L$, $ab \in G_L$
 (ii) $(ab)c = a(bc)$
 (iii) $\exists e$ such that $ae = ea = a$, all $a \in G$
 (iv) $\exists a^{-1}$ for all $a \in G$ such that $a^{-1}a = aa^{-1} = e$.

The Lorentz group is the invariance group for Maxwell's equations, i.e. if the electrodynamic equations are valid in a frame S they will be valid in any frame S' related to S by a transformation belonging to the Lorentz group.

5. Using a Minkowski diagram show that a time interval measured by an observer in a frame moving with respect to the rest frame will appear dilated by the factor γ over that measured in the rest frame. This phenomenon is known as **time dilation**.

6. Two inertial frames S and S' move with velocity $c/3$ with respect to one another. Draw a Minkowski diagram representing these systems, taking Ox and Oct in S to be orthogonal. Draw hyperbolas to indicate unit displacement along the axes Ox, Ox', Oct, Oct' and plot the following world points ($x = 1$, $ct = 1$), ($x' = 1$, $ct' = 1$). Use the diagram to determine the coordinates of these world points in S' and S respectively.

7. Consider the Lorentz transformations.

$$x'' = x' \cosh \alpha' - ct' \sinh \alpha' \qquad ct'' = -x' \sinh \alpha' + ct' \cosh \alpha'$$

$$x' = x \cosh \alpha - ct \sinh \alpha \qquad ct' = -x \sinh \alpha + ct \cosh \alpha$$

and determine the transformation relating (x'', ct'') to (x, ct).

8. Defining the relative velocity of two particles v_r as the velocity of one of them in the rest frame of the other, show that

$$v_r = \left[\frac{(\mathbf{v}_1 - \mathbf{v}_2)^2 - (\mathbf{v}_1 \times \mathbf{v}_2)^2/c^2)}{[1 - (\mathbf{v}_1 \cdot \mathbf{v}_2)/c^2]}\right]^{1/2}$$

Two electron beams are oppositely directed and move with identical velocities $0.8c$ relative to the laboratory frame. What is the relative velocity of the electrons as measured by an observer moving with one of the beams?

9. Cosmic radiation incident on the earth's atmosphere produces π- mesons, both π^+ and π^0. In the rest frame of a π^+ meson its average lifetime (before it decays into a μ-meson and a neutrino) is 2.55×10^{-8} seconds. What is the lifetime of a π^+ meson as measured in the laboratory frame?

10. Write down the general Lorentz transformations for any four-vector **L**.

11. Obtain the transformation formulae for the field components given in (12.36).

12. Using the invariance properties of the antisymmetric four tensor $F_{\mu\nu}$, show that for the electromagnetic field tensor the quantities $(E^2 - c^2B^2)$ and $\mathbf{E} \cdot \mathbf{B}$ are invariant,

13. Using the invariance properties of the fields from question 12 show that, if in a frame S', **E** and **B** are mutually orthogonal then they will be mutually orthogonal in all inertial frames.

14. If $\mathbf{E}\cdot\mathbf{B}=0$ show that it is always possible to find a frame in which $\mathbf{E}=\mathbf{0}$ or $\mathbf{B}=\mathbf{0}$.

15. Show that if $|\mathbf{E}|>|c\mathbf{B}|$ in a frame S then this relation holds in any other inertial frame S' and vice versa.

16. Show that if the angle between **E** and **B** is acute (obtuse) in one frame S it will be acute (obtuse) in any other inertial frame.

17. In a frame of reference S the electric and magnetic fields are mutually orthogonal. Find the velocity of the frame S' relative to S in which there is (a) only an electric field, (b) only a magnetic field.

18. A charge e is at rest in a frame S'. By applying a Lorentz transformation determine the potentials $\mathbf{A}$, ϕ in a frame S moving relative to S' with velocity $\mathbf{v}$. Hence find **E** and **B** (cf. (12.48), (12.49)).

19. In §12.7 we saw that the magnitude of the electric field **E** of a uniformly moving point charge in the direction of motion is reduced compared with the Coulomb field of the particle in its rest frame. In §12.5 we obtained the Lorentz transformations of the field components and found that $\mathbf{E}'_{\parallel}=\mathbf{E}_{\parallel}$. Interpret the apparent anomaly in these two results.

20. Establish the equivalence of (12.42) and (12.50).

21. A charged particle moves in a constant, uniform electric field. Modify the solution given in §12.9 to describe the motion of the particle if, at $t=0$, the particle is at the origin and has momentum $\mathbf{p}_0=p_{0x}\hat{\mathbf{x}}+p_{0z}\hat{\mathbf{z}}$, showing that

$$x(t)=\frac{cp_{0x}}{eE}\ln\left[\frac{\{(p_{0z}+eEt)^2+m_0^2c^2+p_{0zx}^2\}^{1/2}+p_{0z}+eEt}{p_{0z}+\dfrac{W}{c}}\right]$$

$$y(t)=0$$

$$z(t)=\frac{c}{eE}\left[\{(p_{0z}+eEt)^2+m_0^2c^2+p_{0x}^2\}^{1/2}-\frac{W}{c}\right]$$

with W given by (12.71).

Show that the trajectory of the particle is given by

$$z = \frac{W}{eE}\left(\cosh\frac{\omega_0 x}{v_{0x}} - 1\right) + \frac{v_{0z}}{\omega_0}\sinh\frac{\omega_0 x}{v_{0x}}$$

and check that this reduces to (12.83) for $v_{0z} = 0$.

22. Obtain (12.90).

23. Consider the motion of a charged particle in mutually orthogonal **E** and **B** fields as in §12.10. Generalize the solution obtained in §12.10 to the case $I_1 \neq 0$. The motion differs according as $I_1 > 0$ or < 0.

24. The vector potential for a monochromatic plane wave of arbitrary polarization propagating in the x direction is

$$\mathbf{A} = a(\tau)[0, \alpha \cos \omega\tau, (1-\alpha^2)^{1/2} \sin \omega\tau]$$

where $\tau = t - x/c$, $0 \leq \alpha \leq 1$ and $a(\tau)$ is slowly varying. Integrate the appropriate relativistic Lorentz equation to obtain the velocity of an electron in the field of this wave, assuming that the electron starts from rest at the origin. Show that the drift velocity is the same as in the case of a plane polarized wave ($\alpha = 1$).

25. Integrate (12.99), (12.100) ignoring terms of order $\dfrac{1}{a\omega}\dfrac{da}{d\tau}$.

Sketch the trajectory of the particle.

Appendix A

Summary of frequently used results in vector calculus

(i) Vector identities

$$\nabla \cdot (\phi \mathbf{A}) = \mathbf{A} \cdot \nabla \phi + \phi \nabla \cdot \mathbf{A}$$

$$\nabla \times (\phi \mathbf{A}) = \phi \nabla \times \mathbf{A} - \mathbf{A} \times \nabla \phi$$

$$\nabla (\mathbf{A} \cdot \mathbf{B}) = (\mathbf{A} \cdot \nabla)\mathbf{B} + (\mathbf{B} \cdot \nabla)\mathbf{A} + \mathbf{A} \times \nabla \times \mathbf{B} + \mathbf{B} \times \nabla \times \mathbf{A}$$

$$\nabla \cdot (\mathbf{A} \times \mathbf{B}) = \mathbf{B} \cdot \nabla \times \mathbf{A} - \mathbf{A} \cdot \nabla \times \mathbf{B}$$

$$\nabla \times (\mathbf{A} \times \mathbf{B}) = \mathbf{A}(\nabla \cdot \mathbf{B}) - \mathbf{B}(\nabla \cdot \mathbf{A}) + (\mathbf{B} \cdot \nabla)\mathbf{A} - (\mathbf{A} \cdot \nabla)\mathbf{B}$$

$$\nabla \times (\nabla \times \mathbf{A}) = \nabla(\nabla \cdot \mathbf{A}) - \nabla^2 \mathbf{A}$$

(ii) Vector differential operators in curvilinear coordinates

(a) **Cylindrical polar coordinates** (r, θ, z) (Fig. A.1).

$$\nabla \psi = \left[\frac{\partial \psi}{\partial r}, \frac{1}{r}\frac{\partial \psi}{\partial \theta}, \frac{\partial \psi}{\partial z}\right]$$

$$\nabla \cdot \mathbf{A} = \frac{1}{r}\frac{\partial}{\partial r}(rA_r) + \frac{1}{r}\frac{\partial A_\theta}{\partial \theta} + \frac{\partial A_z}{\partial z}$$

$$\nabla \times \mathbf{A} = \left[\frac{1}{r}\frac{\partial A_z}{\partial \theta} - \frac{\partial A_\theta}{\partial z}, \frac{\partial A_r}{\partial z} - \frac{\partial A_z}{\partial r}, \frac{1}{r}\frac{\partial}{\partial r}(rA_\theta) - \frac{1}{r}\frac{\partial A_r}{\partial \theta}\right]$$

$$\nabla^2 \psi = \frac{1}{r}\frac{\partial}{\partial r}\left(r\frac{\partial \psi}{\partial r}\right) + \frac{1}{r^2}\frac{\partial^2 \psi}{\partial \theta^2} + \frac{\partial^2 \psi}{\partial z^2}$$

(b) **Spherical polar coordinates** (r, θ, ϕ) (Fig. A.2).

$$\nabla \psi = \left[\frac{\partial \psi}{\partial r}, \frac{1}{r}\frac{\partial \psi}{\partial \theta}, \frac{1}{r \sin \theta}\frac{\partial \psi}{\partial \phi}\right]$$

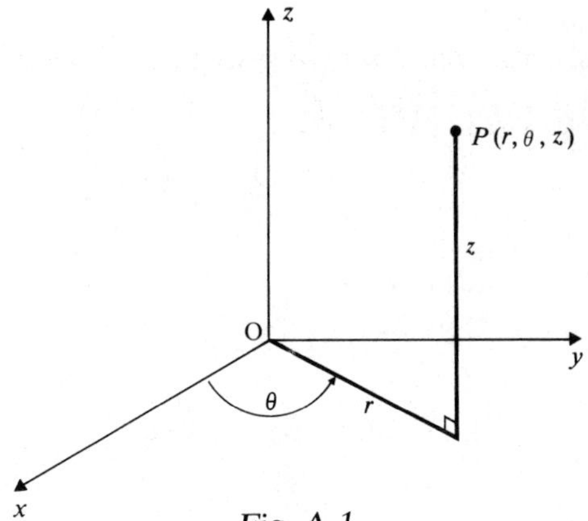

Fig. A.1

$$\nabla \cdot \mathbf{A} = \frac{1}{r^2}\frac{\partial}{\partial r}(r^2 A_r) + \frac{1}{r \sin\theta}\frac{\partial}{\partial \theta}(A_\theta \sin\theta) + \frac{1}{r\sin\theta}\frac{\partial A_\phi}{\partial \phi}$$

$$+ \frac{1}{r\sin\theta}\frac{\partial A_\phi}{\partial \phi}$$

$$\nabla \times \mathbf{A} = \left[\frac{1}{r\sin\theta}\left\{\frac{\partial}{\partial\theta}(A_\phi \sin\theta) - \frac{\partial A_\theta}{\partial\phi}\right\}, \frac{1}{r\sin\theta}\left\{\frac{\partial A_r}{\partial\phi} - \sin\theta\frac{\partial}{\partial r}(rA_\phi)\right\},\right.$$

$$\left.\frac{1}{r}\left\{\frac{\partial}{\partial r}(rA_\theta) - \frac{\partial A_r}{\partial\theta}\right\}\right]$$

$$\nabla^2\psi = \frac{1}{r^2}\frac{\partial}{\partial r}\left(r^2\frac{\partial\psi}{\partial r}\right) + \frac{1}{r^2\sin\theta}\frac{\partial}{\partial\theta}\left(\sin\theta\frac{\partial\psi}{\partial\theta}\right) + \frac{1}{r^2\sin^2\theta}\frac{\partial^2\psi}{\partial\phi^2}$$

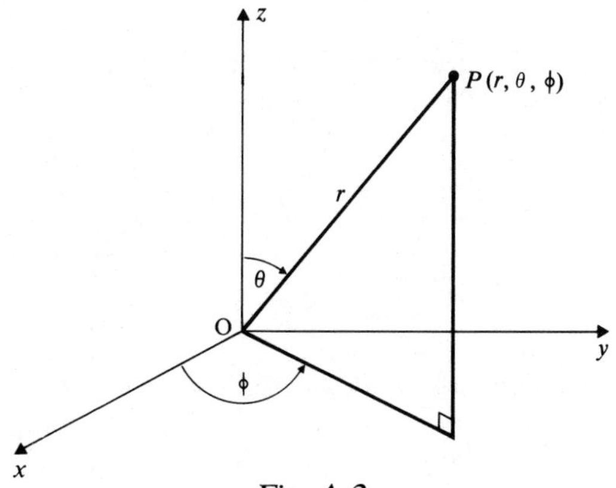

Fig. A.2

(iii) Integral theorems

Gauss' theorem or the **divergence theorem** states that

$$\int_V \nabla \cdot \mathbf{A} \, d\mathbf{r} = \int_S \mathbf{A} \cdot d\mathbf{S}$$

where the region V is bounded by the closed surface S.

Stokes' theorem states that

$$\int_S (\nabla \times \mathbf{A}) \cdot d\mathbf{S} = \oint_C \mathbf{A} \cdot d\mathbf{s}$$

where the open surface S is bounded by the closed contour C.

Some useful related results are:

$$\int_V \nabla\phi \, d\mathbf{r} = \int_S \phi \, d\mathbf{S}$$

$$\int_V \nabla \times \mathbf{A} \, d\mathbf{r} = -\int_S \mathbf{A} \times d\mathbf{S}$$

$$\int_S \nabla\phi \times d\mathbf{S} = -\oint_C \phi \, d\mathbf{s}$$

(iv) Suffix notation

Vector manipulations are often simplified by adopting suffix or tensor notation [10]. The following are equivalent:

Vector notation	Tensor notation	Vector notation	Tensor notation
$\mathbf{A}$	A_i	$\nabla\phi$	$\dfrac{\partial\phi}{\partial r_i}$
$\mathbf{A} \cdot \mathbf{B}$	$A_i B_i$	$\nabla \cdot \mathbf{A}$	$\dfrac{\partial A_i}{\partial r_i}$
$\mathbf{A} \times \mathbf{B}$	$\varepsilon_{ijk} A_j B_k$	$\nabla \times \mathbf{A}$	$\varepsilon_{ijk} \dfrac{\partial A_k}{\partial r_j}$

where

$$\varepsilon_{ijk} = \begin{cases} 1 & \text{if } i, j, k \text{ are a cyclic permutation of } 1, 2, 3 \\ -1 & \text{if } i, j, k \text{ are a non-cyclic permutation of } 1, 2, 3 \\ 0 & \text{if any two subscripts are the same.} \end{cases}$$

The following properties of ε_{ijk} are useful in manipulating tensor formulae:

$$\begin{aligned} \varepsilon_{ijk}\varepsilon_{ilm} &= -\varepsilon_{ijk}\varepsilon_{lim} = \varepsilon_{ijk}\varepsilon_{lmi} \\ &= \delta_{jl}\delta_{km} - \delta_{jm}\delta_{kl} \end{aligned}$$

Appendix B

Units

(i) Gaussian cgs system. Two systems of units are widespread in electrodynamics – SI units and Gaussian cgs units. The former has been used throughout the text but since many physicists use the latter we shall outline in this appendix how quantities in one system may be related to those in the other.

In the Gaussian cgs system length, mass and time are taken as the fundamental units and all others derived from these; in the SI system charge is a fourth fundamental unit. Newton's law allows the unit of force, the dyne, to be defined in terms of centimetres, grammes and seconds, one dyne being that force which produces an acceleration of 1 cm s^{-2} when acting on a mass of 1 gram, i.e.

$$1 \text{ dyn} = 1 \text{ g cm s}^{-2}$$

Coulomb's law is then used to derive the unit of charge:

$$\mathbf{F} = \frac{e_1 e_2}{r^2}\hat{\mathbf{r}}$$

and this unit – the **statcoulomb** or electrostatic unit (esu) – is taken to be that charge which experiences a force of 1 dyn when placed 1 cm from an identical charge i.e.

$$1 \text{ statcoul} = 1 \text{ cm dyn}^{1/2}$$

It follows from

$$V = \frac{e}{r}$$

that the unit of potential

$$1 \text{ statvolt} = 1 \text{ statcoul cm}^{-1}$$

and from

$$\mathbf{E} = \frac{e}{r^2}\hat{\mathbf{r}}$$

that the unit of electric field intensity is the statvolt cm^{-1}.

Current is defined as the rate of change of charge and hence the unit of current is

$$1 \text{ statamp} = 1 \text{ statcoul s}^{-1}$$

The unit of magnetic field **B** is the gauss and in Gaussian cgs units the field at a distance r from a long straight wire bearing a current I is

$$|\mathbf{B}| = \frac{2I}{cr}$$

I being measured in statamps. Thus

$$1 \text{ gauss} = 1 \text{ statcoul cm}^{-2}$$

With these derivations of units for charge, current, electric and magnetic field, Maxwell's equations take the form

$$\nabla \cdot \mathbf{E} = 4\pi\rho \qquad \nabla \cdot \mathbf{B} = 0$$

$$\nabla \times \mathbf{E} = -\frac{1}{c}\frac{\partial \mathbf{B}}{\partial t} \qquad \nabla \times \mathbf{B} = \frac{1}{c}\frac{\partial \mathbf{E}}{\partial t} + \frac{4\pi}{c}\mathbf{j}$$

where ρ, $\mathbf{j}$ are charge and current densities. In a vacuum these combine to give the wave equations

$$\left(\nabla^2 - \frac{1}{c^2}\frac{\partial^2}{\partial t^2}\right)\mathbf{E} = 0; \qquad \left(\nabla^2 - \frac{1}{c^2}\frac{\partial^2}{\partial t^2}\right)\mathbf{B} = 0$$

In the presence of dielectric and magnetic materials we introduce **D** and **H** defined by

$$\mathbf{D} = \mathbf{E} + 4\pi\mathbf{P}; \qquad \mathbf{H} = \mathbf{B} - 4\pi\mathbf{M}$$

so that Maxwell's equations become

$$\nabla \cdot \mathbf{D} = 4\pi\rho_{\text{free}} \qquad \nabla \cdot \mathbf{B} = 0$$

$$\nabla \times \mathbf{E} = -\frac{1}{c}\frac{\partial \mathbf{B}}{\partial t} \qquad \nabla \times \mathbf{H} = \frac{1}{c}\frac{\partial \mathbf{D}}{\partial t} + \frac{4\pi}{c}\mathbf{j}_{\text{free}}$$

For linear, isotropic media

$$\mathbf{D} = \varepsilon \mathbf{E}; \qquad \mathbf{B} = \mu \mathbf{H}$$

The unit of **D** and of **P** is the statvolt cm^{-1}. The unit of **H** is

$$1 \text{ oersted} = 1 \text{ gauss}$$

and that of **M**

$$1 \text{ magnetic moment } cm^{-3} = 1 \text{ gauss}$$

In Gaussian cgs units ε and μ are dimensionless. From Ohm's law $\mathbf{j} = \sigma \mathbf{E}$, it follows that the electrical conductivity is measured in units of s^{-1}.

(ii) Transformations

By comparing the equations of electrodynamics in the two systems one can see how to transform quantities from one to the other. We obtain the following sets of transformations (which are reversible):

Quantity	Gaussian cgs	SI
charge	e	$e/(4\pi\varepsilon_0)^{1/2}$
charge density	ρ	$\rho/(4\pi\varepsilon_0)^{1/2}$
current	I	$I/(4\pi\varepsilon_0)^{1/2}$
current density	$\mathbf{j}$	$\mathbf{j}/(4\pi\varepsilon_0)^{1/2}$
potential	ϕ	$(4\pi\varepsilon_0)^{1/2}\phi$
electric field	$\mathbf{E}$	$(4\pi\varepsilon_0)^{1/2}\mathbf{E}$
magnetic flux density	$\mathbf{B}$	$(4\pi/\mu_0)^{1/2}\mathbf{B}$
vector potential	$\mathbf{A}$	$(4\pi/\mu_0)^{1/2}\mathbf{A}$
polarization	$\mathbf{P}$	$\mathbf{P}/(4\pi\varepsilon_0)^{1/2}$
magnetization	$\mathbf{M}$	$(\mu_0/4\pi)^{1/2}\mathbf{M}$
electric displacement	$\mathbf{D}$	$(4\pi/\varepsilon_0)^{1/2}\mathbf{D}$
magnetic field	$\mathbf{H}$	$(4\pi\mu_0)^{1/2}\mathbf{H}$
dielectric constant	ε	$\varepsilon/\varepsilon_0$
permeability	μ	μ/μ_0
electrical conductivity	σ	$\sigma/4\pi\varepsilon_0$
velocity of light	c	$1/(\varepsilon_0\mu_0)^{1/2}$

To find how a measure of some physical property e.g. polarization is transformed between the two systems we may use the following equivalence table:

Symbol	Quantity	SI	Gaussian cgs
e	charge	1 coul	3×10^9 statcoul
ρ	charge density	1 coul m^{-3}	3×10^3 statcoul cm^{-3}
I	current	1 amp	3×10^9 statamp
$\mathbf{j}$	current density	1 amp m^{-2}	3×10^5 statamp cm^{-2}
ϕ	potential	1 volt	$\frac{1}{300}$ statvolt
$\mathbf{E}$	electric field	1 volt m^{-1}	$\frac{1}{3}\times10^{-4}$ statvolt cm^{-1}
Φ	magnetic flux	1 weber	10^8 maxwell (gauss cm^2)
$\mathbf{B}$	magnetic flux density	1 weber m^{-2} (tesla)	10^4 gauss
$\mathbf{A}$	vector potential	1 weber m^{-1}	$\frac{1}{3}\times10^{-10}$ gauss cm
$\mathbf{P}$	polarization	1 coul m^{-2}	3×10^5 statvolt cm^{-1}
$\mathbf{D}$	electric displacement	1 coul m^{-2}	$12\pi\times10^5$ statvolt cm^{-1}
ε	dielectric permittivity	1 farad m^{-1}	$36\pi\times10^9$ statfarad cm^{-1}
C	capacity	1 farad	9×10^{11} statfarad
$\mathbf{M}$	magnetization	1 amp m^{-1}	$\frac{1}{4\pi}\times10^4$ gauss
$\mathbf{H}$	magnetic field	1 amp m^{-1}	$4\pi\times10^{-3}$ oersted
μ	permeability	1 henry m^{-1}	$\frac{1}{4\pi}\times10^7$ gauss oersted^{-1}
σ	electrical conductivity	1 mho m^{-1}	9×10^9 s^{-1}
R	resistance	1 ohm	$\frac{1}{9}\times10^{-11}$ statohm
$L(M)$	inductance	1 henry	$\frac{1}{9}\times10^{-11}$ stathenry

Appendix C

Physical constants		
Electron charge	e	1.602×10^{-19} coul
Electron mass	m^-	9.109×10^{-31} kg
Electron charge/mass ratio	e/m^-	1.759×10^{11} coul kg^{-1}
Proton mass	m^+	1.672×10^{-27} kg
Proton/electron mass ratio	m^+/m^-	1836
Vacuum permittivity	ε_0	8.854×10^{-12} farad m^{-1}
Vacuum permeability	μ_0	1.257×10^{-6} henry m^{-1}
Velocity of light	c	2.998×10^8 ms^{-1}
Classical electron radius	r_e	2.818×10^{-15} m
Thomson cross-section	σ_T	$6.652 + 10^{-29}$ m^2
Electron volt	eV	1.602×10^{-19} joule
Electron rest energy	$m_0^- c^2$	0.5110 Mev
Proton rest energy	$m_0^+ c^2$	938.28 Mev
Boltzmann's constant	κ	1.381×10^{-23} joule K^{-1}
Avogadro's number	N	6.023×10^{23}
Planck's constant	h	6.626×10^{-34} joule s
Bohr radius	a_0	5.292×10^{-11} m

Appendix D

Notes on the computer programs

Three of the four programs in the text are written for use with a graphics terminal. As there is no standard language available for this purpose at present, we have used machine and system-independent names. Their function is outlined below.

INITAL is used to start the display. Its amin function is to clear the screen and set up space for the graphical information. It may also be necessary to define which portion of the screen is to be used.

GREND ends the display and may reset it to some default values. If this includes erasing the display, insert a READ statement to allow time to examine the picture.

SCAL (XMIN, XMAX, YMIN, YMAX) scales the plotting area so that the bottom left coordinates are (XMIN, YMIN) and the top right (XMAX, YMAX). All subsequent graphical output will first be scaled accordingly. POINT (X, Y) puts a visible dot at (X, Y) using the scale transformation set up by SCAL. (For simplicity we have let all our points be visible; some – such as those before calls to TTEXT – would be better erased.) VECT (DX, DY) draws a line from the position (XO, YO), defined by the last graphics call, to (XO+DX, YO+DY).

NUM (Q) outputs a number Q (real or integer) on to the screen at the current pen position. Some graphics systems require that you first convert Q to a text string. The ENCODE statement is useful for this. TTEXT (NCHAR, IARRAY) outputs the first NCHAR characters stored in array IARRAY onto the screen. It assumes however that they are packed into a word. Using a PDP 11 with, say, 27 characters IARRAY would need 14 elements while for a PDP 10, which stores 5 characters per word, it would need only 6 elements. The extra characters are usually spaces. Using a PDP 10 it would be necessary to change the data statement in PROGRAM LCR to

```
          DIMENSION KMESS(8)
          DATA KMESS /
        1 5HV ACR,5HOSS L,
        2 5HV ACR,5HOSS C,
        3 5HV ACR,5HOSS R,
        4 5HV SUP,5HPLY  /
Also in the DO 30 K=4,2,-1 loop replace the statement
          K1=5*k-4
with
          K1=2*k-1
```

As an example suppose your system has only one routine for plotting MOVE (X, Y, I). I = 3 produces a line from the current pen position to (X, Y) while I = 2 gives a dot at the point (X, Y). Then POINT and VECT would be as follows:

```
          SUBROUTINE POINT(X,Y)
          COMMON /COMPLT/CURRX,CURRY
          CALL MOVE(X,Y,2)
C         UPDATE CURR(ENT) X AND CURR(ENT) Y
          CURRX=X
          CURRY=Y
          RETURN
          END
          SUBROUTINE VECT(DX,DY)
          COMMON/COMPLT/CURRX,CURRY
          X=CURRX+DX
          Y=CURRY+DY
          CALL MOVE(X,Y,3)
C         UPDATE CURR(ENT) X AND CURR(ENT) Y
          CURRX=X
          CURRY=Y
          RETURN
          END
```

Bibliography

1. **Whittaker, E. T.**, *A History of the Theories of Aether and Electricity. Vol. 1. The Classical Theories*, Nelson, London, 1951.
2. **Rohrlich, F.**, *Classical Charged Particles*, Addison-Wesley, Reading, Mass., 1965.
3. **Boyd, T. J. M.** and **Sanderson, J. J.**, *Plasma Dynamics*, Nelson, London, 1969.
4. **Robinson, F. N. H.**, *Macroscopic Electromagnetism*, Pergamon, Oxford, 1973.
5. **Larmor, J. J.,** *Aether and Matter*, Cambridge University Press, London, 1900.
6. **Maxwell, J. C.,** *Treatise on Electricity and Magnetism* (3rd edn) Vols 1 and 2, Clarendon Press, Oxford, London, 1891. Reprinted by Dover, New York, 1954.
7. **Jackson, J. D.**, *Classical Electrodynamics* (2nd edn), Wiley, New York, 1975.
8. **Alfvén, H. and Fälthammar, C-G.**, *Cosmical Electrodynamics* (2nd edn), Clarendon Press, Oxford, 1963.
9 **Abramowitz, M. and Stegun, I. A.** (eds), *Handbook of Mathematical Functions*, Dover, New York, 1965.
10. **Dennery, P. and Krzywicki, A.,** *Mathematics for Physicists*, Harper and Row, New York, 1967.
11. **Pauli, W.,** *Theory of Relativity*, Pergamon, Oxford, 1958.
12. **Ehrlich, R.,** *Physics and Computers*, Houghton Mifflin, Boston, 1973.

Index